The Pauli–Jung Conjecture
and Its Impact Today

I0096164

Edited by
Harald Atmanspacher and Christopher A. Fuchs

Imprint Academic
Exeter 2014

Published in the UK by Imprint Academic
PO Box 200, Exeter EX5 5YX, UK

Distributed in the USA by Ingram Book Company
One Ingram Blvd., La Vergne, TN 37086, USA

ISBN 9781845406684

A CIP catalogue record for this book is available
from the British Library and US Library of Congress.

Table of Contents

Introduction: The Pauli-Jung Conjecture

Harald Atmanspacher and Christopher A. Fuchs

This book is about a meeting of minds in the mid 20th century that was unique in the history of science and philosophy: the encounter of Wolfgang Pauli, one of the architects of modern quantum theory, with Carl Gustav Jung, pioneer of analytical depth psychology. Their common interest was anchored in their search for a worldview better adapted to the extended body of scientific knowledge than what philosophers had offered so far. Their joint target was the "psychophysical problem": How is the interface between the physical and the mental to be understood, and on which idea of reality can it be grounded?

A key place in the history of the psychophysical problem is doubtlessly due to René Descartes and his so-called "substance dualism", which ascribes ontological significance to both the mental (*res cogitans*) and the physical (*res extensa*). Today, Descartes' position does not have many followers any more, and this for a number of reasons. The contribution by Appleby elucidates some of them and contrasts them with Pauli's and Jung's hope for a psychophysically neutral reality beyond mind and matter.

The closest historical precursor of their speculative ideas was the so-called "dual-aspect monism" of Spinoza (see the contributions by Cambray, Seager, Atmanspacher), where the mental and the physical are considered as epistemic aspects of one underlying reality. Ironically, both Pauli and Jung related their thinking much more to Leibniz's psychophysical parallelism than to Spinoza, and Cambray's article identifies possible reasons for this in Jung's oeuvre. He also traces elements of dual-aspect monism in the work of the German idealists from Herder to Schelling.

The physicist Pauli was in the fortunate position to collaborate with a physicist colleague who was educated in philosophy and Jung's psychology as well: Markus Fierz. Karl von Meyenn's masterful edition of Pauli's correspondence includes 336 letters between Pauli and Fierz. From these letters it is obvious how important Fierz's feedback was for Pauli, both for physics problems and for conceptual issues raised with Jung. The article by von Baeyer, whose father was a close friend of Fierz, provides insightful biographical and personal background to the Pauli-Fierz relationship.

The framework of thinking that Pauli and Jung developed, mainly between 1946 and 1954, was both highly speculative and distinctly ahead of its time. This is most clearly visible in the general scientific worldview around the mid 20th century, which then slowly changed from the predominantly reductive physicalist spirit in the 1960s and 1970s to serious explorations of viable alternatives from the 1980s until now. Today versions of dual-aspect monism are among the more promising and prominent ones of those alternatives.[1] This is why we think that the present status of Pauli's and Jung's heritage deserves the notion of something more substantial than a wild speculation: the *Pauli-Jung Conjecture*.

A conjecture is a proposition that is unproven. Karl Popper (1963) pioneered the use of the term "conjecture" in the philosophy of science.[2] But the notion of a conjecture is more familiar in mathematics, as an unproven proposition that appears correct and is hopefully testable on established grounds. Judah Schwartz submits[3]

> that the essence of mathematical creativity lies in the making and exploring of mathematical conjectures. A mathematical conjecture is a proposition about a previously unsuspected relationship thought to hold among mathematical objects.

Famous mathematical conjectures are, e.g., the Beal conjecture, the Collatz conjecture, the Goldbach conjecture, the Maldacena conjecture. Proven conjectures turn into theorems, such as Poincaré's theorem (proven by Perelman in 2003) or Fermat's last theorem (proven by Wiles in 1995).

The *Pauli-Jung Conjecture* is of a different kind. It refers to a philosophical rather than a mathematical proposition. Briefly speaking, it states that the mental (psychological) and the material (physical) are aspects of one underlying reality which itself is psychophysically neutral. A philosophical position like this can arguably not be "proven" in a mathematical sense. However, it can be more or less plausible, and its plausibility may change if it implies consequences which themselves can be corroborated or falsified.

---

[1] A short list of modern dual-aspect approaches (or related neutral monist approaches) in philosophy contains the work of Dave Chalmers, Thomas Nagel, Bertrand Russell, or Kenneth Sayre. But dual-aspect thinking has become interesting for physicists too: David Bohm, Basil Hiley, Bernard d'Espagnat, and Hans Primas are examples.

[2] K. Popper (1963): *Conjectures and Refutations: The Growth of Scientific Knowledge*, Routledge, London.

[3] J.L. Schwartz (1995): Shuttling between the particular and the general: Reflections on the role of conjecture and hypothesis in the generation of knowledge in science and mathematics. In *Software Goes to School*, ed. by D.N. Perkins, K.L. Schwartz, M.M. West, and M.S. Wiske, Oxford University Press, Oxford, p. 95.

Some of the essays in this book – most notably those by Fach, Roesler, and Laveman – indicate how this might play out concretely in our everyday life.

To make things even more ambitious, the *Pauli-Jung Conjecture* is not just a philosophical conjecture: It is a metaphysical conjecture – insofar as it addresses issues that are traditionally considered outside empirical access. For instance, the empirical sciences of today refer either to physical or to mental structures and processes. For the psychophysically neutral reality proposed by Pauli und Jung we do hardly have an idea about which scientific methodology could be used for addressing it. But then, isn't this situation almost characteristic of metaphsyical speculations in previous eras which by now have led to empirical research? Large parts of modern cosmology present clear examples of how pure metaphysics has turned into hard science.

The particular kind of holism inherent in quantum theory is another significant case in point, initiated by the seminal paper by Einstein, Podolsky, and Rosen in 1935 (the famous EPR paper). Shortly after its appearance, Pauli expressed in a letter to Heisenberg that[4]

> it seems to me that for a systematic foundation of quantum mechanics one needs to begin with the composition and decomposition of quantum systems.

This statement describes a key to the more general issue of dual-aspect approaches in philosophy. Neutral monism, as framed by Bertrand Russell,[5] starts with psychophysically neutral elements whose *composition* leads to mental or physical manifestations. Dual-aspect monism according to Pauli and Jung follows the opposite move: starting with a psychophysically neutral whole, an all-embracing one world, or *unus mundus*, the mental and the physical emerge by *decomposing* this whole into parts. In his *Mysterium Coniunctionis* of 1956, Carl Gustav Jung wrote:[6]

> Undoubtedly the idea of the unus mundus is founded on the assumption that the multiplicity of the empirical world rests on an underlying unity, and that not two or more fundamentally different worlds exist side by side or are mingled with one another. Rather, everything divided and different belongs to one and the same world, which is not the world of sense but a postulate ...

---

[4]Letter of 15 June 1935, published in K. von Meyenn, ed. (1985): *Wolfgang Pauli. Wissenschaftlicher Briefwechsel, Band II: 1930–1939*, Springer, Berlin, p. 404.

[5]See B. Russell (1921): *The Analysis of Mind*, George Allen and Unwin, London. However, the pioneering ideas introducing neutral monism go back to Ernst Mach and William James, whom Russell explicitly acknowledges.

[6]C.G. Jung (1970): Mysterium Coniunctionis, *Collected Works, Vol. 14*, Princeton University Press, Princeton, par. 767.

While the composition of elementary systems always leads to a unique result (which can be uniquely reduced back to its components), the decomposition of a whole is generally not unique but depends on contexts. In the *Pauli-Jung Conjecture* these manifold aspects can even be incompatible or complementary, a feature that is not part of any other dual-aspect approach so far. The possibility of incompatible descriptions of parts emerging from wholes clearly derives from Pauli's knowledge of this key insight of quantum theory. It suggests that structural elements of quantum theory may elucidate our understanding of the psychophysical problem.

One almost obvious issue in this direction is the farewell to the classical concept of a detached observer, replaced by the "participating observer" in quantum theory. The observation of a quantum systems always needs to take into account the active role of the observing environment. Any quantum measurement is not simply a reading-off of a measured value but also induces a (generally uncontrollable) change of the state of the system observed. As a consequence, measurement outcomes are probabilistic in principle.[7]

A radical touch has been added to this insight by Fuchs and his collaborators in recent years. They depart from the usual interpretation of probabilities for quantum events by introducing subjective probabilities, explicitly expressing the knowledge acquired by observing subjects (rather than measuring tools). Combined with the idea of measurement as Bayesian updating, the articles by Fuchs and Schack outline this novel kind of "quantum Bayesianism".[8]

It is a basic epistemological point that all kinds of lawful regularities in nature express themselves in observed correlations, which are then – in a second step – interpreted in terms of laws of nature. In Western history of science, the leading candidate for such interpretations is one or another kind of causation – but is this the only possible way of interpreting correlations? A very general expression of causal explanations is Leibniz's principle of sufficient reason: nothing in nature happens without reason.

The sciences of the 19th century constricted this general idea to an almost exclusive pretense to explain any kind of behavior by *efficient causation*. If correlations between events at different times are observed, this could only mean to explain the occurrence of the later event (effect) in terms of the preceding one (cause). Of course, there can be multicausal pathways as well,

---

[7]Max Born found in 1926 that the the square of the state function of a quantum determines the probability that a measurement on that system yields a particular result.

[8]For more background see also C.A. Fuchs (2011): *Coming of Age with Quantum Information. Notes on a Paulian Idea*, Cambridge University Press, Cambridge.

but cause-and-effect relations have become the all-dominant metaphor for scientific explanation – with *chance* as the only alternative.

Quantum theory predicts correlations which cannot be explained by efficient causation, but by decomposing wholes into parts, systems into subsystems. In this process, holistic features become disentangled, and the properties of the emerging parts exhibit holistic, *acausal correlations*. In physics, such correlations are well known due to quantum entanglement (cf. the contribution by Filk).

The *Pauli-Jung Conjecture* goes one step further and posits that a similar principle is at work in psychophysical systems, including both mental and physical properties in correlation with one another. Here it is evidently inconceivable how efficient causation could operate between such categorially distinct entities as mind and matter. Pauli and Jung agreed that *meaning* should be regarded as a proper alternative to efficient causation in this case. Pauli even postulated[9]

> a third type of laws of nature consisting of corrections to chance fluctuations due to meaningful or purposeful coincidences of causally unconnected events.

This third type of laws of nature, beyond efficient causation and blind chance, beyond deterministic and statistical descriptions, is entirely undeveloped in present science. According to the *Pauli-Jung Conjecture* its central interpretive tool would have to be meaning instead of causation. Atmanspacher's and Main's essays elaborate on this point in more detail.

Meaning is the core issue in Jung's concept of synchronicity, an acausal connecting principle correlating mental and physical events. Usually, analytical psychology understands synchronistic events as consequences of ordering factors residing in the collective unconscious, called archetypes. Hogenson's contribution raises the question whether such archetypes are really needed to understand synchronicities and offers an alternative guided by the theory of complex systems. And Gieser's contribution gives an in-depth discussion of the *symbolic* dimension of archetypes from Jung's and Pauli's perspectives.

Ways to use the concept of synchronistic, meaningful coincidences in psychological practice are pointed out by Roesler (synchronicities in psychotherapy) and Fach (synchronicities as special cases of psychophysical phenomena). Laveman's article makes the fascinating proposal to understand the importance of *serendipitous* events in the business world due to synchronistic

---

[9]Attachment to a letter to von Franz of 30 October 1953, published in K. von Meyenn, ed. (1999): *Wolfgang Pauli. Wissenschaftlicher Briefwechsel, Band IV, Teil II: 1953–1954*, Springer, Berlin, p. 336.

occurrences. And Zabriskie's article completes the book with her assessment of synchronicity in the intercultural perspectives of Western literature and Eastern philosophy.

With the metaphysical issues that the *Pauli-Jung Conjecture* raises, it might appear that it just perpetuates the historical divide between ontic and epistemic universes of discourse. However, our intention is different: we propose to seriously explore the interface between them, which is inevitably defined by both the ontic and the epistemic. Since the frontiers of knowledge have never been rigidly fixed, this interface cannot be defined as a static wall behind which empirical access is outright impossible. What may be ontic from one point of view may be epistemic from another. For a future, more refined version of the *Pauli-Jung Conjecture* we envision a metaphysics that distinctly acknowledges participating and committed observers – "both spectators and actors in the great drama of existence".[10]

This collection of essays contains the peer-reviewed contributions to an international interdisciplinary conference on the dialog between Pauli and Jung and its impact for philosophy, science, and society today. Different from a previous conference mainly focusing on Wolfgang Pauli alone,[11] this meeting was specifically designed to include the perspective of Jungian analytical psychology and some of its research-oriented representatives.

The conference would not have been possible without the generous support of the Donald C. Cooper-Fonds (Zürich), the Stefanie und Wolfgang Baumann Stiftung (Basel), and the Collegium Helveticum (Zürich). It was organized by the editors of this volume and took place at Seminarhotel Lihn in Filzbach, a small town in an exciting environment, close to Lake Walensee and the Glarner Alps, from September 23–27, 2012.

We are grateful for the splendid hospitality we experienced at the Lihn and would like to thank Hannes Hochuli and his staff for their support in matters large and small ensuring the success of the conference. Graham Horswell at Imprint Academic, the publisher of this volume, provided helpful advice for its smooth and speedy publication. And our special thanks go to Karin Moos (IGPP Freiburg) for much of the pre-conference arrangements and the preparation of the manuscripts for this book.

---

[10]N. Bohr (1958), *Atomic Theory and the Description of Nature*, Cambridge University Press, Cambridge, p. 81.

[11]The proceedings of this conference were edited by H. Atmanspacher and H. Primas and published as *Recasting Reality. Wolfgang Pauli's Philosophical Ideas and Contemporary Science*, Springer, Berlin 2009.

Mind and Matter:
A Critique of Cartesian Thinking

Marcus Appleby

Abstract

It is argued that the problem of interpreting quantum mechanics, and the philosophical problem of consciousness, both have their roots in the same set of misguided Cartesian assumptions. The confusions underlying those assumptions are analyzed in detail. It is sometimes suggested that quantum mechanics might explain consciousness. That is not the suggestion here. Rather it is suggested that an adequate non-Cartesian philosophy would transform our understanding of both quantum mechanics and consciousness. Consequently it would change our ideas as to just what it is that we are trying to explain.

1 Introduction

Pauli, in a letter to van Franz (Gieser 2005, pp. 243f), wrote:

> Evidently the progress of science must take such a course that the concept "consciousness" will be replaced by a more general or better one.

If one knew that these words were written by a leading 20th century scientist, but did not know that the scientist in question was Pauli, one might think that what is being advocated here is eliminative materialism, or some such similar position (Stich 1983, 1996, Dennett 1993, 2005, Churchland 1988, Churchland 1989, and references therein). Since, however, it is Pauli who is saying this we know he must be thinking along very different lines. Eliminative materialists propose to deal with the mind-body problem by eliminating the mental pole of the duality leaving only the material one. Pauli would reject that proposal because he was looking, not for a materialistic explanation of mental phenomena, but rather for a "psychophysical monism" in which mind and matter are seen as "two aspects of one and the same abstract fact", itself neither physical nor psychological (Meier 2001, pp. 87, 159). It is easy to see why a materialist might want to take an eliminativist attitude to consciousness. The question addressed in this paper is why someone like Pauli, who is not a materialist, would take such an attitude.

What follows is not an exercise in Pauli exegesis. I am not here particularly concerned with Pauli's reasons for taking that view of consciousness. Rather, I am going to give my own reasons for thinking that he might have been basically right.

Before proceeding further, I ought to qualify the notion of consciousness. The meaning of a word like "cat", which can be defined ostensively, is securely anchored. However, the word "consciousness" cannot be defined ostensively, not even by the person whose consciousness it is (it is surely not possible to point one's finger at one's own consciousness). Consequently, if one is not careful, there is a danger that its meaning will float, so that it comes to be used in different ways by different people, or even by the same person at different times. The criticisms of this paper are only directed at *one* of its possible senses.

As an example of a sense of the word which I feel is unlikely to be rendered obsolete by future scientific advance, consider the Glasgow Coma Scale (Teasdale and Jennett 1974, 1976) which is widely used to quantify the level of consciousness in cases of brain damage. It is possible, even likely, that the Glasgow Coma Scale will, in time, come to be replaced by some improved method for quantifying degree of consciousness. It is also likely that scientific advances will lead to a deeper and richer understanding of the phenomenon itself. However, I doubt that this would amount to the kind of development Pauli had in mind when he wrote of the concept of consciousness being "replaced by a more general or better one".

For want of a better term I will refer to the sense in which the word "consciousness" is used in medicine as its "everyday sense". It is true that the medical literature on the subject can be quite technical. However, although medical science has refined the description of states of consciousness, it has done so in a way which remains close to the root meaning. A doctor will understand the statement "the patient is fully conscious" in almost, if not exactly the same sense that the patient's relatives understand it. I take the everyday sense of the word also to include its use in sentences like "she was conscious of the clock ticking," to describe the state of being aware of something.

The critical comments in this essay are not directed at consciousness in the everyday sense, but rather at the concept as it is used in, for example, philosophical discussions of the so-called problem of consciousness. I will refer to this second sense of the word as the Cartesian sense. It is true that nowadays there are not many full-blooded Cartesian dualists left. Nevertheless, a more or less attenuated version of the Cartesian soul continues to be prominent in modern philosophical thinking, and it is this which gives rise to the "problem of consciousness". It is clear from context[1] that it was Cartesian consciousness that Pauli had in mind when he made the statement quoted above.

To see that the everyday and Cartesian senses are different consider the discussion by Chalmers (1996). He begins by saying that consciousness is "intangible" and consequently hard to define (p. 3), which is already an indication that what is in question is something different from consciousness in the everyday sense (consider the likely response of a hospital doctor to the proposition that the state of being non-comatose is intangible, and hard to define). He then goes on to propose the characterization "the subjective quality of experience" (p. 4). Now the meaning of this will be clear enough to someone who has received a certain kind of education. More specifically, it will be clear to someone who has absorbed the basic ideas of the Cartesian philosophy. But it would be unintelligible to anyone who has not had the benefit of such an education (probably the majority of English speakers).

What Chalmers thinks of as the subjective quality of greenness, philosophically unsophisticated people think of simply as greenness. It would take a lot of work to persuade them that they are missing something important. Something that is not taken for granted by the vast majority of speakers cannot be considered to belong to the everyday sense of a word. Of course, one might think that the Cartesian concept of consciousness can be seen to be logically contained in everyday assumptions, if one takes the trouble to think the matter through carefully. However, it is precisely the point of this paper that it is *not* so contained.

Chalmers, like others, thinks that consciousness is hard to define. Why should that be? Searle puts his finger on at least part of the difficulty when he says (Searle 1992, p. 131):

> The reason we find it difficult to distinguish between my description of the objects on the table and and my description of my experience of the objects is that the features of the objects are precisely the conditions of satisfaction of my conscious experiences of them. So the vocabulary I use to describe the table – "there's a lamp on the right and a vase on the left and a small statue in the middle" – is precisely that which I use to describe my conscious visual experiences of the table.

---

[1]In particular, it is clear that Pauli had in mind the so-called privacy of Cartesian consciousness – the property of being undetectable by outside observers.

This provokes the obvious question: If two things have the same description, how does one tell them apart? *Can* one tell them apart? Could it just be that what Searle seeks to convey by the phrase "the contents of my consciousness when I look at my table" is identical to what a less sophisticated person would convey more succinctly, simply by saying "my table"? It seems, however, that that cannot be precisely right, for Searle argues that consciousness is always perspectival. Consequently, he thinks that his visual consciousness of his table only comprises the parts he can directly see. Nevertheless, it is hard to resist the impression that what Searle means by the phrase "the contents of my consciousness" is, if not identical, at any rate close to what an unsophisticated person means by the phrase "the things around me": that the contents of Searle's consciousness, as Searle conceives them to be, can be pictured as something like a film set, convincing when seen from the front, unpainted wood when seen from the back.

This way of thinking is historically important, because it led to idealism. In an amusing critique of idealist philosophy, Stove (1991, p. 116)) asks what is the "product-differentiation": "what are they *selling*, these people who call themselves objective idealists, that a commonsense *materialist* could not consistently buy?" His answer is that there is in fact nothing that a materialist could not consistently buy. In support of this conclusion he cites Bosanquet (one of the more prominent 19th century idealists), who said that "extremes meet", and "a consistent materialist and thorough idealist hold positions which are distinguishable only in name" (Stove 19991, p. 115).

These days idealism has gone out of fashion. However, believers in Cartesian consciousness are still faced with essentially the same problem of differentiating the contents of consciousness (as they conceive them to be) from what commonsense would call the objects around us. It is a difficult problem, and I think that is one of the reasons why "consciousness" is hard to define.

2   The Cartesian Split

Descartes introduced a fundamental split between Cartesian consciousness and Cartesian matter. I am here using the term "Cartesian matter" rather loosely, to refer, not only to the concept of matter originally proposed by Descartes himself, but also to its many descendants. I described the concept of consciousness as it features in, for example, the book by Chalmers (1996) as an attenuated variant of the Cartesian soul. In the same way I would, for example, describe the universal wave function proposed by Everett (DeWitt and Graham 1973) as a (not so attenuated) variant of Cartesian matter. It goes without saying that Chalmers' concept of consciousness differs greatly from Descartes' concept of the soul. However, it shares with the latter the crucial feature of being a receptacle for all the supposedly subjective phenomena which, on a Cartesian view, are excluded from the physical universe. Similarly, Everett's concept of the universal state vector, though obviously very different from Descartes' concept of matter, still shares the crucial feature that it is supposed to be completely describable in purely objective, mathematical terms, without any contamination by the observing subject.

The point to notice is that Cartesian consciousness and Cartesian matter are different aspects of a single conceptual scheme. They are like the two poles of a bar magnet, impossible to isolate. Idealists attempt to cut the bar in two, keeping only the subjective side of the polarity. But, as we saw, when they try to carry that idea through consistently it turns out that the concept of matter has come back in, through the backdoor, so to speak. Materialists attempt to perform the same bisection, keeping only the objective side of the polarity. However, they then face the problem that, no matter how vigorously they attempt to cast doubt on the notion of *qualia* (see, for instance, Section 17 of Lycan (1990)), the fact remains that, to a normally sighted person, green things undeniably do look qualitatively different from red ones. Consequently, if one looks at a green object, while trying to keep in mind that the quality of perceived greenness is not really a feature of the object itself, it is difficult to avoid the thought that the quality of greenness is a feature somehow added by one's own perceptual apparatus. From there it is but a small step to the Cartesian concept of consciousness.

I believe we need to break away from this whole misguided way of thinking: not simply to deny Cartesian consciousness, nor simply to deny Cartesian matter, but to deny both. There are many empirical reasons for taking such a course. Modern neuroscience gives us reasons for being suspicious of Cartesian assumptions about consciousness (see Dennett 1993, 2005, Blackmore 2002, 2004, and references therein), while quantum mechanics gives us equally good reasons for being suspicious of Cartesian assumptions about matter.

The aim of physics, as Descartes conceived it, is to arrive at the one true picture of things, totally objective, and complete in every detail. Before the year 1900 it might have looked as though we were getting steadily closer to that goal.[2] However, quantum mechanics strongly suggests that the goal

---

[2]However, there were 19th century physicists, such as Mach (1959), who did not agree with Descartes about the goal of physics.

is unachievable. In quantum mechanics what you see depends on how you look. Make one kind of measurement on the electromagnetic field and one will obtain results consistent with it being a smoothly varying wave; make another, different kind of measurement and one will obtain results consistent with it being a collection of discrete particles. Similarly, if one observes an atom using a scanning tunneling electron microscope, one will see an apparently solid object. If, on the other hand, one observes it with a γ-ray microscope, one will see a collection of point-like particles separated by empty space.

So which of these pictures is the *true* one? Quantum mechanics declines to say, just as it declines to say what is going on in a physical system when no one is looking. In place of the God-like conspectus of the entire universe, with nothing left out, which Descartes imagined and which continued to inspire physicists for 250 years after him, quantum mechanics merely gives us methods for anticipating what will be observed in this or that particular experimental context. Moreover, the fact, that the outcome depends on the observer's decision as to which measurement to make, casts doubt on the assumption, that physics passively records events that would have happened anyway, in the absence of experimental intervention. This represents a subtle, but important departure from the Cartesian ideal of total objectivity.

Since the 1920s there have been numerous attempts to reconcile quantum mechanics with Cartesian assumptions, as to what the world *ought* to be like (for an overview see Schlosshauer 2011). These attempts have been successful to the extent that it seems there is nothing to logically exclude the possibility that, underlying the observations, there is some universal mathematical mechanism. The difficulty is finding a picture of this kind which is empirically substantiated.

When Einstein embarked on the project of finding an alternative to the Copenhagen interpretation, he doubtless hoped to find a single theory which, like the general theory of relativity, would be uniquely specified by the interplay of various empirical and aesthetic considerations. Doubtless he also hoped for new empirical predictions. Of course, conclusive demonstrations are not to be had in science. So no one can say for sure that Einstein's hopes will not be fulfilled at some time in the future. But it does seem to me that the effect of eighty years of theoretical work has been to make those hopes look increasingly forlorn.

My own feeling is that an adequate understanding of quantum mechanics ultimately depends, not on sophisticated technical developments, but on some simple conceptual shift – something a little like the perceptual shift which occurs when one looks at a diagram like the Necker cube, or the duck-rabbit picture (Wittgenstein 1968, p. 194e, Kihlstrom 2004). Quantum mechanics is not intrinsically weird. It only seems weird because we insist on looking at it through Cartesian spectacles. The problem is that Cartesian assumptions have become so deeply ingrained in our thinking that it is hard to find the right non-Cartesian spectacles.

Turning to the other pole of the Cartesian duality, philosophers are familiar with the privacy of Cartesian consciousness: the fact that the consciousness of another person is, from a Cartesian point of view, just as inescapably hidden as the wave function is in the Bohm interpretation of quantum mechanics (Bell 1987, p. 202). What is less widely appreciated is that there is a problem with ascertaining the contents of one's *own* consciousness.

A particularly striking illustration of this point comes from the study of eye movements in reading (Rayner 1978, 1998). In order to explain it I first need to say something about the physiology of human vision. The region of the retina where the receptors are packed most tightly, and where visual acuity is consequently highest, is called the fovea. The part of the visual field which falls on the fovea subtends an angle of $\sim 1°$ at the center of the lens. Visual acuity falls off rapidly as one moves away from this region, which means that in a single fixation of the eyes one is able to discriminate fine detail in only a very small portion of the visual field (a portion about the size of a thumbnail held at arm's length).

The reason the visual system is nonetheless able to acquire accurate information about the whole environment is that the eyes are continually performing jumps, or saccades. When reading the duration of a single saccade is typically ~ 30 ms, while the duration of the fixation between saccades is typically ~ 200 ms (in other activities the saccades are often bigger, and take correspondingly longer). During a saccade very little information is transmitted to the cortical processing areas (this phenomenon is called saccadic suppression, or saccadic masking). It can consequently be said that most of our visual awareness is based on ~ 4 snapshots per second, each of them covering only a small fraction of the visual field.

These facts already seem very counter-intuitive from a Cartesian point of view: it is surprising (on Cartesian assumptions) that at any moment one sees so little in fine detail, and surprising also that there are so few jumps per second (a movie which ran at 4 frames per second would look jumpy). However, it gets worse (from a Cartesian point of view). The eye muscles give a brief twitch to initiate a saccade, and thereafter the eyeballs move ballistically, subject only to frictional forces. Consequently, a computer attached to an eye-tracking device can calculate where the next fixation is going to be before the eyes actually land there.

This makes possible the following experiment. One takes a page of printed text and projects it onto a screen, replacing all the letters by x's. The experimental subject sits in front of the screen, and his/her eye-movements are monitored. During a saccade the computer calculates where the eyes are going to alight, and puts a handful of letters from the original page just at that point, leaving x's everywhere else. In the next saccade the computer wipes those letters, replacing them by x's, and puts another group of letters at the next fixation point, and so on. To illustrate this, in one experiment the original text was (cf. Rayner 1978):

> By far the single most abundant substance in the biosphere
> is the familiar but unusual inorganic compound called water. In
> nearly all its physical properties water is either unique or at
> the extreme end of the range of a property. It's extraordinary

while what appeared on the screen during one particular fixation was:

> Xx xxx xxx xxxxxx xxxx xxxxxxxx xxxxxxxxx xx xxx xxxxxxxxx
> xx xxx xxxxxxxx xxx xxxsual inorganic coxxxxxx xxxxxx xxxxx. Xx
> xxxxxx xxx xxx xxxxxxxx xxxxxxxxxx xxxxx xx xxxxxx xxxxxx xx xx
> xxx xxxxxxx xxx xx xxx xxxxx xx x xxxxxxxx. Xx'x xxxxxxxxxxxxx

This, and other, similar techniques have been used to acquire a wealth of information about the visual system. However, its relevance to the present discussion is simply this. To an observer whose eye movements are not synchronized with the screen it is obvious (a) that at any moment the screen contains almost nothing but x's and (b) that what is on the screen is constantly changing. However, to the experimental subject, whose eye movements *are* synchronized, the screen looks like a perfectly normal page of text. To convey just how good the illusion is, Grimes (1996) records that one of the first people to conduct an experiment of this kind served as the first experimental subject (see also Dennett 1993, p. 361); after a while he sat back from the apparatus and announced that something must be wrong with the system because the text was not changing – though it was, in fact, working perfectly.

If one reflects on this fact, that it is demonstratively impossible to tell the difference between a normal page of printed text, and a page which at any given moment consists almost entirely of x's, then one becomes genuinely uncertain, as to what precisely *are* the contents of one's own consciousness at any given moment. Looking at the page in front of me I can see that it does not consist almost entirely of x's. I am able to know this because information is integrated across saccades. Consequently, I am aware, not only of the information acquired on this present visual fixation, but also of information acquired on many previous fixations. But *how much* information is integrated across saccades? What precisely is its nature? And precisely how much of that information is contained in my consciousness?

The first two of these questions are empirical questions which can be, and actually are being, investigated by the usual scientific methods. However, the last question is of a different character. At least, it is of a different character if it is consciousness of the Cartesian sort which is in question. On Cartesian principles, consciousness is private. It follows that if I myself cannot tell what exactly are the contents of my own consciousness, then no amount of neuroscientific experimentation can tell either. Like the position of the particle in a two-slit experiment, my consciousness now is indeterminate.

There are numerous other experiments and examples pointing to the same conclusion.[3] I will confine myself to just two other examples. Grimes (1996) used an eye-tracking device coupled to a computer to examine what happened when a picture (as opposed to a page of printed text) was changed in the middle of a saccade. In one such experiment, in a picture of two men wearing differently colored hats, the hats were switched mid-saccade. 100% of the experimental subjects did not notice. Even more dramatically, in another case a parrot, occupying roughly 25% of the picture area, was switched from brilliant green to brilliant red mid-saccade. In this case most of the subjects did notice. But 18% of them did not. 25% of the picture area is a lot, and it raises the question of what exactly is one conscious of, if one does not notice a change as striking as that.

A second illustration is the one given by Dennett (1993, pp. 354–355), of wallpaper in which the pattern consists of a large number of identical images of Marilyn Monroe. If one looks at it, it will only take a second or two to realize that the images are all the same. Since the eye performs only a few saccades per second it is impossible that one has discriminated more than a handful of the images in sufficient detail to be able to identify it. Instead the visual system must essentially be making a guess, based on the small number of cases which it has accurately discriminated. So the question arises again: in a case like this what exactly are the contents of consciousness?

In ordinary life, and in physics before the 20th century, the assumption that a physical object always has a determinate trajectory works well. But when we push our investigations far enough we start to run into difficulties. Similarly with the concept of consciousness: when we start to ask the kind of detailed questions raised above we run into problems not entirely dissimilar to the problems which quantum mechanics reveals for Cartesian matter.

---

[3]For details the reader may consult Dennett (1993, 2005), Blackmore (2002, 2004), Grimes (1996), Simons (2000), and references therein.

It is often thought that quantum indeterminacies are humanly unimaginable. That is to get it exactly the wrong way around. What is impossible to imagine is knowing the position of something to infinitely many decimal places. On other hand, ordinary experience is full of indeterminacies. If someone wants to know what it would be like to perceive an indeterminate position, all they need to do is look at an object in a room, and try to estimate its distance from the walls. It is unlikely that they can achieve even 10% accuracy. Similarly, to know what it is like to perceive a number indeterminacy (such as the indeterminacy of number of photons in a coherent state) all one needs to do is look at a collection of objects on a table. If one is then asked how many objects there are, it is unlikely one will be able to say, without first taking the time to count them up. The fact that one cannot answer straight away (and probably could not answer at all if one did not still have the objects in view) suggests that at the time of asking one was conscious of the objects, but not of their number.

Dennett (1993) has written a book entitled *Consciousness Explained*. Since I agree with Dennett on a number of points I ought to stress that I do not agree with him on this central one. I do not think that he, or anyone else, is close to "explaining consciousness". Like Pauli, I think that a satisfactory understanding of these questions will involve breaking out of the Cartesian mould entirely, and developing a different conceptual framework.

At this stage I should perhaps obviate another potential misunderstanding. There have been a number of attempts to explain consciousness using quantum mechanics (see Atmanspacher 2004 for a review). Since these approaches all depend on adopting non-Copenhagen interpretations of quantum mechanics, and since they take the Cartesian concept of consciousness for granted, it should be apparent, from what I said earlier, that I do not find any of them convincing. If I keep mentioning consciousness and quantum mechanics in the same breath (so to speak) it is not because I think that one of them can be used to explain the other, but because I think that in both cases a clear understanding of the phenomena is obstructed by the same misguided Cartesian philosophy. A second, subsidiary reason is that I cannot help being struck by parallels.[4] What the parallels are worth, I do not know. But I find them interesting.

Here is another parallel. Dennett (1993) argues that in discussions of consciousness it is essential to take careful account of the probe (i.e., the specific

---

[4]For some discussions of this and related points see Meier (2001), Jung (1960), Atmanspacher *et al.* (2002), Atmanspacher *et al.* (2004), Conte *et al.* (2007), Bruza *et al.* (2009), and references therein.

question used to elicit a response at a specific time in a specific experimental context). Furthermore, if one tries to interpret the results obtained using different probes in terms of a single, coherent story – a "trajectory of consciousness" – one runs into difficulties (see, for instance, Dennett's discussion of the color phi and cutaneous rabbit experiments). Also, the probe disturbs the system: it can bring into existence a conscious content which otherwise might not have occurred. This is all reminiscent of the situation in quantum mechanics (there are major differences, but it is reminiscent).

3   Before Descartes

At this stage it will be useful to look at the historical development of Cartesian ideas. In the first place this is a good way to see that the Cartesian concept of consciousness, so far from being a natural intuition (as many people are still inclined to think), actually depends on postulates which, although they have since become second-nature for many people, originally had to be worked out slowly and laboriously. In the second place, it brings out the fact that the Cartesian philosophy was intimately related to the 17th century development of modern science.

The Cartesian concept of consciousness is a 17th century invention. It did not exist before.[5] In order to appreciate just how original a departure it was, one needs to see it in the context of the earlier conceptions it replaced. Concerning classical Graeco-Roman philosophical ideas,[6] Matson (1966) writes:

> Any teaching assistant can set up the mind-body problem so that any freshman will be genuinely worried about it. Yet none of the ancients ever dreamed of it, not even the author of *De Anima*.

And he goes on to observe that "in the whole classical corpus there exists no denial of the view that sensing is a bodily process throughout." Similarly, Caston (2002), discussing the question whether "Aristotle even had a concept

---

[5]Rorty (1979) makes this point in some detail. His discussion is very useful. However, Rorty is not much interested in natural science. In his own words, he tends to "view natural science as in the business of controlling and predicting things, and as largely useless for philosophical purposes" (Saatkamp 1995, p. 32). Consequently he misses a number of points which are crucial for the present discussion. Burtt (2003) is also very relevant.

[6]In the interests of brevity I will here confine myself to the European, Islamic and Jewish philosophical traditions, which are closely related, and which are the ones most relevant to Descartes' intellectual *milieu*. For the bearing of Buddhism on the problem of consciousness see Blackmore (2004).

of consciousness," observes that, although "Aristotle clearly distinguishes being awake and alert from being asleep or knocked out", he "does not use any single word to pick out the phenomena we have in mind," and he "does not share the epistemological concerns distinctive of the Cartesian conception of consciousness, such as privacy or indubitability". In other words, Aristotle had the everyday concept of consciousness, but not the Cartesian one.

3.1 Augustine

There were philosophers in the ancient Graeco-Roman world whose thinking was in *some* ways similar to the Cartesian philosophy. The one who came closest was probably Augustine. It has been suggested, in fact, that Augustine was a significant influence on Descartes (Rorty 1979, Matthews 1992, Menn 1998, Wilson 2008, Matthews 2000, Mann 2000), though opinions differ as to the extent of that influence.[7] Like other philosophers in the Platonic and neo-Platonic tradition (and as one might expect of a Christian theologian), Augustine believed in the existence of an immortal soul. He also thought that one has indubitable knowledge of one's own existence (Augustine 1913, Book XI, Chap. 26):

> In respect of these truths, I am not at all afraid of the arguments of the Academicians, who say, What if you are deceived? For if I am deceived, I am. For he who is not, cannot be deceived; and if I am deceived, by this same token I am. And since I am if I am deceived, how am I deceived in believing that I am? For it is certain that I am if I am deceived.

However, this anticipation of Descartes' *cogito ergo sum* should not be allowed to obscure the considerable differences between Augustine and Descartes. In the first place Augustine, so far from making the indubitability of one's own existence central to his philosophy, only mentions it halfway through the *City of God* (Augustine 1913) – similarly with the argument he gives in *Against the Academics* (Augustine 1950) and *On the Trinity* (Augustine 2002). There is no suggestion that the only thing of which one can be really certain is the existence of one's own consciousness, and that everything else must be deduced from that. On the contrary, he takes it for granted, as something which does not require demonstration, that in most

---

[7]Descartes himself explicitly denied that he had been influenced (though he welcomed what he considered to be the few superficial and purely accidental resemblances as providing useful ammunition in his arguments with Dutch Calvinists). However, as Wilson (2008) points out, that is not, by itself, conclusive since Descartes was in the habit of downplaying, and even outright denying his intellectual debts.

cases sense perceptions convey genuine and reliable information about the external world (O'Daly 1987, p. 95). Concerning this point Matthews (2000) says:

> It is, I should say, a singularly important fact about Descartes's *Meditations* that reading them can put one in the grip of what has come to be called "the problem of the external world." ...There is no similarly desperate ego-isolation in Augustine.

In the second place Augustine's concept of the soul was completely different from the Cartesian one. For Augustine the soul is the "the phenomenon of life in things" (O'Daly 1987, p. 11). On this conception a bird needs a soul in order to fly, quite as much a person needs one in order to think. Finally, Augustine had a different theory of sensation from Descartes. Unlike Descartes, he thought of sensation as an active process, in which "the soul, as agent of sensation, activates the force of sentience through a fine corporeal medium" (O'Daly 1987, p. 82). Thus in vision he thought that rays burst out of the eye and range abroad, "so that seeing becomes a kind of visual touching, just as hearing is, so to speak, aural touching" (O'Daly 1987, p. 82).

In the Cartesian picture the world is conceived as a sort of spectacle, and the observer as a member of the audience, whose role is purely passive. In Augustine's conception, by contrast, it is as if the audience climbs onto the stage and walks around among the actors, touching and feeling them. Given those assumptions, one would not expect him to think, in Cartesian terms, of consciousness as an internal movie show. Unfortunately the obscurities of the texts make it difficult to be sure that he does not. Matthews (2001) argues:

> Although commentators have sometimes suggested otherwise, Augustine's theory of sense perception is not representational, if one understands by "a representational theory of sense perception" one according to which an image or sense-datum is the direct object of perception.

Kenny (2005, p. 215) thinks that judgment is "most likely" correct. Spade (1994), on the other hand, takes a different view. However, it seems to me that the very fact that there is this scope for disagreement is an indication that Augustine cannot really have been thinking in Cartesian terms.[8] If someone has *genuinely* caught the Cartesian bug they tend to make it very obvious.

---

[8]There is some disagreement in the literature, whether Descartes did in fact think that an image or sense-datum is the direct object of perception. See, for example, Yolton (1984), Hatfield (1990), Wilson (1994), and references cited therein.

3.2 Aquinas

It was no different in the medieval period. As one would expect, medieval philosophers had the everyday concept of consciousness. Moreover, Augustine was one of the most widely read philosophers during the medieval period. As a consequence, "it was a commonplace in medieval philosophy that no one can be in doubt about the existence of one's own soul" (Yrjönsuuri 2011, p. 253).

Philosophers were also familiar with Avicenna's argument that it is possible to imagine oneself as a disembodied soul, without sensory experiences. However, they did not have any of the other notions which go to make up the Cartesian concept of consciousness (Rorty 1979, Kenny 2005, Yrjönsuuri 2011, Marenbon 1988, Marenbon 1987, Kenny 1993). The medieval philosopher who is most relevant to the present discussion is Aquinas, since he was the most prominent scholastic philosopher, and consequently the figure most responsible for determining the view which Descartes opposed. Unlike Augustine, who belonged to the Platonic tradition, Aquinas belonged to the Aristotelian one.

Nevertheless they had certain things in common. In the first place Aquinas, like Augustine, considered the soul to be "whatever makes the difference between animate and inanimate objects" (Kenny 1993, p. 129). As Aquinas saw it, a tree, or a beetle, has a soul, just as a person does. Moreover the soul is implicated in *every* manifestation of life: in the act of digesting one's food, or the act of conceiving and bearing a child, no less than in the act of thinking. In the second place Aquinas, like Augustine and like just about every other medieval philosopher, was primarily interested in those aspects of the soul which make people special. It is these which go to make up the medieval concept of mind.

The soul of a beetle is capable of sensation, so sensation was not considered to be something mental. On the other hand neither a beetle nor any other non-human living organism can have abstract thoughts or take rational decisions (or so medieval philosophers assumed). Consequently mind, as medieval philosophers conceived it to be, essentially consists of only two faculties of the soul: intellect and will (see, for example, Kenny 1993, p. 16). The medieval concept of soul was thus much broader than the Cartesian one, while the medieval concept of mind was much narrower (Descartes, by contrast, identified the concepts of mind and soul). From the fact that this was the way in which medieval philosophers parcelled up the phenomena, I think it can already be seen that they were rather unlikely to arrive at anything like the Cartesian concept of consciousness.

For our purposes there are two important differences between Aquinas and Augustine. The first is that Aquinas, following Aristotle, considered that the soul is the form of the body (Augustine, by contrast, was strongly influenced by neo-Platonism). This might be thought a surprising view for someone who, as recently as the last century, could fairly be described as the official philosopher of the Catholic Church (Kenny 2005). How, one might ask, is it to be reconciled with a belief in the immortality of the soul? The answer is, only with difficulty (see Kenny 1993 for a critical discussion). Nevertheless, although Aquinas thought that the soul, like the smile of the Cheshire cat, could survive the death of its body, he also thought that what survives is not the person whose soul it was, and, furthermore, not fully human. As he put it (according to Kenny 1993, p. 138):

> ...but the soul, since it is part of the body of a human being, is not a whole human being, and my soul is not I; so even if a soul gains salvation in another life, that is not I or any human being.

It was therefore essential, as Aquinas saw it, that the soul should be re-united with the body on the day of judgment. It might, perhaps, be said that the fact that Aquinas thought that the soul is detachable from the body makes him in some sense a dualist (though I doubt he would have agreed). However, his dualism (if "dualism" is the right word) is less extreme than that of Descartes (who would not have said that what survives the death of my body is "not I"). It could be said that Aquinas' conception of human nature is *earthier* than the Cartesian one. The second important difference is that Aquinas, unlike Augustine, thought of sensation as a passive process. However, his conception is no closer than Augustine's to the Cartesian one. As Kenny (1993, p. 135) put it:

> In Aquinas' theory there are no intermediaries like sense-data which come between perceiver and perceived. In sensation the sense-faculty does not come into contact with a likeness of the sense-object. Instead, it becomes itself like the sense-object, by taking on the sense-objects form.

My aim in giving this brief historical review was to stress the originality of Descartes' conception of consciousness. If, in over 2000 years of previous philosophical thinking, no one had come up with anything like it, then it follows that, whatever else, the idea cannot be regarded as obvious. The question now arises: what led Descartes to make such a radical break with the philosophical past? It is often suggested that religion, and a consequent belief in the immortality of the soul, is a motive for a dualistic conception of human nature. That may be so, in many cases. However, I do not think it can account for Descartes adopting a much more radical version of dualism than his medieval predecessors. Aquinas, like every other major medieval Latin philosopher, was first and foremost a theologian, whereas Descartes' interests where strongly secular, centered on mathematics, physics and physiology. If religion was the explanation then, of the two, one would expect it to have been Aquinas who had the more ethereal conception of mind. Yet in fact it was just the other way around.

3.3 Galilei

Although it is impossible to establish this point conclusively, there are reasons for believing that Descartes' real motivation came from Galilean physics. Galilei was strongly committed to the Pythagorean[9] idea that the world is fundamentally mathematical in character (Burtt 2003). As he put it in a famous passage from *The Assayer* (Galilei 2008, p. 183):

> Philosophy is written in this all-encompassing book that is constantly open before our eyes, that is the universe; but it cannot be understood unless one first learns to understand the language and knows the characters in which it is written. It is written in mathematical language, and its characters are triangles, circles, and other geometrical figures; without these it is humanly impossible to understand a word of it, and one wanders around pointlessly in a dark labyrinth.

Of course, the universe does not, at first sight, appear to be a book to be written in the language of mathematics. Galilei consequently needed to account for all the seemingly non-mathematical, qualitative features of the world, such as colors, sounds and smells, which do not easily fit in with his mathematizing program. For that purpose he adopted a doctrine of the ancient atomists (Furley 1987), and denied that they are features of objective reality at all, asserting instead that they are somehow produced in the "sensitive body" (Galilei 2008, p. 185):

> Accordingly, I say that as soon as I conceive of a corporeal substance or material, I feel indeed drawn by the necessity of also conceiving that it is bounded and has this or that shape; that it is large or small in relation to other things; that it is in this or that location and exists at this or that time; that it moves or stands still; that it touches or does not touch another body; and that it is one, a few, or many. Nor can I, by any

---

[9]I am using the term "Pythagorean" broadly, and rather loosely, to refer to any belief that the world is in some sense fundamentally mathematical in character. In this sense of the word Einstein (in his later years), Dirac and numerous other theoretical physicists may be said to have had a broadly Pythagorean outlook.

stretch of the imagination, separate it from these conditions. However, my mind does not feel forced to regard it as necessarily accompanied by such conditions as the following: that it is white or red, bitter or sweet, noisy or quiet, and pleasantly or unpleasantly smelling; on the contrary, if we did not have the assistance of our senses, perhaps the intellect and the imagination by themselves would never conceive of them. Thus, from the point of view of the subject in which they seem to inhere, these tastes, odors, colors, etc., are nothing but empty names; rather they inhere only in the sensitive body, such that if one removes the animal, then all these qualities are taken away and annihilated.

I would argue that this passage marks the actual origin of the Cartesian concept of consciousness. It is true that Galilei himself did not go into details, as to the nature of the "sensitive body". But I think that once this step had been taken the subsequent development, though not inevitable,[10] became very natural.

It is worth noting that Galilei did not attempt to justify the distinction between primary qualities,[11] supposed to be objectively real, and secondary qualities, supposed to be in some sense illusory. Descartes did try to justify it, but his justification is not, to my mind, very convincing. I believe that Burtt (2003, p. 311) gets it about right when he says that in its first inception[12] the doctrine of primary and secondary qualities was "buttressed by nothing more than a mathematical apriorism".

Subsequently, of course, the primary-secondary distinction played an important role in science since it allowed physicists to dismiss all the ostensibly qualitative features of the world as a problem for philosophers, and to concentrate on the quantitative, mathematical description of nature. Consciousness, in other words, has been useful to physicists because it has served as a garbage can for all the many things they did not want to have to think about. However, it is time to ask whether it might have outlived its usefulness. If the quantum revolution had never happened, one might still be able to make a case for the primary-secondary distinction. But as it is the quantum revolution did happen, and since then the search for primary qualities consistent with quantum mechanics has been a source of endless difficulties. That being so it is worth asking whether we have any good reason for retaining the notion.

---

[10]Its lack of inevitability can be seen from, for example, the fact (Furley 1987) that the ancient atomists did not develop a concept of consciousness similar to the Cartesian one.

[11]The terms "primary" and "secondary" qualities are actually due to Locke (1975).

[12]At a later date one could appeal to the empirical successes of the classical theories apparently based on the doctrine, but not at the time of its first inception.

In a letter to Mersenne, Descartes (1991, p. 124), after saying that Galilei

> philosophizes much more ably than is usual, in that, so far as he can, he abandons the errors of the Schools and tries to use mathematical methods in the investigation of physical questions

goes on to complain:

> But he continually digresses, and he does not take time to explain matters fully. This, in my view, is a mistake: it shows that he has not investigated matters in an orderly way, and has merely sought the explanations for some particular effects, without going into the primary causes in nature.

It is interesting to observe that Drake (1964) says something a little reminiscent of this, concerning Galilei's failure to give an explicit statement of the law of inertia:

> A modern physicist reading Galilei's writings would share the puzzlement – I might say the frustration – experienced by Ernst Mach a century ago, when he searched those works in vain for the general statement that (he felt) ought to be there. It would become evident to you, as it was to Newton and Mach, that Galilei was in possession of the law of inertia, but you would not then be able to satisfy those historians who demand a clear and complete statement, preferably in print, as a condition of priority.

Drake notes that, as a result, the first statement of the law of inertia "in the form and generality which we accept today" was given by Descartes.[13] I imagine that Descartes would have been equally critical of Galilei's failure to go into details, regarding events inside the "sensitive body." I would suggest that one of his aims in his early works, *The World* (Descartes 1985) and *Treatise on Man* (Descartes 1985), was to rectify that deficiency.

4   The Galilean Core of Descartes' Philosophy

In the mature form of his philosophy, as represented by *Meditations on First Philosophy* (Descartes 1984) and *Principles of Philosophy* (Descartes 1985), Descartes set out to arrive at demonstratively certain knowledge, ultimately resting on the famous proposition *cogito ergo sum*. However, an examination

---

[13]The history of the law of inertia is complicated. For a more recent discussion, and a rather different assessment of Galilei's role in its discovery, see, for example, Hooper (1998).

of the historical record suggests that this may badly obscure the route by which he was originally led to it. In his early works *The World* and *Treatise on Man* there is no mention of the *cogito* argument.[14] Instead, these works are entirely devoted to a mechanistic description of the world, conceived along the lines Galilei had previously suggested, and of our relation to it. Moreover, the treatment is not deductive (as in his subsequent writings) but avowedly hypothetical: he is at pains to stress that he is not saying how the world definitely *is*, but only how it conceivably *might be*.

The World and *Treatise on Man* form part of a larger project, which occupied him during the years 1630 to 1633 (Gaukroger 1995). The other parts were either never written or have been lost; there is also the possibility that parts were included in subsequent publications. At all events the works as we have them now, published posthumously, are incomplete. The reason is that at the end of 1633 Descartes learned of Galilei's condemnation by the inquisition and, not wanting to publish something of which the Church disapproved, he chose to "suppress it rather than to publish it in a mutilated form" (letter to Mersenne of November 1633; Descartes 1991, pp. 40f).

In *The World* Descartes begins by making the same distinction between primary and secondary qualities that Galilei does in *The Assayer*. The fact that he uses one of Galilei's own examples (the tickling sensation produced by a feather) suggests that he was well aware of what Galilei had previously written on the subject. He then goes on to give a mechanistic account of the world framed entirely in terms of the Galilean primary qualities of shape, size, position, motion and time.

In *Treatise on Man* Descartes turns to a description of the human body, particularly the brain, conceived as a mechanism. He ends with a promise to give a description of the "rational soul". Unfortunately this description is one of the parts of the manuscript which was either never written or has been lost. However, since everything he says about the brain is conformable with later accounts (including the status of the pineal gland), we may assume that he intended to give an account of the soul which was similarly conformable. Specifically, we may assume that he intended to describe the soul as a separate, immaterial entity interacting with the brain *via* the pineal gland.

---

[14]There are indications that he had already conceived some of the essentials of the argument he presented in his later works at the time he wrote *The World* and *Treatise on Man*. See the letter to Mersenne dated 27th February 1637 (Descartes 1991, p. 53), and the autobiographical passage in *Discourse on the Method* (Descartes 1985, p. 126). If that were the case it would not necessarily conflict with my speculation that the argument in these early works gives a better idea of his original motivations.

It is fair to say that what Descartes does in these early works is to flesh out, in much greater detail, Galilei's proposal in *The Assayer*. However, Descartes also introduces a significant novelty: in place of Galilei's "sensitive body" Descartes locates the secondary qualities in an immaterial soul. It is impossible to prove, of course, but one may plausibly speculate that it was this – the need to find a home for the secondary qualities – which was the original motivation for the Cartesian soul.

I do not say it was inevitable that Descartes would be led to dualism. Indeed, his contemporary Thomas Hobbes, in the *Third Set of Objections* (published jointly with the *Meditations*), argued for a completely materialistic conception of human nature (Descartes 1984). However, it does seem to me that, given his opinions about primary and secondary qualities, and given the high value he placed on mathematics, it was very natural for Descartes to take such a view. It would offend his Pythagorean sensibilities[15] – his mathematician's sense of system, and harmony – to suppose that, located here and there in the otherwise colorless expanse of mathematical mechanism, there are little brightly painted islands. It would be equally inconsistent to suppose that, dotted around in the mechanism, there are little islands somehow endowed with subjective color experiences. Since he could not locate color perceptions inside the physical universe, what else could he do but locate them outside?

In his later works, beginning with the *Discourse on the Method*, Descartes (1985) presents his ideas in a very different way. In particular, the *cogito* argument, which is not mentioned at all in *The World* and *Treatise on Man*, now becomes central. This argument is another of Descartes' strikingly original departures from previous philosophical thinking. As I mentioned earlier, it was a medieval commonplace, due originally to Augustine, that one cannot doubt the existence of one's own soul (Yrjönsuuri 2011). Moreover, there was a widespread interest in sceptical arguments during the early modern period (Popkin 2003). However, there was no precedent for the way in which Descartes put these ingredients together.

The *cogito* argument begins with what is sometimes called an act of hyperbolic doubt. It is worth asking what motivated this step. As Wittgenstein (1969) has stressed one needs reasons to doubt. One also needs a suitable context. At least, one does if one wants people to listen. Suppose someone expressed doubt, as to whether their head contained sawdust instead of brains

---

[15]Hobbes was not a mathematician, and is unlikely to have shared Descartes' Pythagorean feelings. Perhaps that is the reason he could accept the move to full materialism. Perhaps it is also the reason the ancient atomists (who did not have a Pythagorean vision of the world either) were not led to the Cartesian concept of consciousness.

(Wittgenstein 1969, p. 36e). This would be a much more modest doubt than the global, all-encompassing act of scepticism with which Descartes begins the *cogito* argument. Yet no one would take it seriously, while people *have* taken the Cartesian doubt very seriously indeed. The problem of the external world, and the various philosophical movements to which it has given rise (empiricism, subjective idealism, Kantianism, objective idealism, positivism, pragmatism, phenomenology, etc.) has been the dominant theme in Western philosophy for the last 350 years.

Why is that? I think the answer is that, although in the context of everyday life it would be crazy to doubt the existence of external reality, in the context of the views expressed in *The Assayer*, *The World* and *Treatise on Man* the doubt becomes very reasonable. If one has become convinced that, in sober truth, our senses are radically misleading us as to the existence of colors, sounds, tastes *etc*, then it is surely very natural to wonder if they might also be misleading us as to the existence of shapes, sizes, positions, etc. And if one has got as far as wondering if the senses are to be trusted *at all*, then how does one avoid doubting the existence of external reality? Moreover, I would suggest that that reason for doubting was operative, not only in the mind of Descartes, but also in the minds of his philosophical successors. It was operative precisely because it was widely believed that science had shown that our senses are radically misleading us. Scientists who are scornful of philosophical worries about the existence of the external world miss the point: it was science itself (or what people thought of as science) which originally suggested the worries (Burt 2003).

In short, I would suggest that all the distinctive features of the Cartesian philosophy are consequences[16] of Galilei's original Pythagorean hypothesis that the world is fundamentally mathematical in character, and of the related distinction between primary and secondary qualities. In particular, this whole way of thinking is rooted in the Galilean-Cartesian concept of matter. Cartesian consciousness is a secondary concept, parasitic on that.

5 Toward a New Philosophy of Nature

There is an irony in this story. In the 17th century there was no possibility of finding solid empirical support for the micro-mechanical explanations of such phenomena as color, or heat, on which the Galilean-Cartesian philosophy was based. These explanations remained highly speculative until the 19th

---

[16]I do not mean consequences in a rigorous, deductive logical sense, but in a looser, psychological sense. This is close to Burtt's (2003) conclusion.

century when hard evidence started to accumulate. Even then progress was slow, as can be seen from the fact that in the late 19th century controversy about atomism the two sides were equally matched (Chalmers 2009, Kuhn 1987, Krips 1986, Psillos 2011). A nice illustration of this is the fact that in the 1890s Planck, who was subsequently to inaugurate an atomistic view of electromagnetic radiation, was sceptical about atoms to the extent that Boltzmann could attribute to him the opinion that work on kinetic theory was a "waste of time and effort" (Kuhn 1987, pp. 22f; also see Krips 1986).

It was only in the 20th century that the validity of micro-mechanical explanations of the behavior of matter was established to the satisfaction of every competent physicist. The irony is that the same advances which finally vindicated micro-mechanical explanations also cast serious doubt on Galilean-Cartesian assumptions about what such explanations ought to be like. Indeed, one of the key papers leading to the general acceptance of atomism, Einstein's (1905a) paper on Brownian motion, was published in the same year, by the same person, as one of the key papers casting doubt on Galilean-Cartesian assumptions, Einstein's (1905b) paper on the photo-electric effect.

Quantum mechanics challenges the whole Galilean-Cartesian framework. It is a challenge which has yet to call forth an adequate response. The Copenhagen interpretation provides a way of thinking about quantum experiments which is sufficient for the practical needs of working physicists. But, as its critics point out, it hardly amounts to a coherent philosophy of nature. Yet, instead of taking the hint from experiment, and trying to move forward, the response of those critics has mostly been to fall back on old 17th century modes of thought, and to try to find ways of interpreting quantum phenomena which would be consistent with Cartesian assumptions.

Over half a century ago Pauli described such attempts as "regressive" (see, for instance, the letter to Fierz quoted by Gieser (2005, p. 266)). It seems to me that everything which has happened since tends to confirm that judgment. What we need to do is to dig up the Galilean-Cartesian foundations and replace them with a different conceptual structure, better adjusted to all we have learned since the year 1900.

Cartesian philosophy is built on two key principles: (1) the Pythagorean hypothesis that there is one true, complete description of the world, expressible in mathematical language, and (2) the distinction between primary and secondary qualities. I believe we ought to abandon both those principles.

The idea, naturally suggested by quantum mechanics, that we should dispense with the Pythagorean hypothesis, produces in many people a sense of vertigo. They fear that letting go of this is tantamount to letting go of the concept of physical reality. But that merely shows that they are so fixated on the Galilean-Cartesian way of thinking about physical reality that they are unable to envisage an alternative.

A description is something human. The ability to give descriptions evolved (presumably) in the palaeolithic, for the purpose of communicating such facts as the location of the nearest source of flint-nodules. We have come a long way since then, cognitively speaking. Nevertheless, our modern mathematical descriptions of nature are all expressible in the language of axiomatic set theory, which is a formalization of the naive set theoretic ideas that palaeolithic hunter-gatherers (presumably) used when sorting their stone tools, negotiating their intricate family relationships, etc. Moreover, our mathematical descriptions comprise sequences of propositions, just like the verbal communications of palaeolithic hunter-gatherers. In short, our mathematical descriptions bear a clear human imprint. *Conceivably* the universe splits logically into a collection of sentence-sized morsels, each perfectly adapted to human cognitive capacities.[17] But I see no *a priori* reason for assuming that to be the case.

Our attitude to this question should be empirical. If Einstein had achieved the same stunning success, with his attempt to explain quantum mechanics in terms of classical field theory, that he did with general relativity, then there would be reason to take the Pythagorean hypothesis seriously. But since he did not, and since no one else has either, there are grounds for scepticism. This is not to say that I question the validity of the *partial* descriptions we are able to give. Nor is it to say that I am an anti-realist. It is not even (necessarily) to deny that God is a mathematician. It is only to say that God is, perhaps, a little more subtle and (dare I say?) interesting than Galilei gave him credit for being.

Turning to the primary-secondary distinction, it is obvious that color perceptions are in some sense subjective. The question is, however, whether they are any more subjective than, for example, the statement that the electric field intensity at position $\mathbf{r}$ is $3\mathbf{i} - 4\mathbf{j} + 7\mathbf{k}$ Vm^{-1} – where by "statement" I mean the actual ink marks, or the brain states which occur as one reads them. It is true that a color-blind person will fail to discriminate two colors which a normally sighted person sees to be different: from which it would seem to follow that the color-blind has a different visual experience from the normally sighted person. But then it is equally true that a person who

---

[17]There is *some* overlap here with the discussion in Chapter 1 of Rorty (1989). However, the fact that I agree with Rorty that the universe is not a book should not be taken to imply that I agree with everything else he says in this chapter.

measures the electric field intensity to an accuracy of ± 1 Vm^{-1} will have a different cognitive experience from a person who uses a different instrument to measure it to an accuracy of ± 0.1 Vm^{-1}.

Color perceptions, being perceptions, are subjective by definition. But then, so are quantitative thoughts. Idealists aside, few people are tempted to suppose that, because the belief that carbon has proton number 6 is *only* a belief, therefore carbon does not *really* have proton number 6. No more should one be tempted to suppose that, because the perception of green is *only* a perception, therefore grass is not *really* green.

The function of eyes is to acquire information. Looking at an object is not the same as listening to a verbal description of that object. But what one acquires by looking is still information, and to that extent it may be regarded as a kind of statement.[18] Cartesian-minded classical physicists, like Einstein, supposed that the world is completely describable, in terms of fields (or whatever). Allowing that to be the case, for the sake of the argument, it would not follow that the statements of one's visual system are any more subjective than statements made in the approved mathematical language. What the classical physicist's description says in one way, using the language of fields, the visual system says in another way, using the language of colors.

To be sure, visual statements say less – contain less information – than the classical physics description (supposing that it is valid). But that does not make them subjective. If one takes some data given to 10 significant figures, and rounds everything off to 3 significant figures, one loses a lot of information. But the information which remains is no less objective than it was before. Worrying about the difference between the mathematical description and the description in terms of colors is like worrying about the difference between a description in English and the same description written out in French. Color qualities are no more in the head – and no less in the head – than the electromagnetic field is in the head.

Discussions of *qualia* are often vitiated by the idea that there are two pictures involved: one that is colored (the picture we get from our eyes) and one that is not (the picture we get from physics). This idea goes back to Descartes, of course, with his talk of colors not "resembling" anything in the object. It is based on a confusion, since neither of these pictures exists. There is no picture in the head, as we have seen. Moreover the mathematical descriptions which physics gives us are not pictures either[19] – any more than

---

[18]Descartes makes an analogy between words and colors at the beginning of *The World*. However, he fails to draw what I believe to be the correct conclusion.

[19]It is impossible to imagine the number 3 in the abstract. Similarly, it is impossible to

a verbal description is a picture. Thinking that colors do not exist in reality because there are no colors in the mathematical description is like thinking that a city is colorless because the verbal description in the guidebook is printed in black and white.

Back in the Palaeolithic, when language first developed, abstract, symbolic descriptions conveyed much less information than the descriptions we get from our eyes. It was therefore natural to take the visual description to be the standard, or canonical description, against which verbal descriptions were to be judged. Effectively, reality was identified with the visual description (supplemented with information obtained from the other senses). However, with the development of mathematical physics in the 17th century we found an abstract, symbolic mode of description which, unlike ordinary language, was actually superior to the visual description in terms of informational capacity. It therefore became natural to take the new mathematical description to be the canonical description: in effect, to identify reality with the mathematical description.

It seems to me that the lesson of quantum mechanics is that we should drop the whole idea of there being a canonical description. Galilei's book metaphor is profoundly misleading. There is no mathematical description in the sky. The only descriptions around are the ones we humanly construct and which, being human, are necessarily partial.

6 Conclusion

To say that there is no canonical description with which reality can be identified is not to deny the existence of reality. Supposing there to be a canonical description, we have never known it. Such knowledge of reality as we possess right now is entirely expressed in terms of our ordinary, humanly constructed descriptions. It is not scepticism to suggest that knowledge so expressed is all we ever will possess.

In this paper I have essentially confined myself to a criticism of Cartesian philosophy. To construct an adequate non-Cartesian philosophy would take an enormous amount of work. However, I believe there is reason to think that if we were to undertake that project it might lead to a conceptual revolution equal in magnitude to the 17th century Cartesian one. In particular, it might lead to conceptions of the world, and of human nature, which differ as much from the Cartesian conceptions as the latter did from medieval conceptions. So much so that we would, perhaps, no longer want to use

imagine quantities like vectors, the electric field vector for example.

the words "consciousness" and "matter" (except in their everyday senses, of course).

Acknowledgments

The author is grateful to the Stellenbosch Institute for Advanced Study for their hospitality while carrying out some of the research for this paper. Research at Perimeter Institute is supported by the Government of Canada through Industry Canada and by the Province of Ontario through the Ministry of Research & Innovation.

References

Atmanspacher H., Römer H., and Walach H. (2002): Weak quantum theory: Complementarity and entanglement in physics and beyond. *Foundations of Physics* **32**, 379–406.

Atmanspacher H. (2004): Quantum approaches to consciousness. In *Stanford Encyclopedia of Philosophy*, ed. by E.N. Zalta, accessible at `plato.stanford.edu/entries/qt-consciousness/`.

Atmanspacher H., Filk T., and Römer H. (2004): Quantum Zeno features of bistable perception. *Biological Cybernatics* **90**, 33–40.

Augustine (1913): *The City of God*, transl. by M. Dods, T. & T. Clark, Edinburgh.

Augustine (1950): *Against the Academics*, transl. by J.J. O'Meara, Newman Press, Westminster.

Augustine (2002): *On the Trinity*, ed. by G.B. Matthews, transl. by S. McKenna, Cambridge University Press, Cambridge.

Bell J.S. (1987): *Speakable and Unspeakable in Quantum Mechanics*, Cambridge University Press, Cambridge.

Blackmore S. (2002): There is no stream of consciousness. *Journal of Consciousness Studies* **9**(5), 17–28.

Blackmore S. (2004): *Consciousness: An Introduction*, Oxford University Press, Oxford.

Bruza P., Busemeyer J.R., and Gabora L. (2009): Introduction to the special issue on quantum cognition. *Journal of Mathematical Psychology* **53**(5), 303–305.

Burtt E.A. (2003): *The Metaphysical Foundations of Modern Science*, Dover, New York.

Caston V. (2002): Aristotle on consciousness. *Mind* **111**, 444, 751–815.

Chalmers D.J. (1996): *The Conscious Mind: In Search of a Fundamental Theory*, Oxford University Press, Oxford.

Chalmers A. (2009): *The Scientist's Atom and the Philosopher's Stone*, Springer, Berlin.

Churchland P.M. (1988): *Matter and Consciousness: A Contemporary Introduction*, MIT Press, Cambridge.

Churchland P.S. (1989): *Neurophilosophy: Toward a Unified Science of the Mind-Brain*, MIT Press, Cambridge.

Conte E., Todarello O, Federici A., Vitiello F., Lopane M., Khrennikov A., and Zbilut J.P. (2007): Some remarks on an experiment suggesting quantum-like behavior of cognitive entities and formulation of an abstract quantum mechanical formalism to describe cognitive entity and its dynamics. *Chaos, Solitons and Fractals* **31**, 1076–1088.

Dennett D.C. (1993): *Consciousness Explained*, Penguin, London.

Dennett D.C. (2005): *Sweet Dreams: Philosophical Obstacles to a Science of Consciousness*, MIT Press, Cambridge.

Descartes R. (1985): *The Philosophical Writings of Descartes Vol. 1*, transl. by J. Cottingham, R. Stoothoff and D. Murdoch, Cambridge University Press, Cambridge.

Descartes R. (1984): *The Philosophical Writings of Descartes Vol. 2*, transl. by J. Cottingham, R. Stoothoff and D. Murdoch, Cambridge University Press, Cambridge.

Descartes R. (1991): *The Philosophical Writings of Descartes Vol. 3*, transl. by J. Cottingham, R. Stoothoff, D. Murdoch and A. Kenny, Cambridge University Press, Cambridge.

DeWitt B.S. and Graham N., eds. (1973): *The Many-Worlds Interpretation of Quantum Mechanics*, Princeton University Press, Princeton.

Drake S. (1964): Galilei and the law of inertia. *American Journal of Physics* **32**, 601–608.

Einstein A. (1905a): Über die von der molekularkinetischen Theorie der Wärme geforderte Bewegung von in ruhenden Flüssigkeiten suspendierten Teilchen. *Annalen der Physik* **17**, 549–560.

Einstein A. (1905b): Über einen die Erzeugung und Verwandlung des Lichtes betreffenden heuristischen Gesichtspunkt. *Annalen der Physik* **17**, 132–148.

Furley D. (1987): *The Greek Cosmologists, Volume 1: The Formation of the Atomic Theory and its Earliest Critics*, Cambridge University Press, Cambridge.

Galilei G. (2008): *The Essential Galilei*, ed. and transl. by M.A. Finocchiaro, Hackett, Indianapolis.

Gaukroger S. (1995): *Descartes: An Intellectual Biography*, Oxford University Press, Oxford.

Gieser S. (2005): *The Innermost Kernel: Depth Psychology and Quantum Physics*, Springer, Berlin.

Grimes J. (1996): On the failure to detect changes in scenes across saccades. In *Perception*, ed. by K. Akins, Oxford University Press, Oxford, pp. 89–110.

Hatfield G.C. (1990): *The Natural and the Normative: Theories of Spatial Perception from Kant to Helmholtz*, MIT Press, Cambridge.

Hooper W. (1998): Inertial Problems in Galilei's Preinertial Framework. In *The Cambridge Companion to Galilei*, ed. by P. Machamer, Cambridge University Press, pp. 146–174.

Jung C.G. (1960): *On the Nature of the Psyche*, transl. by R.F.C. Hull, Princeton University Press, Princeton.

Kenny A. (1993): *Aquinas on Mind*, Routledge, London.

Kenny A. (2005): *A New History of Western Philosophy, Volume II: Medieval Philosophy*, Oxford University Press, Oxford.

Kihlstrom J.F. (2004): Joseph Jastrow and His Duck — or is it a Rabbit? See `socrates.berkeley.edu/$\sim$kihlstrm/JastrowDuck.htm`.

Kuhn T.S. (1987): *Black-Body Theory and the Quantum Discontinuity 1894–1912*, University of Chicago Press, Chicago.

Krips H. (1986): Atomism, Poincaré and Planck. *Studies in History and Philosophy of Science* **17**(1), 43–63.

Locke J. (1975): *An Essay Concerning Human Understanding*, ed. by P.H. Nidditch, Oxford University Press, Oxford.

Lycan W.G., ed. (1990): *Mind and Cognition: A Reader*, Blackwell, Oxford.

Mach E. (1959): *The Analysis of Sensations and the Relation of the Physical to the Psychical*, ed. by S. Waterlow, transl. by C.M. Williams, Dover, New York.

Mann W.E. (2000): Review of "Descartes and Augustine" by Stephen Menn. *Philosophical Review* **109**(3), 438–441.

Marenbon J. (1987): *Later Medieval Philosophy (1150-1350): An Introduction*, Routledge, London.

Marenbon J. (1988): *Early Medieval Philosophy (480–1150): An Introduction*, Routledge, London.

Matson W. I. (1966): Why isn't the mind-body problem ancient? In *Mind, Matter and Method: Essays in Philosophy and Science in Honor of Herbert Feigl*, ed. by P.K. Feyerabend and G. Maxwell, University of Minnesota Press, Minneapolis, pp. 92–102.

Matthews G.B. (1992): *Thought's Ego in Augustine and Descartes*, Cornell University Press, New York.

Matthews G.B. (2000): Review of "Descartes and Augustine" by Stephen Menn. *Philosophy and Phenomenological Research* **61**(3), 721–723.

Matthews G.B. (2001): Knowledge and illumination. In *The Cambridge Companion to Augustine*, ed. by E. Stump and N. Kretzmann, Cambridge University Press, Cambridge, pp. 205–233.

Meier C.A., ed. (2001): *Atom and Archetype: The Pauli/Jung Letters 1932–1958*, Princeton University Press, Princeton.

Menn S. (1998): *Descartes and Augustine*, Cambridge University Press, Cambridge.

O'Daly G. (1987): *Augustine's Philosophy of Mind*, University of California Press, Berkeley.

Popkin R.H. (2003): *The History of Scepticism from Savonarola to Bayle*, Oxford University Press, Oxford.

Psillos S. (2011): Moving molecules above the scientific horizon: On Perrin's case for realism. *Journal for General Philosophy and Science* **42**(2), 339–363.

Rayner K. (1978): Eye movements in reading and information processing. *Psychological Bulletin* **85**(3), 618–660.

Rayner K. (1998): Eye movements in reading and information processing: 20 years of research. *Psychological Bulletin* **124**, 372–422.

Rorty R. (1979): *Philosophy and the Mirror of Nature*, Princeton University Press, Princeton.

Rorty R. (1989): *Consistency, Irony, and Solidarity*, Cambridge University Press, Cambridge.

Saatkamp H.G., ed. (1995): *Rorty and Pragmatism: The Philosopher Responds to His Critics*, Vanderbilt University Press, Nashville.

Schlosshauer M., ed. (2011): *Elegance and Enigma: The Quantum Interviews*, Springer, Berlin.

Searle J.R. (1992): *The Rediscovery of the Mind*, MIT Press, Cambridge.

Simons D.J. (2000): Current approaches to change blindness. *Visual Cognition* **7**, 1–15.

Spade P.V. (1994): Medieval Philosophy. In *The Oxford Illustrated History of Western Philosophy*, ed. by A. Kenny, Oxford University Press, Oxford, pp. 55–106.

Stich S.P. (1983): *From Folk Psychology to Cognitive Science: The Case Against Belief*, MIT Press, Cambridge.

Stich S.P. (1996): *Deconstructing the Mind*, Oxford University Press, Oxford.

Stove D.C. (1991): *The Plato Cult and other Philosophical Follies*, Blackwell, Oxford.

Teasdale G. and Jennett B. (1974): Assessment of coma and impaired consciousness: A practical scale. *Lancet* **2**, 81–84.

Teasdale G. and Jennett B. (1976): Assessment and prognosis of coma after head injury. *Acta Neurochirurgica* **34**, 45–55.

Wilson C. (2008): Descartes and Augustine. In *A Companion to Descartes*, ed. by J. Broughton and J. Carriero, Blackwell, Oxford, pp. 33–51.

Wilson M.D. (1994): Descartes on sense and resemblance. In *Reason, Will and Sensation: Studies in Descartes' Metaphysics*, ed. by J. Cottingham, Oxford University Press, Oxford, pp. 209–228.

Wittgenstein L. (1968): *Philosophical Investigations*, Blackwell, Oxford.

Wittgenstein L. (1969): *On Certainty*, ed. by G.E.M. Anscombe and G.H. von Wright, Blackwell, Oxford.

Yolton J.W. (1984): *Perceptual Acquaintance from Descartes to Reid*, University of Minnesota Press, Minneapolis.

Yrjönsuuri M. (2011): Consciousness. In *Encyclopedia of Medieval Philosophy*, ed. by H. Lagerlund, Springer, Berlin, pp. 227–229.

The Influence of German Romantic Science on Jung and Pauli

Joe Cambray

Abstract

The past two decades have witnessed a renewed interest in the scientific tradition associated with German Romanticism. In this paper, some select influences from this tradition on the philosophical positions of C.G. Jung and Wolfgang Pauli will be presented, much of which was not overtly acknowledged by either author. Extending this inquiry, the importance of the philosophical ideas of Baruch Spinoza, especially his views on the mind-body relationship (dual-aspect monism), will be reassessed both for his impact on German Romanticism and Idealism, and on Pauli and Jung's synchronicity hypothesis, including the notion of the psychoid archetype. Possible reasons for the omission of Spinoza's views from the Pauli-Jung hypothesis will be explored.

1 Introduction

Both Carl Gustav Jung (1875–1961) and Wolfgang Pauli (1900–1958) received their primary and higher educations in German speaking cultures. In this they were exposed to the scientific thinking and traditions of Germanic culture even as it was evolving. Jung's focus was in medicine, specializing in psychiatry, Pauli's in theoretical physics. Though much had changed in science during the 25 years separating their educations, they did share linguistic and cultural histories, which included knowledge of the German Romantic and Classical period. This is an era that has been lost to many modern educated individuals, perhaps due to the stain cast by the shadow of National Socialism's misuse of the figures and ideas from this earlier period in Germanic history.

Over the past several decades some scholars of the history of science have been turning increased attention to alternative traditions of scientific research outside the standard Western canon. In this process there has been a growing interest in the scientific work done by natural philosophers from the German Romantic and Classical period.[1] Links have been made between the research projects of this era and emerging areas in contemporary science such as epigenetics (see e.g., Amundson 2005, Laubichler 2007, Gissis and Jablonka 2011).

In this paper I will explore select strands of German Romanticism that appear to have influenced Jung (Cambray 2011a,b, 2013, 2014a,b) and include parallel influences on Pauli. As I have written elsewhere, the impact of these precursors on Jung were often not overtly acknowledged by him. At times they were implicitly present, at others the omissions may have been related to emotional conflicts and anxiety of influence. Lack of adequate references to the thought of the philosopher Baruch Spinoza (1632–1677) will also be discussed, especially in light of his impact on the German Romantics.

2   German Romantic Influences on Jung

While Jung's debt to the philosophical and artistic traditions associated with the German Romantic and Classical periods has been well documented (see, e.g., Ellenberger 1970, pp. 199–228, Woodman 2005, Bishop 2008, 2009), his borrowings from the scientific work of this era has remained largely unstudied. In a series of papers stemming from Jung's recently published *Red Book* (Jung 2009) I have discussed these influences (Cambray 2013). The incorporation of images from Ernst Haeckel's scientific artwork into the mandala paintings in the *Red Book* served as my point of entry (Cambray 2011b). Haeckel (1834–1919) was in the second generation of Romantic scientists in Germany and one of the most renowned biologists of his day. His evolutionary theories drew upon Goethe, Darwin and Lamarck; he was a friend and correspondent of Darwin. Haeckel's biogenetic law that "ontogeny recapitulates phylogeny" was embraced by both Freud and Jung in their theorizing.[2]

---

[1]Compare the extensive literature by, e.g., Cunningham and Jardine (1990), Poggi and Bossi (1994), Müller-Siever (1997), Chaouli (2002), Richards (2002), Holmes (2008), Holland (2009, 2010). In addition a number of high quality biographies of key figures in these movements have also appeared (e.g., Richards 2008, Rupke 2008).

[2]While the biogenetic law in the form Haeckel presented it (recapitulation of adult forms during embryological development) has been disproved, modified versions (recapitulation of embryological forms) does have some validity. Furthermore, the burgeoning field of epigenetics which "may be defined as the study of any potentially stable and, ideally, heritable change in gene expression or cellular phenotype that occurs without changes in Watson-Crick base-pairing of DNA" (Goldberg *et al.* 2007), treats Haeckel as an important forerunner (Laubichler 2007, Churchill 2007) and sheds light on multigenerational transmission of trauma (Cambray 2014a,b).

Tracing these roots further back to the first generation of the German Romantic scientists I next turned to Alexander von Humboldt rather than exploring the complex mixture of influences from Goethe, whose orientation is classical rather than romantic and whom Jung readily acknowledges – but primarily for his literary works, in particular his *Faust*, rather than his scientific studies. Humboldt befriended Jung's grandfather when he was a young man in exile and helped him move to Basel to become the director of the medical school there; this was in fact how the Jung family became Swiss, a story the Jung family proudly acknowledges (Jung 2011).

A scientific traveler and discoverer, Humboldt's published personal narratives of his journeys through Latin America were major sources of influence on Charles Darwin when years later he took his epic voyage on the Beagle. They even bear important resemblances to Jung's parallel journey into his psychological interior as documented in the *Red Book*. Though Jung only cited Humboldt's final publication *Cosmos* in his writings, the personal influence on his theories and methods is unmistakable (Cambray 2014a).

Several other scientists from this tradition whose work contains ideas that Jung incorporated without reference are Hans Christian Ørsted (1777–1851) and Johann Wilhelm Ritter (1776–1810). Ritter founded the discipline of electrochemistry and demonstrated the equivalence of galvanic (bio-electric) and voltaic (inorganic) phenomena (Wetzel 1990, pp. 201f). His work was integral to a change in cosmological perspectives from the Newtonian clockwork universe to a more organic vision which brought back the notion of a "world soul" into scientific discourse (an excellent bilingual edition of Ritter's work is now available by Holland (2010)). Jung would later speak of the *anima mundi* of the alchemists though without noting Ritter's original work.[3]

Ørsted's serendipitous discovery of the link between electricity and magnetism served as the basis for the later experiments of Michael Faraday. He first formulated a field theory of electromagnetism in preparation of James Clerk Maxwell's theoretical formulation of the laws governing these fields. Many features of Jung's model of therapeutic action ultimately derive from the field models of the 19th century, imported into psychology by William James. While Jung readily acknowledges James' influence on him, he does not overtly trace his field model to Ørsted, Faraday, or Maxwell (Cambray 2011c). In his last book *The Soul in Nature*, Ørsted, through Humboldt's in-

---

[3]Ritter also discovered the ultraviolet (uv) end of the visible light spectrum based on Romantic science notions of opposites by exploiting Herschel's discovery of the infrared (ir) end of the spectrum. Jung employed the ir-uv polarity within the light spectrum as a metaphor for archetypes having somatic (ir) and spiritual or imagistic (uv) poles.

fluence, also echoed Spinoza's dual-aspect monism, noting "soul and nature are one, seen from two different sides" (Ørsted 1852, p. 384). The relevance of this perspective for the psychoid as Jung and Pauli formulated this radical notion will be explored later.

3 German Romantic Influences on Pauli (as Seen from His Dreams)

Pauli's dream series that comprises Part II of Jung's *Psychology and Alchemy* includes a snippet (#17): "The dreamer goes for a long walk, and finds a blue flower on the way." In his reflections on this fragment Jung sees a symbol of the self in the blue flower harkening back to (Jung 1953/1968, par. 101)

> a more romantic and lyrical age ... when the scientific view of the world had not yet broken away from the world of actual experience – or rather when this was only just beginning.

As he continues his analysis Jung sees this "numinous emanation from the unconscious showing the dreamer ... the historical place where he can meet friends and brothers of like mind, where he can find the seed that wants to sprout in him too", ultimately arriving at an identification of this image with one from alchemy: "The sapphire blue flower of the hermaphrodite".

Here Jung's portentous intuition spots a "historical regression" back to the romantic period and beyond to the world of alchemy as places where the dreamer (Pauli) can discover precursors for his own strivings that are as yet dimly seen. However, explicit reference to Novalis is surprisingly missing. As the iconic image for German Romanticism, the blue flower is traceable to Novalis (1772–1801), from his novel *Heinrich von Ofterdingen* (first published in 1802, the year after his death) where it symbolizes the unattainable, barely but hauntingly glimpsed by the hero of the tale, an idealistic young poet. Further, this is not solely a literary conceit as Novalis was well read in the science of his day, especially geology, mining and biology. He sought to combine the experimental with the aesthetic (Cunningham and Jardine 1990, pp. 4-6) in a manner similar to the scientific approach of von Humboldt where precise measurement and aesthetic response were recorded together as the full human description of experimentation. Emotional, imaginal impressions of the soul were considered formative aspects of the world itself in an attempt to reunite the subjective and objective aspects of experience. In this view the blue flower is symbolic of a holistic vision of a romantic science, a vision largely unattainable as a verifiable theory at that point in history.

Later, when he amplified Pauli's famous "world-clock" dream with its vertical, blue disc, Jung (1953/1968, par. 320) conjectures[4]

> that blue, standing for the vertical, means height and depth (the blue sky above, the blue sea below) ... [and] the vertical would correspond to the unconscious. But the unconscious in man has feminine characteristics, and blue is the traditional color of the Virgin's celestial cloak.

According to analyst Remo Roth, Pauli wrote to Markus Fierz[5] about his distress with this view, stating that he would emphasize (Roth 2004, chap. 3.3.9)

> that the blue color, associated with the female, is of *pagan and chthonic origin*. It is the cornflower of the Greek fertility goddess Demeter. For him, exactly this fertility is the positive aspect of "mother Earth" which is constellated in his unconscious ... This is the deepest reason why he has such a strong aversion against the blue coat of the "disinfected" Heavenly Queen.

However, the issue here is not only which feminine aspects of Pauli's psyche are linked to the symbolic associations around the color blue, but the metaphysical worldview of each man. In a letter of February 1953, Pauli writes to Jung about his understanding of the Assumption (Meier 2001, p. 87):

> But as a symbol of the *monistic union of matter and soul*, this assumption has an even deeper meaning for me ... In the empirical world of phenomena there must always be the difference between "physical" and "psychic", and it was the mistake of the alchemist to apply a monist (neutral) language to concrete chemical processes. But now that matter has also become an abstract *invisible reality* for the modern physicist, the prospects for a psychophysical monism have become much more favorable. Inasmuch as I now believe in the possibility of a simultaneous religious and scientific function of the appearance of archetypal symbols, the fact of the declaration of the new dogma was and is for me a *clear sign* that the *psychophysical problem* is also now constellated anew in the scientific sphere.

Pauli envisions this monistic perspective as the fundament for Jung's concept of synchronicity. In addition, Pauli is hereby linking a cultural, religious

---

[4] Jung also discusses this dream in his Terry lectures with a similar understanding of the blue, see Jung (1969, par. 111ff).

[5] Markus Fierz was a physicist colleague of Pauli and brother of Jungian analyst and psychiatrist Heinrich Fierz. See the contribution by von Baeyer in this volume.

event with a parallel (synchronistic?) paradigm shift in science, itself a fascinating, original insight.[6] What comes next in the letter is highly significant; Pauli expresses his distress to Jung about his handling of the Assumption in his recently published *Answer to Job*, especially Pauli's anticipation (Meier 2001, p. 87)

> ... on the subject of matter and on the psychophysical problem when you came to the new dogma. To my disappointment, however, I found that there was no mention of the latter, and matter itself was alluded to only briefly in the expressions "creaturely man" and "incarnation of God", otherwise being basically ignored.

Although Pauli tries to justify Jung's silence by seeing *Answer to Job* in relation to the essay on synchronicity, there is an important question as to the degree of overlap or divergence of their worldviews.

In Jung's ginger reply, he takes up the criticism though only speaking about how he (Meier 2001, p. 98)

> attempted to open up a new path to the "state of spiritualization" (German: "Beseeltheit") of matter by making the assumption that "being is endowed with meaning" (i.e., extension of the archetype in the object).

Pauli accepts Jung's answer but makes clear he has "no use at all for the 'Being' definition you assign to metaphysical judgments" (Meier 2001, p. 102). They continue to explore and debate the issues around the identity of the archetypal world with *physis*, which results in Jung bringing forward the concept of the psychoid "which represents an approach to neutral language in that it suggests the presence of a non-psychic essence" (Meier 2001, p. 111). While this forms a resting place for their metaphysical arguments, I believe there remains something of a philosophical gap between their views. To bring this into greater relief, I will start with the position of Spinoza's philosophy in German Romanticism, together with Pauli and Jung's reactions to him before looking more closely at their formulation of the psychoid.

4 Spinoza's Influence on the German Romantics and Idealists

From the time of his death in 1677 until late into the 18th century Spinoza was a figure of disrepute in German academic and ecclesiastical circles. Read

---

[6]This could be seen as one of the first statements of the idea of cultural synchronicity; see chapter 4 of Cambray (2009) for a discussion of this topic.

as an atheist, freethinker and political radical his works were treated as dangerous to the established order of princes and clergy. However, debate in the mid 1780s between Moses Mendelssohn and Friedrich Heinrich Jacobi about Lessing's embrace of Spinoza brought much attention and renewed interest to Spinoza's ideas. For a valuable discussion of the significance of the debate on the transformation of opinion about Spinoza's value see Beiser (1987, especially chap. 2).

The fact that Spinoza questioned the Bible as the source of revelation, but rather saw it as a historical, cultural document which could be subjected to critical appraisal, fit well with the rising pietist movement in Germany. By removing the Bible from its unique, central position as a document of unquestioned truth, Spinoza's view strongly appealed to that strand of Lutheranism wishing to focus on direct, personal experience of the divine, removing it further from mediation by the clergy. Further, Spinoza's equating God and Nature was to deeply influence many of the German Romantics in their seeking a mystical, holistic experience of the natural world.

In this vein, the well known high esteem in which Goethe held Spinoza had a powerful shaping influence on many of the German Romantics of the time of Goethe. According to Richards, Goethe in his 1785 essay *Study after Spinoza* "endorsed the basic Spinozistic thesis that God and nature were one, so that the world must be both divine and natural simultaneously" (Richards 2002, p. 379). Similarly the holistic features of Spinoza's philosophy apparently impressed Goethe to conceive of the interrelatedness of all individuals within a larger whole. Applied to anatomy this meant "one had to examine the range of animal skeletons in comparative fashion in order to come to an adequate idea, or archetype, of 'the' animal skeleton" (Richards 2002, p. 379), the archetype here being the holistic, ideal form from which the individual expressions derived. Before tracing this line of influence to Pauli and Jung, I would like to look a bit more at Goethe's embrace of Spinoza, and the impact of this on Schelling.[7]

Germanist Michael Forster, among others, has persuasively argued that philosopher Johann Gottfried Herder is the central figure in introducing and inspiring Goethe's reading of Spinoza. He retells the story of Herder meeting and befriending "the young Goethe at an inn named 'Zum Geist'" in 1770, and how both men went on to write "*Tractatus*-inspired" works under Herder's guidance (Forster 2012, p. 64). Their friendship and interest in

---

[7]Bishop (2008, p. 3) briefly looks at the influence of Spinoza on Goethe, Schiller and Nietzsche and from them on to Jung. He cites the same seven passages used here (see Sect. 5 below), but primarily to suggest that "Jung seems to have enjoyed at least a passing familiarity with some of the main ideas found in Spinoza".

Spinoza recurred as "Goethe would later continue to follow Herder's lead in the interpretation of Spinoza when they re-read Spinoza together in Weimar in the early 1780s" (Forster 2012, p. 61). Goethe's persistent defense of Spinoza's philosophy against charges of atheism was crucial to his works being embraced by the younger Romantics who followed him. This does, however, create a dilemma for our understanding of Jung's relative disregard of Spinoza, as we will come to shortly.

The list of the German Romantics and Idealists considered to have been strongly influenced in a positive manner by Spinoza's writings is impressive. Forster notes that "in addition to Lessing and Herder, further neo-Spinozists included Goethe, Schelling, Hegel, Schleiermacher, Hölderlin, Novalis, and Friedrich Schlegel" (Forster 2010, p. 47). By extension through these philosophers, various other major figures including the Romantic scientists incorporated much of Spinoza's views on God and Nature. One example follows from Michael Mack's interpretation of Noyes' tracing out the post-colonial and ecological themes in Herder's and Goethe's writings, which Mack (2010, p. 53) directly links to their "Spinozist approach". These concerns are exactly what can be found in the scientific observations of Alexander von Humboldt as previously discussed (Cambray 2014a,b).

Schelling as the chief architect of *Naturphilosophie*, which formed the philosophical backing for the scientific efforts of the Romantics, drew heavily on Spinoza. Early in his career Schelling incorporated aspects of Spinoza's ideas on intellectual intuition, which Nassar sees as guiding his privileging ontology over epistemology in his philosophy of nature (Nassar 2012). This is exemplified in a passage quoted by Bishop wherein Schelling differentiates intellectual intuition from a more basic enthusiasm[8] and turns to Spinoza: "As he thought of himself as submerged in the absolute object, he could also think of his self; he could think of his self as annihilated, without at the same time having to think of it as existing" (Bishop 2000, pp. 208f).

Through these channels dual-aspect monism enters into the fundamental perspective of *Naturphilosophie* and undergirds the research program of the German Romantic scientists. In the next generation of scientists these ideas reappear, for example in the writings of Haeckel, whose influence on Jung was mentioned before. Haeckel's own brand of monism, which he saw as nature religion was derived from his reading of Goethe and Spinoza. As he aged this became an increasingly important concern for him. In 1892 he published a monograph *Der Monismus als Band zwischen Religion und Wissenschaft* (English: monism as a link between religion and science) which

---

[8] "Schwärmerei" is the German term for what the enlightenment scorned as enthusiasm.

experienced remarkable popularity (seventeen editions were produced) and led to the founding of the *Monisten-Bund* (monist league, cf. Richards 2002, Chap. 9).

The complexity of feelings which Jung had about the *Monisten-Bund* are revealed by his letters of protest to the *Neue Zürcher Zeitung* in January 1912 about a lecture attacking psychoanalysis given at the Kepler-Bund (an organization directly antithetical to Haeckel's Monisten-Bund). Moreover, he had expressed concern about the Monisten-Bund in a letter to Freud in 1910 for its inclusion of non-professionals (McGuire 1974, letter 217J). At the time of his letters to the newspaper, opposition to psychoanalysis and Haeckel's Monisten-Bund were closely linked (see Ellenberger 1970, pp. 810–815, for discussion).

5   Jung's Minimal Recognition of Spinoza

A survey of Jung's collected works, published seminars, and letters reveals scant reference to Spinoza. In total there are only seven short references to Spinoza in the whole of Jung's writings (and none in his correspondence with Pauli). To assess Jung's limited view of Spinoza in detail, his statements about the philosopher are presented in the following:

1. In *Psychiatric Studies*, Jung (1957/1970, par. 100, n. 49) disparagingly mentions a hypnopompic vision of Spinoza as an example of the way in which "imaginative people are particularly subject to them" (hypnagogic hallucinations) and underscores this by noting that hypnopompic hallucinations "are the essentially the same as the hypnagogic ones". The irony of this can best be seen against Jung's own exploration of his waking visions, i.e., his own hypnopompic hallucinations out of which his *Red Book* emerged. This leads to the suggestion that Jung's avoidance of Spinoza may have had unconscious psychological factors at play, including vulnerability to highly activated unconscious contents. The comment cited above comes from a very early period in Jung's career, well before his own explorations of the unconscious and even before his encounters with Freud. Based on his subsequent explorations the remark seems to point to a concealed anxiety which in turn may have contributed to his choice of profession – psychiatry as a way of gaining purchase on his own unconscious processes. He was an assiduous student, reading all of the psychiatric literature then available on the subject during his first years at the *Burghölzli*.

2. In *Psychological Types*, Jung (1971, par. 770) states in the definitions section: "Intuitive knowledge possesses an intrinsic certainty and conviction,

which enabled Spinoza (and Bergson) to uphold the *scientia intuitiva* as the highest form of knowledge." The identification of scientific (or intellectual) intuition as the apex of knowledge does not accord well with Jung's own gnostic orientation. It moved him to label Spinoza reductively as a rationalist rather than to explore the significance of Spinoza's views on intuition, which have a profundity beyond the mere scientific formulation of ideas.

In fairness to Jung, Spinoza's form of presenting his arguments in the *Ethics* as quasi-geometric axioms makes the text difficult to penetrate and has struck some readers as overly rational. However, many of the German writers who embraced Spinoza, e.g., Goethe, Hamann, Herder, Jacobi, Novalis, and so forth, asserted the value of feelings, subjectivity, and emotions against an idealization of reason. They acknowledged that the importance of emotional life derives in part from their reading of Spinoza on affects in his various text including the *Ethics*. His views on the passions as the basis of human life and of the mind mitigate against any view of him simply as a rationalist. Thus we are left with an unresolved problem of Jung's dismissal of Spinoza on these grounds.

3. In *Instinct and the Unconscious*, Jung (1960/1969b, par. 276) cites Spinoza as one of the philosophers who, he feels, diminished the notion of the archetype:

> From Descartes and Malebranche onward, the metaphysical value of the "idea" or archetype steadily deteriorated. It became a "thought", an internal condition of cognition, as clearly formulated by Spinoza: "By 'idea' I understand a conception of the mind which the mind forms by reason of its being a thinking thing." Finally Kant reduced the archetypes to a limited number of categories of understanding. Schopenhauer carried the process of simplification still further, while at the same time endowing the archetypes with an almost Platonic significance."

4. In *The Phenomenology of the Spirit in Fairy Tales*, Jung (1959/1969, par. 385) explores the term "spirit": First he raises the notion of the spirit as the principle standing in opposition to matter, and as the vehicle of psychic phenomena, or even of life itself. Then he remarks:

> In contradiction to this view there stands the antithesis: spirit and nature. Here the concept of spirit is restricted to the supernatural or anti-natural, and has lost its substantial connection with psyche and life. A similar restriction is implied in Spinoza's view that spirit is an attribute of the One Substance. Hylozoism goes even further, taking spirit to be a quality of matter.

This passage suggests that Jung does not grasp the dual-aspect monism of Spinoza's philosophy, rather the holistic background is misread as a reductive collapse into a fundamental materialism. Perhaps this is due to the complexity of Spinoza's theism combining his dual-aspect theory ("the mental and the physical are distinct modes of a single substance, God"; Audi 1999, p. 686) with his identification of God with Nature ("Deus, sive Natura"; Audi 1999, p. 870). Jung apparently misconstrues this (along with various critics of Spinoza) as a kind of nominalism regarding the spirit, reducing "God" to nature taken solely as what is physically manifest.[9] However, Spinoza was no Hobbesian materialist, nor even a neutral monist like Hume, but his dual-aspect theory is in truth an infinite-aspect theory, where the mental and physical are just two among "infinitely many modes of this one substance" (Audi 1999, p. 686) – though the only two accessible by humans.

We are again left with the dilemma of how to explain Jung's facile disregard of Spinoza, now concerning his contribution to the mind-body problem. From his synchronicity essay, Jung's stated preference for Leibniz as a precusor to his own stance on this is clear, including Meier's raising the issue of the mind-body relationship as a form of synchronicity (Jung 1960/1969a, par. 938, n. 70). However, Spinoza's dual-aspect monism is an unacknowledged precursor of Jung's notion of the psychoid realm, much more appropriate than Leibniz's parallelism.[10] Similarly, Spinoza's view on the motivational aspects of human affects and desires as a part of nature is in close accord with Jung's implicit theory of motivation through what he termed psychic energy. Yet this link is not stated – as if it were unconsciously blocked.

5. Again in *The Phenomenology of the Spirit in Fairy Tales*, Jung (1959/1969, par. 390f) continues to dismiss Spinoza's monism as rational materialism:

> The transcendent spirit became the supranatural and transmundane cosmic principle of order and as such was given the name of "God", or at least it became an attribute of the One Substance (as in Spinoza) or one

---

[9]Guilherme (2008, p. 24) has recently shown that the great Spinozist Edwin Curley "reads Spinoza's metaphysics in the light of modern physics" and thus distorts his view of the one substance. While Guilherme does not attempt to completely define Spinoza's understanding of substance, he does show how Curley seriously truncates Spinoza's position in a way that has some resemblance to Jung's reading of him.

[10]As Beiser (1987, p. 53) points out: "Mendelssohn reveals that there are many points of similarity between Leibniz and Spinoza, and argues that Leibniz had taken some of his characteristic doctrines from Spinoza. Leibniz's notion of the preestablished harmony, for example, is said to have its source in Spinoza's idea that the mind and the body are independent attributes of one and the same substance."

Person of the Godhead (as in Christianity)... One had only to give the One Substance another name and call it "matter" to produce the idea of a spirit which was entirely dependent on nutrition and environment, and whose highest form was the intellect or reason.

However, here are Spinoza's own words (according to Bennett 2010):

God's nature doesn't involve either intellect or will. I know of course that many think they can demonstrate that a supreme intellect and a free will pertain to God's nature; for, they say, they know nothing they can ascribe to God more perfect than what is the highest perfection in us.

It seems that Spinoza is reaching for more than Jung would credit him. Similarly his view of intellectual intuition goes beyond rational thought: "the highest stages of knowledge consist in a form of intuitive insight, which transcends mere reasoning or conceptual knowledge in that it enables us to grasp the essence of individual things" (Mander 2012). His pantheism, so embraced by Goethe, Schelling and a host of romantics remains in the shadows for Jung.

6. In *The Role of the Unconscious*, Jung (1964/1970, par. 27) discusses the compensatory function of the unconscious and its symbol-creating function as paradoxical. In this context, he tells a story of a young rabbi who was a pupil of Kant's and was threatened with being cursed as a heretic "as happened to Spinoza". Here Jung recounts a story about the *shofar* "that is blown at the cursing of heretic" and how it functions as object and symbol for the rabbi but only as mere object for the heretic. This story offers a valuable clue to Jung's distancing himself from Spinoza. It seems as if Jung's own anxieties along with his experiences of being seen and treated as a heretic for his forays into mysticism – by his Freudian colleagues, the theologians he wrestled with, and even the scientific community – are operative in his facile dismissal of Spinoza.

At the level of politics, Jung's conservativism is well known despite his avant-garde psychology (Sherry 2010). This is in sharp contrast to Spinoza's radical political stance, especially as delineated in his *Tractatus Theologico-Politicus* and filtered through German Romantic tradition. These differences in political orientation were likely a further source of alienation for Jung.

7. In *The Love Problem of a Student*, Jung (1964/1970, par. 199) offers a hierarchy of mystical love: "Beginning with the highest mystery of the Christian religion, we encounter, on the next-lower stages, the *amor Dei* of Origen, the *amor intellectualis Dei* of Spinoza, Plato's love of the Idea, and the *Gottesminne* of the mystics ..." The greatest of Jewish and Pagan

philosophers are placed in rungs beneath Christian mysteries, precisely the matters Jung's own father had failed to adequately explain to him as a boy (Jung 1961, pp. 52–59). Perhaps reducing the importance of Spinoza compensated for his distress over what he viewed as paternal weakness.

From these passages, combined with the absence of citations of Spinoza as a precursor to the ideas of the psychoid and synchronicity, I would suggest a complex at play, a Spinoza complex?! The fact that neither Pauli nor Jung thought to include Spinoza, either in their discussions of the psychoid quality of the archetype or in their exchanges on synchronicity, suggests they shared this aversion. A clue to this may be found in various compendia of philosophy such as the *Cambridge Dictionary of Philosophy* (Audi 1999, p. 874): "Spinoza has affected the philosophical outlook of such diverse twentieth-century thinkers as Freud and Einstein". Curiously Jung and Pauli may have felt allergic to the philosophical views of the men whom they most strove to differentiate themselves from, so that any embrace of Spinoza may have felt like self-betrayal.[11] To assess the full impact of this complex on Jung's theorizing a fuller exploration should be undertaken.

6   The Psychoid Realm and Dual-Aspect Monism

Around the time that Jung composed his essay on synchronicity, he also worked on various drafts of the essay that we now know as "On the Nature of the Psyche" (Jung 1954/1969, pars. 343–442). This essay is the one in which Jung first puts forward his ideas on the psychoid. During this period he was in active, regular correspondence with Pauli and various ideas in this essay were discussed in their correspondence at the time. I will not review the detailed development of the concept but simply look to a representative statement that captures the implicit dual-aspect monism (Jung 1954/1969, par. 418):

> Since psyche and matter are contained in one and the same world, and moreover are in continuous contact with one another and ultimately rest on irrepresentable, transcendental factors, it is not only possible but fairly probable, even, that psyche and matter are two different aspects of one and the same thing. The synchronicity phenomena point, it seems to me, in this direction, for they show that the nonpsychic can behave like the psychic, and vice versa, without there being any

---

[11]Pauli's disputes with Einstein, who strongly embraced Spinoza in his field metaphysics, may have further contributed to his omission from their considerations (Atmanspacher, private communication; see also Seager 2009, p. 88).

> causal connection between them. Our present knowledge does not allow us to do much more than compare the relation of the psychic to the material world with two cones, whose apices, meeting in a point without extension – a real zero-point – touch and do not touch.

Jung obviously struggles with formulating the relationship of psyche and matter, which in its microcosmic (human) form is the mind-body problem. His acknowledgment of the likely fundamental unity is cautious, almost to the point of vanishing, something that will trouble Pauli. Though Jung's worldview has been deeply altered by his understanding of quantum physics, especially from his interaction with Pauli, I do not believe he truly grasped the concept of quantum entanglement and its implications for his hypotheses of the psychoid and of synchronicity. A greater familiarity with Spinoza's understanding of the equivalence of "God" and "Nature" would have helped him move further in this direction.[12]

In contradistinction, Atmanspacher (2012) has cogently argued that the Jung-Pauli collaboration resulted in the conceptualization of synchronicity as involving the psychoid layer of the unconscious as a modern form of dual-aspect monism, emerging from the interface of quantum field theory with depth psychology. Consistent with this view, Pauli introduced the notion of complementarity into his discussions with Jung,[13] for example, in his plea for a "monistic union of matter and soul" (Meier 2001, p. 87). And, later in his correspondence with Jung, Pauli again insists that the concept of the archetype must not be limited to psychic factors but includes physical aspects as well. Pauli quotes Jung, reminding him that he (Jung) has "to question the solely psychic nature of the archetypes". Then Pauli continues in a letter of March 1953 (Meier 2001, p. 106):

---

[12]Ironically the Jung-Pauli relationship is filled with entanglements of a relational nature, as Jung initially refused to analyze Pauli and sent him to a junior colleague. This way he imagined to be capable of collecting the dreams of this remarkable young man without contaminating them by his own thought. (Jung later used a dream series by Pauli in his *Psychology and Alchemy*.) Shortly after he relented and did work with Pauli, especially with his dreams. They went on to have an extensive correspondence, even a friendship, and published the book *The Interpretation of Nature and the Psyche* together.

[13]Elsewhere I have pointed out how Jung confused various terms derived from the Copenhagen interpretation of quantum mechanics (Cambray 2009, p. 24). Jung referred to "correspondence" when in fact he was borrowing the concept of "complementarity" as Bohr had applied it when he had borrowed it from William James. (Bohr also used the notion of correspondence but for purposes different from Jung's). Furthermore, in writings on dreams Jung favored the term "compensation" for the unconscious response to conscious attitude as frequently detected in dreams, while having a more limited view of "complementation". His struggle with this concept may provide an additional hint at a complex activated around formulating the psychoid (the Spinoza complex).

> I feel that you should certainly *take these doubts seriously* and not once again make too much of the *psychic factor*. When you say that "the psyche is partly of a material nature", then for me as a physicist this takes on the form of a metaphysical statement. I prefer to say that psyche and matter are governed by common, neutral, "not in themselves ascertainable" ordering principles.

Common, neutral, ordering principles do point to a strongly monistic orientation coming from Pauli, more than Jung's vanishing point of contact/origin mentioned above, though it is difficult in Pauli's letter to differentiate neutral monism from dual-aspect monism (for a discussion of the differences in these approaches see Section 1.1 of Atmanspacher 2012). That the ordering principles are "not in themselves ascertainable", however, gives a hint into the quantum logic Pauli is drawing on, as in the feature of nonlocality inherent in the notion of entanglement.

Exploring the differences between neutral and dual-aspect monism, Atmanspacher also provides an important insight into the significance and consequences of Pauli's argument by contextualizing his statements in terms of quantum field theory and thereby highlighting the role of measurement in collapsing nonlocal (holistic) quantum states.[14] He employs the properties of quantum systems to draw the conclusion that they emerge not from neutral monism, but a type of dual-aspect monism comprised of (Atmanspacher 2012, p. 107)

> ontic states and associated intrinsic properties [which] refer to the holistic concept of reality and are operationally inaccessible, [together with] epistemic states and associated contextual properties [which] refer to a local concept of an operationally accessible reality.

---

[14] A pertinent website at the University of Oregon (abyss.uoregon.edu/~js/glossary/holism.html) says the following about quantum holism:
"The emergence of a quantum entity's previously indeterminate properties in the context of a given experimental situation is an example of relational holism. We cannot say that a photon is a wave or a particle until it is measured, and how we measure it determines what we will see. The quantum entity acquires a certain new property – position, momentum, polarization – only in relation to its measuring apparatus. The property did not exist prior to this relationship. It was indeterminate.
Quantum relational holism, resting on the nonlocal entanglement of potentialities, is a kind of holism not previously defined. Because each related entity has some characteristics – mass, charge, spin – before its emergent properties are evoked, each can be reduced to some extent to atomistic parts, as in classical physics. The holism is not the extreme holism of Parmenides or Spinoza, where everything is an aspect of the One. Yet because some of their properties emerge only through relationship, quantum entities are not wholly subject to reduction either. The truth is somewhere between Newton and Spinoza. A quantum system may also vary between being more atomistic at some times and more holistic at others; the degree of entanglement may vary."

In this model the act of measurement (not the subjectivity of the scientist) is the operation creating the transition between states (Atmanspacher 2012, p. 108): "Measurement suppresses the connectedness constituting a holistic reality and generates approximately separate local objects constituting a local reality". Applied to the mind-matter relationship this logic points to the correlation that Meier first suggested: there is a synchronistic link between the mental and the physical which are indeterminate in the holistic world.

Observations as of states of mind or body break the symmetry of the indeterminate holistic states of being. Emergence can follow, as in waking from a dream. It is at this epistemic level that actual synchronistic experience can manifest, and do so as emergent phenomena as I have previously described (e.g., Cambray 2002, 2009). Furthermore as the emergent forms themselves can increase in levels of complexity through self-organization that derives from enhanced interconnectedness, there is the potential for increased inclusion and a movement towards more global structures. The mind-matter relation is thus nonlinear with a tendency towards a renewed holism, even if this cannot be fully achieved. In short, the proposed dual-aspect monistic theory leads to an individuation model in which emergence from an initial undifferentiated holism sets off a striving for wholeness that can never be completed, though increasingly complex structures evolve from the effort.

7 Conclusion

In this paper I have tried to identify some of the historical precursors to the concepts of the psychoid and of synchronicity that have been overlooked, first and foremost by those who formulated these ideas, Jung and Pauli. While the attempt to uncover reasons for these omissions remains speculative, recovering links to the scientific work of the German Romantics may help us re-examine some of the dichotomies that have crept into and characterized our modern cosmology, such as the tendency to split art from science, or religion from science.

The holistic perspective that can emerge from this recovery offers essential elements missing from our canonical worldview. Means for describing the interdependence of subjective and objective aspects of reality can potentially arise from such studies, in accordance with the inquiries of contemporary philosophers such as Nagel (2012) and scientists such as Abrams and Primack (2011). Should this bear fruit, the Jung-Pauli relationship may be more fully appreciated for their profoundly pioneering cosmological perspective.

References

Abrams N.E. and Primack J.R. (2011): *The New Universe and the Human Future*, Yale University Press, New Haven.

Amundson R. (2005): *The Changing Role of the Embryo in Evolutionary Thought: Roots of Evo-Devo*, Cambrdige University Press, Cambridge.

Atmanspacher H. (2012): Dual-aspect monism à la Pauli and Jung. *Journal of Consciousness Studies* **19**(9-10), 96–120.

Audi R., ed. (1999): *The Cambridge Dictionary of Philosophy*, Cambridge University Press, Cambridge.

Beiser F.C. (1987): *The Fate of Reason: German Philosophy from Kant to Fichte*, Harvard University Press, Cambridge.

Bennett J. (2010): Ethics. Manuscript available at www.earlymoderntexts.com/pdfbits/spinoza1.pdf.

Bishop P. (2000): *Synchronicity and Intellectual Intuition in Kant, Swedenborg, and Jung*, Edwin Mellen Press, Lewiston.

Bishop P. (2008): *Analytical Psychology and German Classical Aesthetics: Goethe, Schiller, and Jung, Volume 1: The Development of the Personality*, Routledge, London.

Bishop P. (2009): *Analytical Psychology and German Classical Aesthetics: Goethe, Schiller, and Jung, Volume 2: The Constellation of the Self*, Routledge, London.

Cambray J. (2002): Synchronicity and emergence. *American Imago* **50**(4), 304–409.

Cambray J. (2009): *Synchronicity: Nature and Psyche in an Interconnected Universe*, Texas A&M University Press, College Station.

Cambray J. (2011a): Jung, science, and his legacy. *International Journal of Jungian Studies* **3**(2), 110–124.

Cambray J. (2011b): L'Influence d'Ernst Haeckel dans le Livre Rouge de Carl Gustav Jung. In [*Recherches Germaniques, Revue Annuelle Hors Serie* **8**, 41–59.

Cambray J. (2011c): Moments of complexity and enigmatic action: A Jungian view of the therapeutic field. *Journal of Analytical Psychology* **56**(2), 296–309.

Cambray J. (2013): The Red Book: Entrances and Exits. In *The Red Book: Reflections on C. G. Jung's Liber Novus*, ed. by T. Kirsch and G. Hogenson, Routledge, London, pp. 36–53.

Cambray J. (2014a): Romanticism and revolution in Jung's science. In *Jung and the Question of Science*, ed. by R. Jones, Routledge, London, in press.

Cambray J. (2014b): Jung, science, German romanticism: A contemporary perspective. In *Jung in the Academy and Beyond: The Fordham Lectures – 100 Years Later*, Spring Journal Books, New Orleans, in press.

Churchill F.B. (2007): Living with the biogenetic law: A reappraisal. In *From Embryology to Evo-Devo*, ed. by M.D. Laubichler and J. Maienschein, MIT Press, Cambridge.

Cunningham A. and Jardine N., eds. (1990): *Romanticism and the Sciences*, Cambridge University Press, Cambridge.

Chaouli M. (2002): *The Laboratory of Poetry: Chemistry and Poetics in the Work of Friedrich Schlegel*, Johns Hopkins University Press, Baltimore.

Ellenberger H.F. (1970): *The Discovery of the Unconscious: The History and Evolution of Dynamic Psychiatry*, Basic Books, New York.

Forster M.N. (2010): *After Herder: Philosophy of Language in the German Tradition*, Oxford University Press, Oxford.

Forster M.N. (2012): Herder and Spinoza. In *Spinoza and German Idealism*, ed. by E. Förster and Y.Y. Melamed, Cambridge University Press, Cambridge, pp. 59–84.

Gissis S.B. and Jablonka E. (2011): *Transformations of Lamarckism: From Subtle Fluids to Molecular Biology*, MIT Press, Cambridge.

Goldberg A C., Allis D., and Bernstein E. (2007): Epigenetics: A landscape takes shape. *Cell* **128**, 635–638.

Guilherme A. (2008): Spinoza's substance: A reply to Curley. *Conatus* **2**(4), 19–24.

Holland J. (2009): *German Romanticism and Science: The Procreative Poetics of Goethe, Novalis, and Ritter*, Routledge, London.

Holland J. (2010): *Johann Wilhelm Ritter: Key Texts on the Science and Art of Nature*, Brill, Leiden.

Holmes R. (2008): *The Age of Wonder*, Vintage Books, New York.

Jung A. (2011): The grandfather. *Journal of Analytical Psychology* **56**(5), 653–673.

Jung C.G. (1953/1968): *Psychology and Alchemy, Collected Works 12*, Princeton University Pres, Princeton.

Jung C.G. (1954/1969): On the nature of the psyche. In *The Structure and Dynamics of the Psyche, Collected Works 8*, Princeton University Press, Princeton.

Jung C.G. (1957/1970): *Psychiatric Studies, Collected Works 1*, Princeton University Press, Princeton.

Jung C.G. (1959/1969): The phenomenology of the spirit in fairy tales. In *The Archetypes and the Collective Unconscious, Collected Works 9*, Princeton University Press, Princeton.

Jung C.G. (1960/1969a): Synchronicity: An acausal connecting principle. In *The Structure and Dynamics of the Psyche, Collected Works 8*, Princeton University Press, Princeton.

Jung C.G. (1960/1969b): Instinct and the unconscious. In *The Structure and Dynamics of the Psyche, Collected Works 8*, Princeton University Press, Princeton.

Jung C.G. (1961): *Memories, Dreams, Reflections*, Vintage Books, New York.

Jung C.G. (1964/1970): *Civilization in Transition, Collected Works 10*, Princeton University Press, Princeton.

Jung C.G. (1969): Psychology and religion. In *Psychology and Religion: West and East, Collected Works 11*, Princeton University Press, Princeton, pp. 3–105.

Jung C.G. (1971): Psychological types. In *Collected Works 6*, Princeton University Press, Princeton.

Jung C.G. (2009): *The Red Book. Liver Novus*, ed. by S. Shamdasani, W.W. Norton, New York.

Laubichler M.D. (2007): Does history recapitulate itself? Epistemological reflections on the origins of evolutionary developmental biology. In *From Embryology to Evo-Devo*, ed. by M.D. Laubichler and J. Maienschein, MIT Press, Cambridge.

Mack M. (2010): *Spinoza and the Specters of Modernity: The Hidden Enlightenment of Diversity from Spinoza to Freud*, Continuum, New York.

Mander W. (2012): Pantheism. In *Stanford Encyclopedia of Philosophy*, ed. by E.N. Zalta, available at `plato.stanford.edu/entries/pantheism/`.

McGuire W., ed. (1974): *The Freud/Jung Letters: The Correspondence between Sigmund Freud and C.G. Jung*, transl. by R. Manheim and R.F.C. Hull, Princeton University Press, Princeton.

Meier C.A., ed. (2001): *Atom and Archetype: The Pauli/Jung Letters 1932–1958*, Princeton University Press, Princeton.

Müller-Siever H. (1997): *Self-Generation: Biology, Philosophy, and Literature Around 1800*, Stanford University Press, Stanford.

Nagel T. (2012): *Mind and Cosmos*, Oxford University Press, Oxford.

Nassar D. (2012): Spinoza in Schelling's early conception of intellectual intuition. In *Spinoza and German Idealism*, ed. by E. Förster and Y.Y. Melamed, Cambridge University Press, Cambridge, pp. 136–155.

Ørsted, H.C. (1852): *The Soul in Nature*, translated by L. and J.B. Horner, Henry G. Bohn, London.

Poggi S. and Bossi M., eds. (1994): *Romanticism in Science: Science in Europe 1790–1840*, Kluwer, Dordrecht.

Richards R.J. (2002): *The Romantic Conception of Life: Science and Philosophy in the Age of Goethe*, University of Chicago Press, Chicago.

Richards R.J. (2008): *The Tragic Sense of Life: Ernst Haeckel and the Struggle over Evolutionary Thought*, University of Chicago Press, Chicago.

Roth R. (2004): *The Return of the World Soul*, manuscript available at paulijungunusmundus.eu/synw/jungneoplatonismaristotlep4.htm.

Rupke N. (2008): *Alexander von Humboldt: A Metabiography*, University of Chicago Press, Chicago.

Seager W. (2009): A new idea of reality: Pauli on the unity of mind and matter. In *Recasting Reality: Wolfgang Pauli's Philosophical Ideas and Contemporary Science*, ed. by H. Atmanspacher and H. Primas, Springer, Berlin, pp. 83–97.

Sherry J. (2010): *Carl Gustav Jung: Avant-Garde Conservative*, Palgrave Macmillan, New York.

Wetzel W.D. (1990): Johann Wilhelm Ritter: Romantic physics in Germany. In *Romanticism and the Sciences*, ed. by A. Cunningham and N. Jardine, Cambridge University Press, Cambridge, pp. 199–212.

Markus Fierz:
His Character and His Worldview

Hans Christian von Baeyer

Abstract

Markus Fierz (1912–2006) was Wolfgang Pauli's assistant, friend, prolific correspondent, and eventual successor. In this lecture I briefly review his biography, including my own interactions with him, before turning to some of his thoughts on physics, psychology, and quantum mechanics. His views overlapped, complemented, or clashed with many of Pauli's. My purpose is not so much to celebrate Fierz's contributions to physics as to propose him as a sober, well informed and astute mediator who can help to throw light on the insights of the strange, demonic, and often obscure genius who was Pauli. Fierz's influence was particularly evident in Pauli's ambitious but unsuccessful project of psycho-physical unification.

1 Introduction

The twentieth of June 2012 marked the centenary of the birth of Markus Fierz, assistant, friend, and successor of Wolfgang Pauli. Most physicists have heard of him since he wrote four important papers with Pauli, plays a supporting role in every book about Pauli, and, with the exception of Heisenberg, exchanged more letters with Pauli than any other physicist. But a centenary celebration? Aren't those reserved for the big stars, not their satellites?

Well, Fierz was unique in belonging to the inner circles not only of Pauli, but of Carl Jung as well. He orbited a double star, as it were. Besides, centenaries are not for the dead – they couldn't care less. For us, the living, birthdays represent an excuse and opportunity for reflecting on the past. A case in point: It was a centenary that provided the impetus for Markus Fierz to branch out beyond theoretical physics. It was on the occasion of a tercentenary that he started writing about history and philosophy. I own a flimsy carbon copy, personally corrected in ink, of a manuscript entitled (originally in German): "Isaac Newton, his character and his worldview, by Markus Fierz – a lecture given on 21 December 1942, in memory of Newton's 300th birthday." Gieser (2005) lists this lecture among the major influences on Pauli's famous Kepler study. So never underestimate the potential power of a centenary!

Here I will briefly sketch Fierz's biography, emphasizing its points of intersection with the lives of my father and myself, and then turn to some reflections on his thoughts about foundational issues. My sources include, besides the voluminous Pauli literature, a packet of unpublished letters, a couple of private memoirs, and a glittering anthology of Fierz's lectures and essays (Fierz 1988) produced during the second half of his career. That book was published in 1988 but not, as far as I know, translated into English.

2   Character

Markus Fierz was born in Switzerland, where he lived until his death in 2006.[1] His father was an industrial chemist and professor in Zürich, where Markus himself studied and later taught. His mother, Linda Fierz-David, was a prominent analytical psychologist, writer, and close collaborator of Jung. In fact, she built a vacation home next door to Jung's residence and decorated it in a style steeped in colorful Jungian symbolism. The myths and fairy tales she told her two sons nourished the imagination of young Markus, whose prodigious memory allowed him to draw on them for the rest of his long life.

If Markus took after his scientist father, his twin brother Heiner followed in their mother's footsteps by becoming a Jungian analyst. Thus, through his mother and his brother, Markus was exposed to the Jungian mystique from birth – long before Pauli ever encountered it. So when Pauli took Fierz as his assistant, he hired an unusual young man who could hold his own in discussions of psychology as well as physics. Furthermore I suspect that the idea of the Fierz twins, Markus and Heiner, the physicist and the psychiatrist, must have resonated with Pauli's love of mirror symmetry when he began speculating about the unification of physics with psychology. Heiner, incidentally, was my godfather. So theoretically, though not in fact, he was my spiritual guardian.

In 1931 Markus began studying in Göttingen, then the mecca of European physics. In the university orchestra, where he played viola, he met another

---

[1]See von Meyenn (2007) for an obituary for Fierz with interesting biographical material and a complete bibliography of his publications.

violist and physicist: my father. The two students shared similar family backgrounds in the European intellectual bourgeoisie, and hit it off at once.

It is amusing to compare what the two friends wrote about each other in their respective memoirs, written half a century later. According to my father, Markus was "the essence of Swiss independence in thought and deed. His face appeared as though carved from wood, hardwood, and combined complete honesty and utter impassivity" as well as a "clean, lively intelligence, which is open to almost anything." He relates an anecdote about Fierz:

> Once, for a wager, he said he could take a couple of bars of gold into Germany and back without being detected by the border police. He did it. The bars were heavy and weighed down the pockets of his raincoat, but such was his look of probity, of irreproachability, that not one of the guards could suspect him.

Though the story sounds unlikely, it illustrates the impression of rock-solid imperturbability and self-confidence that Fierz projected.

Fierz's description of my father is different:

> Like me, [von Baeyer] sat one night in the viola section, where I noticed him immediately. Finally I met a person whose aura seemed comprehensible to me: from there a friendship for life developed, even though we ended up living in very different places.

Notice the reversed perspective. For Fierz a friendship is justified not so much by the positive qualities of the other as by his own positive reaction. Apparently he not only understood, but approved of my father's character. And that was sufficient to take him on as a friend.

In the summer of 1932 Adolf Hitler came through Göttingen on a speaking tour. Fierz and my father attended the event together, and, with the help of hindsight, described it in their respective memoirs. Fierz recalled that the venue was decorated with small flags, but since swastikas were still in short supply at that time, the organizers had to resort to alternating them with little Swiss flags. Hitler, Fierz wrote,

> reminded me of a superannuated Boy Scout leader. His allusions to sex and homosexuality were unmistakable: a strong, hysterical and frightening personality. This, then, was the man whom the German people would soon follow heedlessly into slavery and shame. Of course I could not know that at the time, but that it was shameful to follow this Führer, that was clear.

In the following January Hitler assumes power, and the golden years of physics in Göttingen end abruptly. Fierz retreats immediately home to

Zürich, to get his PhD with Gregor Wentzel, the W in the WKB approximation. My father returns to his home in Heidelberg to earn his PhD with the future Nobel Laureate Walter Bothe.

In Zürich Fierz is fascinated by Pauli's lectures. He was not so much lecturing as talking to himself, Fierz recalled, in an unclear, nasal voice, accompanied by tiny, indecipherable writing on the blackboard. Furthermore, Pauli occasionally lost his thread, and stared into space, as though doubting what he had just said. When he subsequently continued, all the while mumbling incomprehensible words, nodding, or shaking his head, nobody knew what had suddenly bothered him. Fierz found all this very mysterious, and it emphasized the demonic aura that surrounded this unusual man.

Fierz was captivated by the beauty and mystery of theoretical physics, but contrary to his customary self-confidence, he felt inadequate and afraid to commit himself to it. Eventually it was a dream that gave him the courage to proceed. Wandering behind Pauli and Wentzel along Lake Zürich he comes to a wonderful tree with blue and golden leaves, which he identifies as the tree of life. He interprets this scene as a sign that he is on the right path. Fierz remembered his dreams, analyzed them diligently, and paid attention to their messages.

I think this anecdote reveals something important. According to Victor Weisskopf, Fierz was the only assistant with whom Pauli felt sufficiently comfortable to share his innermost thoughts on psychology and mysticism. In Fierz, Pauli found himself a soulmate whose honesty assured that he would never stoop to humoring his famous boss.

After obtaining his degree, Fierz briefly visited Heisenberg in Leipzig, and then Pauli asked him to become his assistant at Zürich. I quote from Fierz's memoirs:

> Sure, I wasn't as experienced as Weisskopf, but he wanted to try me anyhow. I was very frightened by the offer. Not only did I doubt that I was up to the task, but I was afraid of Pauli. Not of his infamous sharp tongue, because his mean and even insulting assaults seemed quite harmless to me. No, I was afraid of the eerie, hidden qualities of the man. These sinister traits were a natural phenomenon that caused his body to sway in strange, irregular motions, so he resembled a man in a trance. All the other physicists felt this too, and experienced it as the "Pauli effect", which everyone, including Pauli, believed in. Personally I never experienced a Pauli effect, probably because I was able to recognize the demon even without material evidence.

How like Fierz – his claim to understand Pauli's remarkable aura without requiring dramatic, visible proof, the way everyone else did.

In any case, he would have preferred to reject Pauli's offer, but felt that fate was calling him, so he accepted. Pauli proved to be an excellent teacher. He came to Fierz's office almost every day to ask about progress. When, inevitably, he began to quibble, Fierz bravely held his own and contradicted him. On the other hand, when Fierz grew discouraged, Pauli supported him with intelligence and sensitivity. Fierz concluded that "he understood people very well – when he wanted to."

For the next 22 years Fierz had a close relationship with Pauli – you might call it a friendship, Fierz thought – and they exchanged hundreds of letters. But in all that time they never used each other's first names or the familiar German "Du".

After his postdoc time with Pauli, Fierz became professor in Basel. In the 1950s I studied there myself one semester, and took his course on introductory quantum mechanics. What a contrast his lectures presented to Pauli's. He spoke without notes, but what appeared on the board in precise calligraphy was text-book perfect. Once a week he took me home for lunch. After the meal with his wife Menga, he dazzled me with stories about the history of physics and mathematics.

After my family emigrated to Canada, the correspondence between Markus and my father withered. But they did have a memorable final reunion in 1977. The last chapter of Fierz's memoirs describes the week he and Menga spent in Greece with my father and my stepmother. The account is a fascinating travelogue, but it doesn't report anything about the conversations or the feelings between the two old friends.

For me, that's too bad, because Fierz himself understood people very well – when he wanted to. In his lecture about Newton he begins by drawing a sensitive and enlightening sketch of the man's human qualities before proceeding to his ideas. This tactic served him well in other excursions into the history of ideas, such as his splendid little biography of the Renaissance polymath Girolamo Cardano. In this essay I am following the same plan: I began by talking about Fierz's life. Now I turn to his thoughts.

3 Worldview

Although Fierz was a distinguished physicist in his own right, what interests us especially here is his role as a sounding board for Pauli. Since Fierz was less enigmatic than Pauli, and had nothing of the mysterious or demonic about him, he is more accessible to us. It seems instructive, therefore, to study Fierz in order to shine light upon Pauli. At the very least, to the extent that Fierz shared ideas with Pauli, and occasionally even anticipated them, he helps to make Pauli appear less singular and more approachable.

Consider the unique joint publication by Jung and Pauli (1952). Two questions reverberate through Pauli's contribution, which is entitled "The influence of archetypical ideas on the formation of scientific theories in Kepler." Since scientific theories obviously do *not* follow logically from empirical observations (Einstein called them "free inventions") the question arises: Where *do* new scientific concepts come from? Pauli answers that they come, in part, from philosophical and religious preconceptions in our unconscious. They bubble up from there through dreams and daydreams whose symbolic language is then transformed, with great effort, into the mathematical language of science. Kepler's writings serve to illustrate these irrational roots of rational thinking, in Fierz's apt formulation.

Pauli's second question is this: All but the most stubborn materialists concede that science alone does *not* explain everything in the world. How, Pauli asks, can we regain a more comprehensive worldview, what the Germans call a *Weltanschauung*? A worldview covers not only the external, physical universe, but the inner, psychological world as well. Pauli was not able to answer this second question, but believed that Kepler's view of the world was more unified than ours, and that quantum mechanics may contain hints for a modern solution to the challenge of merging psychology with physics.

The Jung-Pauli book was published in 1952.[2] Five years earlier, Pauli had given a lecture to the Psychological Club of Zürich on the same subject, under an identical title, but without the elaborate illustrations of the published text. Five years before *that* – in other words, a decade before the celebrated Jung-Pauli book – Fierz had given his tercentenary lecture on Newton, published in Fierz (1943). It turns out that Pauli's two questions also occupied Fierz. Just as Newton's physics is closer to ours than Kepler's, Fierz is more accessible than Pauli.

The sources of Newton's deepest ideas, Fierz found, were his religious beliefs. Absolute space, for example, is a manifestation of God's ubiquity. Absolute time is an expression of God's eternity. This suggestion is not introduced idly or speculatively, but explained explicitly and in detail in the *Principia*. In Newton's words, translated from the Latin: "[God] lasts forever and he is present everywhere, and by existing forever and everywhere he has established duration and space, eternity and infinity."

---

[2]Fierz (1979) published a commentary to this book, based on a lecture in the philosophy-of-science colloquium at Zürich in 1978.

What's more, the idea is supported by Biblical references. For example, Newton quotes Saint Paul in Acts 18, 28: "For in him we live and move, in him we exist." God is space – what could be plainer? This is actually an old Jewish tradition. The word *maqom*, for space, is one of God's names. While God's *nature* appears as space and time, his *dominion* appears in the form of natural laws. And the most obvious effect of his actions is gravity.

It seems that for Newton the purpose of physics was not at all to reveal the mechanism of the world, as his followers, down to our own time, imagined. Rather, it was to demonstrate God's influence on the world. Since God is utterly incomprehensible to us, the laws of nature – such as universal gravitation – are incomprehensible too. Anything we can say about God is merely symbolic – and therefore the laws of nature are symbolic too.

Where Pauli had to dig deeply into Kepler's writings to extract their lessons, Fierz is merely reporting what Newton himself writes clearly and explicitly. Furthermore, where Kepler related the three-dimensionality of the world to the holy trinity, Newton, as a Deist, rejected the trinity in favor of a single God. Regardless of its theological implications, this difference simplifies the entire discussion, because the trinity is a notion steeped in mystery. The tone of these passages of the *Principia*, even in their references to religion, is rational and polemical, and not in the least bit mystical. Fierz's lecture concluded:

> Newton's contributions to physics are among the pillars of the mecha-nistic/rationalist trends of the subsequent centuries. However, he him-self never succumbed to the intellectual optimism of his followers, who believed to have in their hands the keys to the solution of the problems of the world. His deeper understanding of nature, and of the human heart, saved him from this error. Furthermore, he knew that if all our knowledge comes from experience, then it can never transcend our experience.

In short, in his essay about Newton, Fierz made the same point that Pauli made in his Kepler study, but he made it more transparently – at least for me.

While he was delving into Kepler and his critics in 1947, Pauli read Fierz's essay and thanked him for it. Of course he could not pass up the opportunity to add some witty glosses of his own. He wrote to Fierz, who was in Basel at the time, that Newton had installed space and time at the right hand of God because that position had just been vacated by Jesus Christ after his demotion by the Deists. Later, Pauli went on, it took truly extraordinary efforts to pull space and time back down again from this exalted spot. Those efforts, of course, were a reference to relativity, a subject Pauli knew so well.

But Pauli does not comment on Fierz's multiple uses of the word "symbolic" which would play such a central role in their discussions in subsequent years.

In the following year Pauli acknowledged reading another lecture by Fierz. This one was given at the annual Eranos Conference in Ascona, Switzerland, near the Italian border.[3] Eranos has been meeting from 1933 until today, and brings together scholars from East and West for a weeklong, interdisciplinary discussion of some specific theme. Schrödinger had spoken there, and Jung was a key participant. The list of Eranos lectures includes one by Hermann Weyl on "Science as a symbolic human construct," but none by Pauli. In 1948 Fierz was invited to speak about a humanistic topic from the point of view of a physicist. The title he chose was "Zur physikalischen Erkenntnis" which means something like "On knowledge in physics" or "On the epistemology of physics". This lecture (Fierz 1949) opens Fierz's collection of essays which I mentioned earlier.

After Pauli read Fierz's Eranos lecture, he decided to reach out. On 12 August 1948 (von Meyenn 1993, pp. 558–562) he reported to Fierz a long conversation with Jung about the program he called psycho-physical unification. At the end Jung asked Pauli what his fellow physicists thought of the idea. Pauli didn't know – he had never revealed it. Although he realized that it would be a daring experiment – his very reputation might be at risk – he decided to start discussing such matters with Fierz. The conversation would continue for the remaining decade of Pauli's life.

Just as Fierz's views on the first question (where do novel physical concepts come from?) seems simpler and more straightforward than Pauli's, his take on the second question – psycho-physical unification – is illuminating. In his Eranos lecture he begins by pointing out that the worldview of classical physics depended on the strict separation of the observing subject from the observed objective world. This neat division condemns physics to be powerless in the face of human phenomena such as cognition and consciousness. The physical worldview is powerful, but it is unbalanced, one-sided. "Fortunately", writes Fierz, "quantum mechanics shows a way out."

Quantum mechanics demonstrates emphatically that individual events are not subject to the venerable, grand law of cause and effect that underlies classical physics. According to Fierz, only the totality of many similar events, a highly abstract concept, is amenable to theoretical investigation. The language of this description is mathematical, and thus symbolic. Of course mathematics has been the principal tool of physics since the time of Galilei,

---

[3]Pauli, in his letter of 7 January 1948 (von Meyenn 1993, pp. 495–498), implies that this talk had originally been presented to the Psychological Club of Zürich.

but Fierz contends that quantum mechanics brings out the symbolic nature of physics with special clarity.

He goes to considerable trouble to flesh out how mathematics is symbolic. One is tempted to see in mathematics the epitome of rationality. This, he claims, is not altogether wrong, but it is insufficient: Mathematics shows aspects that go far beyond mere logic. At its most primitive level there are triangles, squares, and five Platonic solids which exert a fascination that is certainly not rational. Even the integers are the basis of symbolic systems like that of Pythagoras. But beyond that Fierz points out that all integers have individual, qualitative characters – that they are no mere quantities.

For illustration he recalls that algebraic equations up to fourth order can be solved by explicit formulas, which do not exist for fifth order and higher. Thus, the numbers 2, 3, and 4 have attributes not shared by any other integers. Mathematical constructs, Fierz concludes, can be compared to archetypical symbols.

At this point Fierz introduces a compelling simile. He asks: Why can't physical science be extended to cover the immaterial universe? The answer is that physical science has developed a very specific, well-defined technique called the scientific method, which applies only to reproducible phenomena. Accordingly, what is unique and individual cannot be a subject for physics. So the inner, spiritual world of dreams and feelings, which is by nature irreproducible, escapes the understanding of physical science. "The physicist resembles King Midas, in whose hands everything turned to gold", Fierz concluded. "The method of physics seems to be such that nature, or the world, contemplated physically, turns out to be physical." What a pessimistic comment on our profession of physics! By our own method we are condemned to miss half of the world we live in.

Fierz goes on to suggest that the way to study the inner, irrational world is not physics, but psychology: "One can expect that a sufficiently developed psychology will make every problem appear to be psychological. And again this royal claim will turn out to be the gift of King Midas."

To reconcile the two contrasting world views of physics and psychology, Fierz proposes – independently of Pauli, I believe – that they are complementary, and might one day be combined into a highly abstract, symbolic, unified representation of the world. But he has no idea what this novel science would look like, except that in comparison, today's theoretical physics would look like mere child's play.

On June 2, 1949, Pauli wrote to Fierz (von Meyenn 1993, p. 657) that he was satisfied with everything in the Eranos lecture, except the end. He didn't believe that a unified psycho-physical theory would necessarily be so incredibly difficult. But from then until Pauli's death nine years later the two friends made no progress toward that Holy Grail. It remains a challenge for the future.

I would like to end my remarks by examining in more detail the concluding passage of Fierz's Eranos lecture. I have read it many times, and find tantalizing hints of profound insights, but I have a hard time organizing them into a rigorous, logical argument. In this volume, dedicated to the effort of distilling meaning from obscure texts, I'll give it a try.

The world, Fierz writes, can be seen under two aspects, which he chooses to call, in admittedly oversimplified fashion, the physical and the psychological. This division, he feels, is more convincing than splitting the world into the two separate categories of matter and spirit. It is seductive to regard physics and psychology as complementary, but this suggestion is, of course, only an analogy.

There follows a painstaking illustration of the concept of complementarity, as applied to the example of the wave-particle duality of light. I will skip this passage and, using a lawyerly phrase, stipulate its contents. But the example does not exhaust the meaning of complementarity. For that, Fierz cautions his lay audience, a course on quantum mechanics would be required. This brings me to the final paragraphs of the Eranos lecture (my translation):

> At least it should be clear by now that a physical phenomenon cannot be described in an intuitively accessible way [anschaulich in German] without seriously taking into account the manner in which the phenomenon is visualized. But this is not achieved by subjecting the effect of the measuring apparatus on the object under observation to a causal analysis. Rather, the measuring apparatus determines a certain "aspect" of the phenomenon, which can be interpreted as a consequence of the effect of the apparatus on the object. This effect cannot be quantitatively controlled. It can be shown that this failure is related to the fact that the measuring apparatus and the object under observation must be rigorously distinguished from each other.
>
> More generally, it seems to be the case that the perception of any object presupposes a subject, which must be distinguished from the former. But there must always be a relationship between object and subject. But a real relationship implies effects of subject and object on each other. These cannot be accounted for during the act of perception. For if they were, the separation of object from subject would be suspended, and then the distinction would be a mere verbal one, an unjustified formulation of our thinking.
>
> For this reason every act of cognition [Erkenntnis] appears as an act of creation [ein schöpferischer Akt], in which subject and object, by act-

ing on each other, experience a transformation [*Wandlung*]. In contrast, classical physics corresponds to the position that an act of perception causes changes unilaterally in the subject, which is why objects appear unaffected and absolute. To me it appears essential that it remains to a certain extent arbitrary what is counted as belonging to the sphere of the object, and what to the sphere of the subject. From this emerge the different aspects under which reality reveals itself.

Physics and psychology seem to me to be complementary ways to view the world, each corresponding to a certain attitude of consciousness [*Einstellung des Bewusstseins*]. The aspects of the world gained with the help of physics and psychology are images of the same world, which cannot be united in an intuitively accessible way. This could only become possible in the framework of a symbolic representation, which would, to be sure, have a highly abstract character. It would be comprehensible only to the scientifically educated, and in comparison with this new science, theoretical physics would appear as a simple and introductory subject. What this science, which leads to such a comprehensive symbolic understanding of the world, looks like, about that, it must be said, we have not the faintest idea.

We recognize in this passage yet another reference to John Bell's infamous "shifty split" – in this case the movable boundary between object and subject. Fierz implies that by suitably moving the location of this split we can apprehend the world either physically or psychologically. But I don't understand that last step of his argument – how he jumps from the mobility of the split to the complementarity of physics and psychology.

However, I salute David Mermin's recent declaration (Mermin 2012) that Quantum Bayesianists have, at long last, fixed the shifty split.[4] They have pushed it right up to the boundary between the objective world and my own personal, conscious perception of the world. But I suspect that Fierz and Pauli wanted to push even further, right into the territory that Mermin carefully avoided, the realm of dreams and hallucinations. There is work to be done, so let's roll up our sleeves.

References

Fierz M. (1943): Isaac Newton. Sein Charakter und seine Weltansicht. *Vierteljahresschrift der Naturforschenden Gesellschaft Zürich* **88**, 198–216.

Fierz M. (1949): Zur physikalischen Erkenntnis. In *Naturwissenschaft und Geschichte* by M. Fierz, Birkhäuser, Basel, pp. 9–30.

---

[4]Compare the contributions by Fuchs and by Fuchs and Schack in this volume.

Fierz M. (1979): Naturerklärung und Psyche. Ein Kommentar zu dem Buch von C.G. Jung und W. Pauli. In *Naturwissenschaft und Geschichte* by M. Fierz, Birkhäuser, Basel, pp. 181–191.

Fierz M. (1988): *Naturwissenschaft und Geschichte*, Birkhäuser, Basel.

Gieser S. (2005): *The Innermost Kernel*, Springer, Berlin.

Jung C.G. and Pauli W. (1952): *Naturerklärung und Psyche*, Rascher, Zürich. English translation: *The Interpretation of Nature and the Psyche*, Routledge and Kegan Paul, London 1955.

Mermin N.D. (2012): Commentary: Quantum mechanics: Fixing the shifty split. *Physics Today* **65**(7), 8.

von Meyenn K., ed. (1993): *Wolfgang Pauli. Wissenschaftlicher Briefwechsel Band III: 1940–1949*, Springer, Berlin.

von Meyenn K. (2007): Markus Eduard Fierz (1912–2006). *Mind and Matter* **5**, 241–267.

Quantum Bayesianism
for the Uninoculated

Christopher A. Fuchs

Abstract

This article summarizes the quantum Bayesian point of view of quantum mechanics, with special emphasis on the view's most radical variant – quantum Bayesianism, briefly QBism.[1] QBism has its roots in personalist Bayesian probability theory, is crucially dependent upon insights from modern quantum information theory, and most recently, has set out to investigate whether the world might be of a type sketched by some early 20th century philosophies: pragmatism, pluralism, nonreductionism, and meliorism.

1   A Feared Disease

The end of the last decade saw a media frenzy over the possibility of an H1N1 flu pandemic. The frenzy turned out to be misplaced, but it did serve to remind us of a basic truth: That a healthy body can be stricken with a fatal disease which to outward appearances is nearly identical to a common yearly annoyance. There are lessons here for quantum mechanics. In the history of physics, there has never been a healthier body than quantum theory. No theory has ever been more all-encompassing or more powerful. Its calculations are relevant at every scale, from subnuclear particles to table-top lasers, to the cores of neutron stars and even the first three minutes of the universe. Yet since its founding days, many physicists have feared that quantum theory's common annoyance – the continuing feeling that something at the bottom of it does not make sense – may one day turn out to be the symptom of something fatal.

There is something about quantum theory that is different in character from any physical theory before. To put a finger on it, the issue is this:

---

[1] The present article represents a much condensed version of my manuscript "QBism, the perimeter of quantum Bayesianism", available at arxiv.org/abs/1003.5209.

The basic statement of the theory – the one we have all learned from our textbooks – seems to rely on terms our intuitions balk at as having any place in a fundamental description of reality. The notions of "observer" and "measurement" are taken as primitives, the very starting point of the theory. This is an unsettling situation! Shouldn't physics be talking about *what is* before it starts talking about *what will be seen* and who will see it? Few have expressed this more forcefully than John Bell (1990):

> What exactly qualifies some physical systems to play the role of "measurer"? Was the wavefunction of the world waiting to jump for thousands of millions of years until a single-celled living creature appeared? Or did it have to wait a little longer, for some better qualified system ... with a PhD?

One sometimes feels that until this issue is settled, fundamental physical theory has no right to move on.

But what constitutes "progress" in quantum foundations? How would one know progress if one saw it? Through the years, it seems that the most popular strategy has taken its cue (even if only subliminally) from the tenor of Bell's quote: The idea has been to remove the observer from the theory just as quickly as possible, and with surgical precision. In practice this has generally meant to keep the *mathematical structure* of quantum theory as it stands (complex Hilbert spaces, operators, tensor products, etc.), but, by hook or by crook, to invent a story about the mathematical symbols that involves no observers at all.

In short, the strategy has been to reify or objectify all the mathematical symbols of the theory and then explore whatever comes of the move. Three examples suffice to give a feel: In the de Broglie-Bohm "pilot wave" version of quantum theory on N particles, there are no fundamental measurements, only "particles" flying around in a $3N$-dimensional configuration space, pushed around by a wave function regarded as a real physical field in that space. In "spontaneous collapse" versions, systems are endowed with quantum states that generally evolve unitarily, but occasionally collapse without any need for measurement. In Everettian or "many-worlds" quantum mechanics, it is only the world as a whole – they call it a multiverse – that is really endowed with an intrinsic quantum state, and that quantum state evolves deterministically, with only an *illusion from the inside* of probabilistic "branching".

The trouble with all these interpretations as quick fixes for Bell's vivid remark is that they look to be just that, *really quick fixes*. They look to be interpretive strategies hardly compelled by the particular details of the quantum formalism, adding only more or less arbitrary appendages to it. This already explains in part why we have been able to exhibit three such different strategies, but it is worse: Each of these strategies gives rise to its own set of incredibilities – ones which, if one had Bell's gift for words, one could make look just as silly. Pilot-wave theories, for instance, give instantaneous action at a distance, but not actions that can be harnessed to send detectable signals. If so, then what a delicately balanced high-wire act nature presents us with? Or the Everettians: Their world purports to have no observers, but then it has no probabilities either. What are we then to do with the Born rule for calculating quantum probabilities? Throw it away and say it never mattered?

2 Quantum States Do Not Exist

There is another lesson from the H1N1 virus. It is that sometimes immunities can be found in unexpected populations. To some perplexity at the beginning, it seemed that people over 65 – a population usually more susceptible to fatalities with seasonal flu – fared much better than younger folk with H1N1. The leading theory was that the older population, in its years of other exposures, has developed various latent antibodies. The antibodies were not perfect, but they were a start. And so it may be for quantum foundations.

Here, the latent antibody is the concept of *information*, and the perfected vaccine, we believe, will arise in part from the theory of single-case, personal probabilities – the branch of probability theory called Bayesianism. Metaphorically, the older population corresponds to some of the founders of quantum theory (Heisenberg, Pauli, Einstein) and some of the younger disciples of the Copenhagen school (Peierls, Wheeler, Peres), who, though they disagreed on many details of the vision – *whose information? information about what?* – were unified on one point: That quantum states are not something out there, in the external world, but instead are expressions of information.

Before there were people using quantum *theory* as a branch of physics, before they were *calculating* neutron-capture cross-sections for uranium and working on all the other practical problems the theory suggests, there were no quantum states. The world may be full of stuff and things of all kinds, but among all the stuff and all the things, there is no unique, observer-independent, *quantum-state kind of stuff*. The immediate payoff of this strategy is that it eliminates the conundrums arising in the various objectified-state interpretations. A paraphrase of a quote by Hartle (1968) makes the point decisively:

> A quantum-mechanical state being a summary of the observer's information about an individual physical system changes both by dynamical laws, and whenever the observer acquires new information about the system through the process of measurement. The existence of two laws for the evolution of the state vector becomes problematical only if it is believed that the state vector is an objective property of the system. If, however, the state of a system is defined as a list of ... [experimental] propositions together with their ... [probabilities of occurrence], it is not surprising that after a measurement the state must be changed to be in accord with ... new information. The "reduction of the wave packet" does take place in the consciousness of the observer, not because of any unique physical process which takes place there, but only because the state is a construct of the observer and not an objective property of the physical system.

It says that the real substance of Bell's fear is just that, the fear itself. To succumb to it is to block the way to understanding the theory on its own terms. Moreover, the shriller notes of Bell's rhetoric are the least of the worries: The universe did not have to wait billions of years to collapse its first wave function – wave functions are not part of the observer-independent world.

But this much of the solution is an elderly and somewhat ineffective antibody. It can be significantly strengthened by lessons learned from the field of quantum information theory – the multidisciplinary field that brought about quantum cryptography, quantum teleportation, and will one day bring about quantum computation. What the protocols and theorems of quantum information pound home is the idea that quantum states look, act, and feel like information in the technical sense of the word – the sense provided by probability theory and Shannon's information theory.

There is no more beautiful demonstration of this than Robert Spekkens's "toy model" for mimicking various features of quantum mechanics (Spekkens 2007). In that model, the "toys" are each equipped with four possible mechanical configurations, but the players, the manipulators of the toys, are consistently impeded – for whatever reason – from having more than one bit of information about each toy's actual configuration (or a total of two bits for each two toys, three bits for each three toys, and so on). The only things the players can know are their states of uncertainty about the configurations. The wonderful thing is that these states of uncertainty exhibit many of the characteristics of quantum information: from the no-cloning theorem to analogues of quantum teleportation, quantum key distribution, entanglement monogamy, and even interference in a Mach-Zehnder interferometer. More than two dozen quantum phenomena are reproduced *qualitatively*, and all the while one can identify the underlying cause of the occurrence: The phenomena arise in the uncertainties, never in the mechanical configurations. It is the states of uncertainty that mimic the formal apparatus of quantum theory, not the toys' so-called *ontic states* (states of reality).

What considerations like this tell the ψ-ontologists – i.e., those who attempt to remove the observer too quickly from quantum mechanics by giving quantum states an unfounded ontic status – was well put by Spekkens (2007):

> [A] proponent of the ontic view might argue that the phenomena in question are not mysterious if one abandons certain preconceived notions about physical reality. The challenge we offer to such a person is to present a few simple physical principles by the light of which all of these phenomena become conceptually intuitive (and not merely mathematical consequences of the formalism) within a framework wherein the quantum state is an ontic state. Our impression is that this challenge cannot be met. By contrast, a single information-theoretic principle, which imposes a constraint on the amount of knowledge one can have about any system, is sufficient to derive all of these phenomena in the context of a simple toy theory ...

The point is that, far from being an appendage cheaply tacked on to the theory, the idea of quantum states as information has a simple unifying power that goes some way toward explaining why the theory has the particular mathematical structure it actually does. By contrast, who could take the many-worlds idea and derive any of the structure of quantum theory out of it? This would be a bit like trying to regrow a lizard from the tip of its chopped-off tail: The Everettian conception never purported to be more than a reaction to the formalism in the first place.

There are, however, aspects of Bell's challenge (or the mindset behind it), that remain a worry. And upon these, all could still topple. There are the old questions of *whose information?* and *information about what?* which must be addressed before any vaccination can be declared a success. It must also be settled whether quantum theory is obligated to give a criterion for what counts as an observer. Finally, because no one wants to give up on physics, we must tackle head-on the most crucial question of all: If quantum states are not part of the stuff of the world, then what is? What sort of stuff does quantum mechanics say the world *is* made of?

Good immunology does not come easily. But this much is sure: The glaringly obvious (that the central part of quantum theory is about information) should not be abandoned rashly: To do so is to lose grip of the theory as it is applied in practice, with no better grasp of reality in return. If on the other hand, one holds fast to the central point about information, initially frightening though it may be, one may still be able to reconstruct a picture of reality from the unfocused edge of vision. Often the best stories come from there anyway.

3 Quantum Bayesianism

Every area of human endeavor has its bold extremes – ones that say: "If this is going to be done right, we must go this far. Nothing less will do." In probability theory, the bold extreme is the personalist Bayesian account of it (Bernardo and Smith 1994). It says that probability theory resembles the character of formal logic – both provide a set of criteria for testing consistency.

In the case of formal logic, the consistency is between truth values of propositions. However, logic itself does not have the power to *set* the truth values it manipulates. It can only say if various truth values are consistent or inconsistent. The actual values come from another source. Whenever logic reveals a set of truth values to be inconsistent, one must dip back into the source to find a way to alleviate the discord. But precisely in which way to alleviate it, logic gives no guidance. "Is the truth value for this one isolated proposition correct?" Logic itself is powerless to say.

The key idea of personalist Bayesian probability theory is that it too is a calculus of consistency (or "coherence", as the practitioners call it), but in this case for one's decision-making degrees of belief. Probability theory can only determine whether various degrees of belief are consistent or inconsistent with each other. The actual beliefs come from another source, and there is nowhere to pin their responsibility but on the agent who holds them. As Lindley (2006) put it:

> The Bayesian, subjectivist, or coherent, paradigm is egocentric. It is a tale of one person contemplating the world and not wishing to be stupid (technically, incoherent). He realizes that to do this his statements of uncertainty must be probabilistic.

A probability *assignment* is a tool an agent uses to make gambles and decisions – it is a tool he uses for navigating life and responding to his environment. Probability *theory* as a whole, on the other hand, is not about a single isolated belief, but about a whole mesh of them. When a belief in the mesh is found to be incoherent with the others, the theory flags the inconsistency. However, it gives no guidance for how to mend any incoherences it finds. To alleviate the discord, one can only dip back into the source of the assignments – specifically, the agent who attempted to sum up all his history, experience, and expectations with those assignments in the first place. This is the reason for the terminology that a probability is a "degree of belief" rather than a "degree of truth" or "degree of facticity."

Where personalist Bayesianism breaks away the most from other developments of probability theory is that it says there are no *external* criteria for declaring an isolated probability assignment right or wrong. The only basis for a judgment of adequacy comes from the *inside*, from the greater mesh of beliefs the agent may have the time or energy to access when appraising coherence.

It was not an arbitrary choice of words to title the previous section

QUANTUM STATES DO NOT EXIST,

but a hint of the direction we must take to develop a perfected vaccine. This is because the phrase has a precursor in a slogan Bruno de Finetti, the founder of personalist Bayesianism, used to vaccinate probability theory itself. In the preface of his seminal book, de Finetti (1990) writes, centered in the page and in all capital letters,

PROBABILITY DOES NOT EXIST.

It is a powerful statement, constructed to put a finger on the single most significant cause of conceptual problems in pre-Bayesian probability theory. A probability is not a solid object, like a rock or a tree that the agent might bump into, but a feeling, an estimate inside oneself.

Previous to Bayesianism, probability was often thought to be a physical property – something objective and having nothing to do with decision-making or agents at all. But when thought so, it could be thought only inconsistently so. And hell hath no fury like an inconsistency scorned. The trouble is always the same in all its varied and complicated forms: If probability is to be a physical property, it had better be a rather ghostly one – one that can be told of in campfire stories, but never quite prodded out of the shadows. Here's a sample dialogue:

> **Pre-Bayesian:** Ridiculous, probabilities are without doubt objective. They can be seen in the relative frequencies they cause.
>
> **Bayesian:** So if $p = 0.75$ for some event, after 1000 trials we'll see exactly 750 such events?

Pre-Bayesian: You might, but most likely you won't see that exactly. You're just likely to see something close to it.

Bayesian: Likely? Close? How do you define or quantify these things without making reference to your degrees of belief for what will happen?

Pre-Bayesian: Well, in any case, in the infinite limit the correct frequency will definitely occur.

Bayesian: How would I know? Are you saying that in one billion trials I could not possibly see an "incorrect" frequency? In one trillion?

Pre-Bayesian: OK, you can in principle see an *incorrect* frequency, but it'd be ever less *likely*!

Bayesian: Tell me once again, what does "likely" mean?

This is a cartoon of course, but it captures the essence and the futility of every such debate. It is better to admit at the outset that probability is a degree of belief, and deal with the world on its own terms as it coughs up its objects and events. What do we gain for our theoretical conceptions by saying that along with each actual event there is a ghostly spirit (its "objective probability," its "propensity," its "objective chance") gently nudging it to happen just as it did? Objects and events are enough by themselves.

In quantum mechanics, too, if ghostly spirits are imagined behind the actual events produced in quantum measurements, one is left with conceptual troubles to no end. The defining feature of qantum Bayesianism (Caves *et al.* 2002, Fuchs 2002, Fuchs and Schack 2004, Caves *et al.* 2007, Fuchs 2010, Fuchs and Schack 2013) is that it says along the lines of de Finetti: "If this is going to be done right, we must go this far." Specifically, there can be no such thing as a right and true quantum state, if such is thought of as defined by criteria *external* to the agent making the assignment. Quantum states must instead be like personalist, Bayesian probabilities.

The direct connection between the two foundational issues is this. Quantum states, through the Born rule, can be used to calculate probabilities. Conversely, if one assigns probabilities for the outcomes of a well-selected set of measurements, then this is mathematically equivalent to making the quantum-state assignment itself. The two kinds of assignments determine each other uniquely. Just think of a spin-$\frac{1}{2}$ system. If one has elicited one's degrees of belief for the outcomes of a measurement of the spin σ_x in direction x, and similarly one's degrees of belief for the outcomes of σ_y and σ_z measurements, then this is the same as specifying a quantum state itself: For if one knows the quantum state's projections onto three independent axes, then that uniquely determines a Bloch vector, and hence a quantum state. Something similar is true of all quantum systems of all sizes and dimensionality. There is no mathematical fact embedded in a quantum state ρ that is not embedded in an appropriately chosen set of probabilities. Thus generally, if probabilities are personal in the Bayesian sense, then so too must be quantum states.

What this buys interpretatively, beside airtight consistency with the best understanding of probability theory, is that it gives each quantum state a home. Indeed, a home localized in space and time – namely, the physical site of the agent who assigns it! By this method, one expels once and for all the fear that quantum mechanics leads to "spooky action at a distance," and expels as well any hint of a problem with "Wigner's friend". It does this because it removes the very last trace of confusion over whether quantum states might still be objective, agent-independent, physical properties.

The innovation here is that, for most of the history of efforts to take an informational point of view about quantum states, the supporters of the idea have tried to have it both ways: on the one hand quantum states are not real physical properties, yet on the other there is a right quantum state independent of the agent after all. For instance, one hears things like: "The *right* quantum state is the one the agent should adopt if he had all the information." The tension in these two desires leaves their holders open to attack on both flanks and general confusion all around.

Take first instantaneous action at a distance – the distaste of this idea is often one of the strongest motivations for those seeking to take an informational stance on quantum states. Without the protection of truly personal quantum-state assignments, action at a distance is there as doggedly as it ever was. And things only get worse with "Wigner's friend" if one insists there be a *right* quantum state. As it turns out, the method of mending this conundrum displays one of the most crucial ingredients of QBism. Let us put it in plain sight.

"Wigner's friend" is the story of two agents, Wigner and his friend, and one quantum system – the only deviation we make from a more common presentation is that we put the story in informational terms. It starts off with the friend and Wigner having a conversation. Suppose they both agree that some quantum state $|\psi\rangle$ captures their common beliefs about the quantum system. Furthermore suppose they agree that at a specified time the friend will make a measurement on the system of some observable (outcomes $i = 1, \ldots, d$). Finally, they both note that if the friend gets outcome i, he will (and should) update his beliefs about the system to some new quantum state $|i\rangle$. There the conversation ends and the action begins: Wigner walks away and turns his back to his friend and the supposed measurement. Time passes to some point beyond when the measurement should have taken place.

What now is the "correct" quantum state each agent should have assigned to the quantum system? We have already concurred that the friend will and should assign some $|i\rangle$. But what about Wigner? If he were to consistently dip into his mesh of beliefs, he would very likely treat his friend as a quantum system like any other: one with some initial quantum state ρ capturing his (Wigner's) beliefs of *it* (the friend), along with a linear evolution operator U to adjust those beliefs with the flow of time.[2] Suppose this quantum state includes Wigner's beliefs about everything he assesses to be interacting with his friend – in old parlance, suppose Wigner treats his friend as an isolated system. From this perspective, before any further interaction between himself and the friend or the other system, the quantum state Wigner would assign for the two together would be $U(\rho \otimes |\psi\rangle\langle\psi|)U^\dagger$ – most generally an entangled quantum state. The state of the system itself for Wigner would be gotten from this larger state by a partial trace operation. In any case, it will not be an $|i\rangle$.

Does this make Wigner's new state assignment incorrect? After all, "if he had all the information" (i.e., all the facts of the world), wouldn't that include knowing the friend's measurement outcome? Since the friend should assign some $|i\rangle$, shouldn't Wigner assign the same (if he had all the information)? Or is it the friend who is incorrect? For if the friend had "all the information", wouldn't he say that he is neglecting that Wigner could put the system and himself into the quantum computational equivalent of an iron lung and forcefully reverse the so-called measurement? That is, Wigner, if he were sufficiently sophisticated, should be able to force

$$U(\rho \otimes |\psi\rangle\langle\psi|)U^\dagger \longrightarrow \rho \otimes |\psi\rangle\langle\psi| . \tag{1}$$

And so the back and forth goes. Who has the *right* state of information? The conundrums simply gets too heavy if one tries to hold to an agent-independent notion of correctness for otherwise personalistic quantum states. QBism dispels these and similar difficulties of the "aha, caught you!" variety by being conscientiously forthright. *Whose information?* "Mine!" *Information about what?* "The consequences (for *me*) of *my* actions upon the physical system!" It's all "I-I-me-me mine," as the Beatles sang.

The answer to the first question surely comes as no surprise by now, but why on earth the answer to the second? "It's like watching a QBist

---

[2]For an explanation of the status of unitary operations from the QBist perspective, as personal judgments directly analogous to quantum states themselves, see Fuchs (2002) and Fuchs and Schack (2004).

shoot himself in the foot", a friend once said. Why something so egocentric, anthropocentric, psychology-laden, myopic, and positivistic (we have heard any number of expletives) as *the consequences (for me) of my actions upon the system?* Why not simply say something neutral like "the outcomes of measurements"? To the uninitiated, our answer for *information about what?* surely appears to be a cowardly, unnecessary retreat from realism. But it is the opposite. The answer we give is the very injunction that keeps the potentially conflicting statements of Wigner and his friend in check, at the same time as giving each agent a hook to the external world in spite of QBism's egocentric quantum states.

For QBists, the real world, the one both agents are embedded in – with its objects and events – is taken for granted. What is not taken for granted is each agent's access to the parts of it he has not touched. Wigner holds two thoughts in his head: (1) that his friend interacted with a quantum system, eliciting some consequence of the interaction for himself, and (2) after the specified time, for any of Wigner's own further interactions with his friend or system or both, he ought to gamble upon their consequences according to $U(\rho \otimes |\psi\rangle\langle\psi|)U^\dagger$. One statement refers to the friend's potential experiences, and one refers to Wigner's own. So long as it is kept clear that $U(\rho \otimes |\psi\rangle\langle\psi|)U^\dagger$ refers to the latter – how Wigner should gamble upon the things that might happen to him – making no statement whatsoever about the former, there is no conflict.

The world is filled with all the same things it was before quantum theory came along, like each of our experiences, that rock and that tree, and all the other things under the sun. It is just that quantum theory provides a calculus for gambling on each agent's own experiences – it does not give anything else than that. It certainly does not give one agent the ability to conceptually pierce the other agent's personal experience. It is true that with enough effort Wigner could enact Eq. (**??**), causing him to predict that his friend will have amnesia to any future questions on his old measurement results. But we always knew Wigner could do that – a mallet to the head would have been good enough.

The key point is that quantum theory, in this light, takes nothing away from the usual world of common experience we already know. It only *adds*. At the very least it gives each agent an extra tool with which to navigate the world. More than that, the tool is here for a reason. QBism says when an agent reaches out and touches a quantum system – when he performs a *quantum measurement* – that process gives rise to birth in a nearly literal sense. With the action of the agent upon the system, the no-go theorems of Bell and Kochen-Specker assert that something new comes into the world that was not there previously. It is the "outcome," the unpredictable consequence for the very agent who took the action. John Wheeler (1982) said it this way, and we follow suit: "Each elementary quantum phenomenon is an elementary act of 'fact creation'."

With this much, QBism has a story to tell on both quantum *states* and quantum *measurements*, but what of quantum *theory* as a whole? The answer is found in taking it as a *universal* single-user theory in much the same way that Bayesian probability theory is. It is a user's manual that *any* agent can pick up and use to help make wiser decisions in this world of inherent uncertainty.[3] To say it in a more poignant way: In my case, it is a world in which *I* am forced to be uncertain about the consequences of most of *my* actions; and in your case, it is a world in which *you* are forced to be uncertain about the consequences of most of *your* actions.

"And what of God's case? What is it for him?" Trying to give *him* a quantum state was what caused this trouble in the first place! In a quantum mechanics with the understanding that each instance of its use is strictly single-user – "my measurement outcomes happen right here, to me, and I am talking about my uncertainty of them" – there is no room for most of the standard, year-after-year quantum mysteries.

The only substantive *conceptual* issue left before synthesizing a final vaccine is whether quantum mechanics is obligated to derive the notion of agent for whose aid the theory was built in the first place. The answer comes from turning the tables. Thinking of probability theory in the personalist Bayesian way, as an extension of formal logic, would one ever imagine that the notion of an agent, the user of the theory, could be derived out of its conceptual apparatus? Clearly not. How could you possibly get flesh and bones out of a calculus for making wise decisions? The logician and the logic he uses are two different substances – they live in conceptual categories worlds apart. One is in the stuff of the physical world, and the other is somewhere

---

[3] Most of the time one sees Bayesian probabilities characterized as measures of ignorance or imperfect knowledge. But that description carries with it a metaphysical commitment that is not at all necessary for the personalist Bayesian, where probability theory is an extension of logic. Imperfect knowledge? It sounds like something that, at least in imagination, could be perfected, making all probabilities zero or one – one uses probabilities only because one does not know the true, pre-existing state of affairs. Language like this, the reader will notice, is never used in this paper. All that matters for a personalist Bayesian is that there is *uncertainty* for whatever reason. There might be uncertainty because there is ignorance of a true state of affairs, but there might be uncertainty because the world itself does not yet know what it will give – i.e., there is an objective indeterminism. As will be argued in later sections, QBism finds its happiest spot in an unflinching combination of "subjective probability" with "objective indeterminism".

nearer to Plato's heaven of ideal forms. Look as one might in a probability textbook for the ingredients to reconstruct the reader himself, one will never find them. So too, QBism says of quantum theory.

With this we finally pin down the precise way in which quantum theory is "different in character from any physical theory posed before". For QBism, quantum theory is not something *outside* probability theory – it is not a picture of the world as it is, as say Einstein's program of a unified field theory hoped to be – but rather it is an *addition* to probability theory itself. As probability theory is a *normative* theory, not saying what one *must* believe, but offering rules of consistency an agent should strive to satisfy within his overall mesh of beliefs, so it is the case with quantum theory.

To take this substance into one's mindset is all the vaccination one needs against the threat that quantum theory carries something viral for theoretical physics as a whole. A healthy body is made healthier still. For with this protection, we are for the first time in a position to ask, with eyes wide open to what the answer could not be, *just what after all is the world made of?* Far from being the last word on quantum theory, QBism is the start of a great adventure.

4   Hilbert-Space Dimension as a Universal Capacity

A common accusation against QBism is that it leads straight away to solipsism, "the belief that all reality is just one's imagining of reality, and that one's self is the only thing that exists". The accusation goes that, if a quantum state only represents the degrees of belief held by some agent then the agent's beliefs must be the source of the universe. The universe could not exist without him. This being such a ridiculous idea, QBism is dismissed out of hand, *reductio ad absurdum*. It is so hard for the QBist to understand how anyone could think this (it being the antithesis of everything in his worldview) that another Latin phrase comes to mind: *non sequitur*.

A fairer-minded assessment is that the accusation springs from our opponents "hearing" much of what we do say, but interpreting it in terms drawn from a particular conception of what physical theories *always ought to be*: Attempts to directly represent (map, picture, copy, correspond to, correlate with) the *universe* – with "universe" here thought of as a static, timeless block that just *is*. From such a "representationalist" point of view, *if* (a) quantum theory is a proper physical theory, (b) its essential theoretical objects are quantum states, and (c) quantum states are states of belief, *then* the universe that "just is" corresponds to a state of belief. What else is this but a kind of solipsism?

QBism sidesteps the poisoned dart by asserting that quantum theory is just not a physical theory in the sense the accusers want it to be. Rather it is an addition to personal, Bayesian, normative probability theory. Its normative rules for connecting probabilities (personal judgments) were developed in light of the *character of the world*, but there is no sense in which the quantum state itself represents (pictures, copies, corresponds to, correlates with) a part or a whole of the external world, much less a world that *just is*. In fact the very character of the theory points to the inadequacy of the representationalist program when attempted on the particular world we live in.

QBism does not argue that representationalism must be wrong always and in all possible worlds (perhaps because of some internal inconsistency). Representationalism may well be true in this or that setting – we take no stand on the matter. We only know that for nearly 90 years quantum theory has been actively resistant to representationalist efforts on *its* behalf. This suggests that it might be worth exploring some philosophies upon which physics rarely sets foot. Physics of course should never be constrained by any one philosophy, but it does not hurt to get ideas and insights from every source one can. If one were to scan the philosophical literature for schools of thought representative of what QBism actually is about, it is not solipsism one will find, but nonreductionism (Dupré 1993, Cartwright 1999), metaphysical pluralism (James 1996a, Wahl 1925), empiricism (James 1940, 1996b), indeterminism and meliorism[4] (James 1884), and above all pragmatism (Thayer 1981).

A form of nonreductionism can already be seen in play in our answer to whether the notion of agent should be derivable from the quantum formalism itself. We say that it cannot be and it should not be, and to believe otherwise is to misunderstand the subject matter of quantum theory. But nonreductionism also goes hand in hand with the idea that there is real particularity and "interiority" in the world. Think again of the "I-I-me-me mine" feature that shields QBism from inconsistency in the "Wigner's friend" scenario. When Wigner turns his back to his friend's interaction with the system, that

---

[4]Strictly speaking, meliorism is the doctrine "that humans can, through their interference with processes that would otherwise be natural, produce an outcome which is an improvement over the aforementioned natural one". But we would be reluctant to take a stand on what "improvement" really means. So said, all we mean in the present essay by meliorism is that the world before the agent is malleable to some extent – that his actions really can change it.

piece of reality – Bohr might call it a "phenomenon" – is hermetically sealed from him. It has an inside, a vitality that he takes no part in until he again interacts with one or both relevant pieces of it. *With respect to Wigner*, it is a bit like a universe unto itself.

If one seeks the essence of indeterminism in quantum mechanics, there may be no example more directly illustrative of it than "Wigner's friend". For it expresses to a tee William James's notion of indeterminism (James 1884, p. 145):

> [Chance] is a purely negative and relative term, giving us no information about that of which it is predicated, except that it happens to be disconnected with something else – not controlled, secured, or necessitated by other things in advance of its own actual presence. ... What I say is that it tells us nothing about what a thing may be in itself to call it "chance". ... All you mean by calling it "chance" is that this is not guaranteed, that it may also fall out otherwise. For the system of other things has no positive hold on the chance-thing. Its origin is in a certain fashion negative: it escapes, and says, Hands off! coming, when it comes, as a free gift, or not at all.
>
> This negativeness, however, and this opacity of the chance-thing when thus considered *ab extra*, or from the point of view of previous things or distant things, do not preclude its having any amount of positiveness and luminosity from within, and at its own place and moment. All that its chance-character asserts about it is that there is something in it really of its own, something that is not the unconditional property of the whole. If the whole wants this property, the whole must wait till it can get it, if it be a matter of chance. That the universe may actually be a sort of joint-stock society of this sort, in which the sharers have both limited liabilities and limited powers, is of course a simple and conceivable notion.

The train of logic back to QBism is this. If James and our analysis of "Wigner's friend" are right, the universe is not *one* in a very rigid sense, but rather more truly a pluriverse.[5] To get some sense of what this can mean, it is useful to start by thinking about what it is not. A good example can be found by taking a solution to the vacuum Maxwell equations in some extended region of spacetime. Focus on a compact subregion and try to conceptually delete the solution within it, reconstructing it with some new set of values. It cannot be done. The fields outside the region (including the boundary) uniquely determine the fields inside it. The interior of the

___

[5]The term "pluriverse" is again a Jamesian one. He used it interchangeably with the word "multiverse", which he also invented. The latter, however, has been coopted by the Everettians, so we will strictly use only the term pluriverse.

region has no identity but that dictated by the rest of the world – it has no "interiority" of its own. The pluriverse conception says we will have none of that. And so, for any agent immersed in this world there will always be uncertainty for what will happen upon his encounters with it.

What all this hints at is that for QBism the proper way to think of our world is as the empiricist or radical metaphysical pluralist does. Let us launch into making this clearer, for that process more than anything will explain how QBism hopes to interpret Hilbert-space dimension.

The metaphysics of empiricism can be put like this. Everything experienced, everything experienceable, has no less an ontological status than anything else. You tell me of your experience, and I will say it is real, even a distinguished part of reality. A child awakens in the middle of the night frightened that there is a monster under her bed, one soon to reach up and steal her arm – that *we-would-call-imaginary* experience has no less a hold on onticity than a Higgs-boson detection event at the large hadron collider at CERN. They are of equal status from this point of view – they are equal elements in the filling out and making of reality. This is because the world of the empiricist is not a sparse world like the world of Democritus (*nothing but* atom and void) or Einstein (*nothing but* unchanging spacetime manifold equipped with this or that field), but a world overflowingly full of variety – a world whose details are beyond anything grammatical (rule-bound) expression can articulate.

Yet this is no statement that physics should give up, or that physics has no real role in coming to grips with the world. It is only a statement that physics should better understand its function. What is being aimed for here finds its crispest, clearest contrast in a statement by Feynman (1965):

> If, in some cataclysm, all of scientific knowledge were to be destroyed, and only one sentence passed on to the next generation of creatures, what statement would contain the most information in the fewest words? I believe it is the atomic hypothesis (or the atomic fact) that all things are made of atoms – little particles that move around in perpetual motion, attracting each other when they are a little distance apart, but repelling upon being squeezed into one another. ... Everything is made of atoms. That is the key hypothesis.

The problem is the imagery that usually lies behind the phrase "everything is made of". William James called it the great original sin of the rationalistic mind (James 1997, p. 246):

> Let me give the name of "vicious abstractionism" to a way of using concepts which may be thus described: We conceive a concrete situation

by singling out some salient or important feature in it, and classing it under that; then, instead of adding to its previous characters all the positive consequences which the new way of conceiving it may bring, we proceed to use our concept privatively; reducing the originally rich phenomenon to the naked suggestions of that name abstractly taken, treating it as a case of "nothing but" that, concept, and acting as if all the other characters from out of which the concept is abstracted were expunged. Abstraction, functioning in this way, becomes a means of arrest far more than a means of advance in thought. It mutilates things; it creates difficulties and finds impossibilities; and more than half the trouble that metaphysicians and logicians give themselves over the paradoxes and dialectic puzzles of the universe may, I am convinced, be traced to this relatively simple source. *The viciously privative employment of abstract characters and class names* is, I am persuaded, one of the great original sins of the rationalistic mind.

QBism's peculiar way of looking at things realizes that physics *actually can be done* without any accompanying vicious abstractionism. You do physics as you have always done it, but you throw away the idea that "everything is made of essence X" before even starting.

Physics – in the right mindset – is not about identifying the bricks with which nature is made, but about identifying what is *common to* the largest range of phenomena it can get its hands on. The idea is not difficult once one gets used to thinking in these terms. Carbon? The old answer would go that it is *nothing but* a building block that combines with other elements according to the following rules, blah, blah, blah. The new answer is that carbon is a *characteristic* common to diamonds, pencil leads, desoxyribonucleic acid, burnt pancakes, the space between stars, the emissions of Ford pick-up trucks, and so on – the list is as unending as the world is itself. For, carbon is also a characteristic common to this diamond and this diamond and this diamond and this. But a flawless diamond and a purified zirconium crystal, no matter how carefully crafted, have no such characteristic in common: Carbon is not a *universal* characteristic of all phenomena. The aim of physics is to find characteristics that apply to as much of the world in its varied fullness as possible. However, those common characteristics are hardly what the world is made of – the world instead is made of this and this and this. The world is constructed of every particular there is and every way of carving up every particular there is.

An unparalleled example of how physics operates in such a world can be found by looking at Newton's law of universal gravitation. What did Newton really find? Would he be considered a great physicist in this day when every news magazine presents the most cherished goal of physics to be a "theory of everything"? For the law of universal gravitation is hardly that! Instead, it *merely* says that every body in the universe tries to accelerate every other body toward itself at a rate proportional to its own mass and inversely proportional to the squared distance between them. Beyond that, the law says nothing else particular of objects, and it would have been a rare thinker in Newton's time, if any at all, who would have imagined that all the complexities of the world could be derived from that limited law. Yet there is no doubt that Newton was one of the greatest physicists of all time. He did not give a "theory of everything", but a "theory of one aspect of everything". And only the tiniest fraction of physicists of any variety have ever worn a badge of that more modest kind. It is as von Baeyer (2009) wrote:

> Great revolutionaries don't stop at half measures if they can go all the way. For Newton this meant an almost unimaginable widening of the scope of his new-found law. Not only Earth, Sun, and planets attract objects in their vicinity, he conjectured, but all objects, no matter how large or small, attract all other objects, no matter how far distant. It was a proposition of almost reckless boldness, and it changed the way we perceive the world.

Finding a theory of "merely" one aspect of everything is hardly something to be ashamed of. It is the loftiest achievement physics can have in a living, breathing nonreductionist world.

Which leads us back to Hilbert space. Quantum theory – that user's manual for decision-making agents immersed in a world of *some* yet to be fully identified character – makes a statement about the world to the extent that it identifies a quality common to all the world's pieces. QBism says the quantum state is not one of those qualities. But against Hilbert spaces themselves, particularly their distinguishing characteristic one from the other, *dimension*, QBism carries no such grudge. Dimension is something one posits for a body or a piece of the world, much like one posits a mass for it in the Newtonian theory. Dimension is something a body holds all by itself, regardless of what an agent thinks of it.

The claim here is that quantum mechanics, when it came into existence, implicitly recognized a previously unnoticed capacity inherent in all matter – call it *quantum dimension*. In one manifestation, it is the fuel upon which quantum computation runs (Fuchs 2004, Blume-Kohout *et al.* 2002). In another it is the raw irritability of a quantum system to being eavesdropped upon (Fuchs 2004).

When quantum mechanics was discovered, something was *added* to matter in our conception of it. Think of the apple that inspired Newton to his law. With its discovery the color, taste, and texture of the apple did not disappear, the law of universal gravitation did not reduce the apple privatively to *just* gravitational mass. Instead, the apple was at least everything it was before, but afterward even more – for instance, it became known to have something in common with the moon. A modern-day Cavendish would be able to literally measure the further attraction an apple imparts to a child already hungry to pick it from the tree.

Something similar is the case with Hilbert-space dimension. Those diamonds we have used to illustrate the idea of nonreductionism, in very careful conditions, could be used as components in a quantum computer (Prawer and Greentree 2008). Diamonds have among their many properties something not envisioned before quantum mechanics – that they could be a source of relatively accessible Hilbert-space dimension and as such have this much in common with any number of other proposed implementations of quantum computing. Diamonds not only have something in common with the moon, but now also with the ion-trap quantum-computer prototypes around the world.

Diamondness is not something to be derived from quantum mechanics. Quantum mechanics is something we *add* to the repertoire of things we already say of diamonds, to the things we do with them and the ways we admire them. This is a very powerful realization: For diamonds already valuable, become ever more so as their qualities compound. And saying more of them, not less (as is the goal of all reductionism), has the power to suggest all kinds of variations on the theme. For instance, thinking in quantum mechanical terms might suggest a technique for making "purer diamonds". Though to an empiricist this phrase means not at all what it means to a reductionist. It means that these similar things called diamonds can suggest exotic variations of the original objects with various pinpointed properties this way or that. Purer diamond is not *more* of what it already was in nature. It is a new species, with traits of its parents to be sure, but nonetheless stand-alone, like a new breed of dog.

To the reductionist, of course, this seems exactly backwards. But then, it is the reductionist who must live with a seemingly infinite supply of conundrums arising from quantum mechanics. It is the reductionist who must live in a state of arrest, rather than moving on to the next stage of physics. Take a problem that has been a large theme of the quantum foundations meetings for the last 30 years. To put it in a commonly heard question: "Why does the world look classical if it actually operates according to quantum mechanics?" The touted mystery is that we never "see" quantum superposition and entanglement in our everyday experience.

The real issue is this. The expectation of the quantum-to-classical transitionists is that quantum theory is at the bottom of things, and "the classical world of our experience" is something to be derived out of it. QBism says: "No. Experience is neither classical nor quantum. Experience is experience with a richness that classical physics of any variety could not remotely grasp." Quantum mechanics is something put on top of raw, unreflected experience. It is additive to it, suggesting wholly new types of experience, while never invalidating the old.

To the question of why no one has ever *seen* superposition or entanglement in diamond before, the QBist replies: It is simply because before recent technologies and very controlled conditions, as well as lots of refined analysis and thinking, no one had ever mustered a mesh of beliefs relevant to such a range of interactions (factual and counterfactual) with diamonds. No one had ever been in a position to adopt the extra normative constraints required by the Born rule. For QBism, it is not the emergence of classicality that needs to be explained, but the emergence of our new ways of manipulating, controlling, and interacting with matter that do.

In this sense, QBism declares the quantum-to-classical research program unnecessary (and actually obstructive) in a way not so dissimilar to the way Bohr's 1913 model of the hydrogen atom declared another research program unnecessary (and actually obstructive). Bohr's great achievement above all the other physicists of his day was in being the first to say: "Enough! I shall not give a mechanistic explanation for these spectra we see. Here is a way to think of them with no mechanism." The important question is how matter can be coaxed to do new things. It is in the ways the world yields to our desires, and the ways it refuses to, that we learn the depths of its character.

5   The Future

There is so much still to do with QBism. So far we have only given the faintest hint of how QBism should be mounted onto a larger empiricism. It will be noticed that QBism has been quite generous in treating agents as physical objects when needed. "I contemplate you as an agent when discussing your experience, but I contemplate you as a physical system before me when discussing my own." Our solution to "Wigner's friend" is the great example of this. Precisely because of this, however, QBism knows that its story cannot end as a story of gambling agents – that is only where it starts.

Agency, for sure, is not a derivable concept as the reductionists and vicious abstractionists would have it, but QBism, like all of science, should strive for a Copernican principle whenever possible. We have learned so far from quantum theory that before an agent the world is really malleable and ready through their intercourse to give birth. Why would it not be so for every two parts of the world? And this newly defined valence, quantum dimension, might it not be a measure of a system's potential for creation when it comes into relationship with those other parts?

It is a large research program whose outline is just taking shape. It hints at a world, a pluriverse, that consists of an all-pervading "pure experience", as William James called it.[6] Or, as John Wheeler (1982) put it in the form of a question:

> It is difficult to escape asking a challenging question. Is the entirety of existence, rather than being built on particles or fields of force or multidimensional geometry, built upon billions upon billions of elementary quantum phenomena, those elementary acts of "observer-participancy", those most ethereal of all the entities that have been forced upon us by the progress of science?

Expanding this notion, making it technical, and trying to weave its insights into a worldview is the better part of future work. Quantum states, QBism declares, are not the stuff of the world, but quantum *measurement* might be. Might a one-day future Shakespeare write with honesty:

> Our revels are now ended. These our actors,
> As I foretold you, were all spirits and
> Are melted into air, into thin air ...
> We are such stuff as quantum measurement is made on.

References

Banks E.C. (2003): *Ernst Mach's World Elements: A Study in Natural Philosophy*, Kluwer, Dordrecht.

Bell J.S. (1990): Against measurement. *Physics World* **3**, 33–41.

Bernardo J.M. and Smith A.F.M. (1994): *Bayesian Theory*, Wiley, Chichester.

---

[6]Aside from James's own publications (James 1996a, 1996b), further reading on this concept and related subjects is due to Lamberth (1999), Taylor (1996), Wild (1969), Banks (2003).

Blume-Kohout R., Caves C.M., and Deutsch I.H. (2002): Climbing mount scalable: Physical-resource Requirements for a scalable quantum computer. *Foundations of Physics* **32**, 1641–1670.

Cartwright N. (1999): *The Dappled World: A Study of the Boundaries of Science*, Cambridge University Press, Cambridge.

Caves C.M., Fuchs C.A., and Schack R. (2002): Quantum probabilities as Bayesian probabilities. *Physical Review A* **65**(2) 022305.

Caves C.M., Fuchs C.A., and Schack R. (2007): Subjective probability and quantum certainty. *Studies in History and Philosophy of Modern Physics*, **38**(2), 255–274.

de Finetti B. (1990): *Theory of Probability*, Wiley, New York.

Dupré J. (1993): *The Disorder of Things: Metaphysical Foundations of the Disunity of Science*, Harvard University Press, Cambridge.

Feynman R.P. (1965): *The Character of Physical Law*, MIT Press, Cambridge.

Fuchs C.A. (2002): Quantum mechanics as quantum information (and only a little more). In *Quantum Theory: Reconsideration of Foundations*, ed. by A. Khrennikov, Växjö University Press, Växjö, pp. 463–543.

Fuchs C.A. (2004): On the quantumness of a Hilbert space. *Quantum Information and Computation* **4**, 467–478.

Fuchs C.A. (2010): *Coming of Age with Quantum Information*, Cambridge University Press, Cambridge.

Fuchs C.A. and Schack R. (2004): Unknown quantum states and operations, a Bayesian View. In *Quantum Estimation Theory*, ed. by M.G.A. Paris and J. Řeháček, Springer, Berlin, pp. 151–190.

Fuchs C.A. and Schack R. (2013): Quantum-Bayesian coherence. *Reviews of Modern Physics*, in press. Manuscript available at `arXiv:1301.3274v1`.

Hartle J.B. (1968): Quantum mechanics of individual systems. *American Journal of Physics* **36**(8), 704–712.

James W. (1884): *The Will to Believe and Other Essays in Popular Philosophy; Human Immortality*, Dover, New York.

James W. (1940): *Some Problems of Philosophy*, Longmans, Green, and Co., London.

James W. (1996a): *A Pluralistic Universe*, University of Nebraska Press, Lincoln.

James W. (1996b): *Essays in Radical Empiricism*, University of Nebraska Press, Lincoln.

James W. (1997): *The Meaning of Truth*, Prometheus Books, Amherst.

Lamberth D.C. (1999): *William James and the Metaphysics of Experience*, Cambridge University Press, Cambridge.

Lindley D.V. (2006): *Understanding Uncertainty*, Wiley, Hoboken.

Prawer S. and Greentree A.D. (2008): Diamond for quantum computing. *Science* **320**, 1601–1602.

Spekkens R.W. (2007): Evidence for the epistemic view of quantum states: A toy theory. *Physical Review A* **75**, 032110.

Taylor E. and Wozniak R.H., eds. (1996): *Pure Experience: The Response to William James*, Thoemmes Press, Bristol.

Thayer H.S. (1981): *Meaning and Action: A Critical History of Pragmatism*, Hackett, Indianapolis.

von Baeyer H.C. (2009): *Petites Leçons de Physique dans les Jardins de Paris*, Dunod, Paris.

Wahl J. (1925): *The Pluralist Philosophies of England and America*, Open Court, London.

Wheeler J.A. (1982): Bohr, Einstein, and the strange lesson of the quantum. In *Mind in Nature*, ed. by R.Q. Elvee, Harper & Row, San Francisco, pp. 1–23.

Wild J. (1969): *The Radical Empiricism of William James*, Doubleday, Garden City.

Quantum Measurement
and the Paulian Idea

Christopher A. Fuchs and Rüdiger Schack

Abstract

In the quantum Bayesian (or QBist) conception of quantum theory, "quantum measurement" is understood not as a comparison of something pre-existent with a standard, but instead indicative of the creation of something new in the universe: Namely, the fresh experience any agent receives upon taking an action on the world. We explore the implications of this for any would-be ontology underlying QBism. The concept that presently stands out as a candidate "material for our universe's composition" is "experience" itself, or what John Wheeler called "observer-particpancy".

Of every would be describer of the universe one has a right to ask immediately two general questions. The first is: "What are the materials of your universe's composition?" And the second: "In what manner or manners do you represent them to be connected?"

William James 1988

1   Introduction

John Bell famously wrote that the word "measurement" should be banished from fundamental discussions of quantum theory (Bell 1990). In this paper we look at quantum measurement from the perspective of quantum Bayesianism, or "QBism" (Fuchs 2002a, 2004, 2010a, 2013, Caves *et al.* 2007), and argue that the word "measurement" is indeed problematic, even from our perspective. However, the reason it is problematic is not that the word is "unprofessionally vague and ambiguous", as Bell (1987) said. Rather, it is because the word's usage engenders a misunderstanding of the subject matter of quantum theory. We say this because from the view of QBism quantum theory is a smaller theory than one might think – it is smaller because it indicates the world to be a bigger, more varied place than the usual forms of the philosophy of science allow for.

Crucial to the QBist conception of measurement is the slogan – inspired by Peres' (1978) more famous one – that "unperformed measurements have no outcomes". Mindful of James' injunction preceding this article, however, we believe that making precise the intuition behind this slogan is the first step toward characterizing "the materials of our universe's composition".

2 Bayesian Probabilities

Let us put quantum theory to the side for a moment, and consider instead basic Bayesian probability theory (Savage 1954, de Finetti 1990, Bernardo and Smith 1994, Jeffrey 2004). There the subject matter is an agent's expectations for this and that. For instance, an agent might write down a joint probability distribution $P(h_i, d_j)$ for various mutually exclusive hypotheses h_i, $i = 1, \ldots, n$, and data values d_j, $j = 1, \ldots, m$, appropriate to some phenomenon.

A major role of Bayesian theory is that it provides a scheme (Dutch-book coherence, see Vineberg 2011) for how these probabilities should be related to other probabilities, $P(h_i)$ and $P(d_j)$ say, as well as to any other degrees of belief the agent has for other phenomena. The theory also prescribes that if the agent is given a specific data value d_j, he should update his expectations for everything else within his interest. For instance, under the right conditions (Diaconis and Zabell 1982, Skyrms 1987), he should reassess his probabilities for the h_i by conditionalizing:

$$P_{\text{new}}(h_i) = \frac{P(h_i, d_j)}{P(d_j)} \tag{1}$$

But what is this phrase "given a specific data value"? What does it really mean in detail? Shouldn't one specify a mechanism or at least a chain of logical or physical connectives for how the raw fact signified by d_j comes into the field of the agent's consciousness? And who is this "agent" reassessing his probabilities anyway? Indeed, what is the precise definition of an agent? How would one know one when one sees one? Can a dog be an agent? Or must it be a person? Maybe it should be a person with a PhD?[1]

We are thus led to ask: Should probability theory really be held accountable for giving answers to all these questions? In other words, should a book like *The Foundations of Statistics* (Savage 1954) spend some of its pages

---

[1] This is a tongue-in-cheek reference to Bell (1990) again.

demonstrating how the axioms of probability – by way of their own power – give rise, at least in principle, to agents and data acquisition itself? Otherwise, should probability theory be charged with being "unprofessionally vague and ambiguous"?

Probability theory has no chance of answering these questions because they are not questions within the subject matter of the theory. Within probability theory, the notions of "agent" and "given a data value" are primitive and irreducible. Guiding agents' decisions based on data is what the whole theory is constructed for – just like primitive forces and masses are what the whole theory of classical mechanics is constructed for. As such, agents and data are the highest elements within the structure of probability theory – they are not to be constructed from it, but rather agents are there to receive the theory's guidance, and the data are there to designate the world external to the agent.

3   Quantum Bayesianism

QBism says if all of this is true of Bayesian probability theory in general, it is true of quantum theory as well. As the foundations of probability theory dismiss the questions of where data come from and what constitutes an agent, so can the foundations of quantum theory dismiss them too.

There will surely be a protest from some readers at this point: "It is one thing to say all this of probability theory, but quantum theory is a wholly different story." Or: "Quantum mechanics is no simple branch of mathematics, be it probability or statistics. Nor can it plausibly be a theory about the insignificant specks of life in our vast universe making gambles and decisions. Quantum mechanics is one of our best theories of the world! It is one of the best maps we have drawn yet of what is actually out there." But this is where these readers err. We hold fast: Quantum theory is simply *not* a "theory of the world". Just like probability theory is not a theory of the world, quantum theory is not as well. It is a theory for the use of agents immersed in and interacting with a world of a particular character, the quantum world.

By declaring this, we certainly do not want to dispense with the idea of a world external to the agent. Indeed it must be as Gardner (1983) says:

> The hypothesis that there is an external world, not dependent on human minds, made of *something*, is so obviously useful and so strongly confirmed by experience down through the ages that we can say without exaggerating that it is better confirmed than any other empirical

hypothesis. So useful is the posit that it is almost impossible for anyone except a madman or a professional metaphysician to comprehend
a reason for doubting it.

Yet there is no implication in these words that quantum theory, for all its
success in chemistry, physical astronomy, laser making, and so much else,
must be read off as a theory of the world. There is room for a significantly
more interesting form of dependence: Quantum theory is conditioned by the
character of the world, but yet is not a theory directly of it. Confusion on
this point, we believe, is what has caused most of the discomfort in quantum
foundations in the 86 years since the theory's coming to a relatively stable
form.

4 Measurement

Returning to our discussion of Bell and the word "measurement," it is not
because we think it unprofessionally vague and ambiguous that we regard
"measurement" as problematic. It is because the word subliminally whispers
the philosophy of its birth – that quantum mechanics *should* be conceived
in a way that makes no ultimate reference to agency, and that agents are
constructed out of the theory, rather than taken as the primitive entities the
theory is meant to aid. In a nutshell, the word deviously carries forward the
impression that quantum mechanics should be viewed as a theory directly of
the world.

One gets a sense of the boundaries the word "measure" places upon
our interpretive thoughts by turning to any English dictionary. Here is a
sampling from `dictionary.com/`:

- to ascertain the extent, dimensions, quantity, capacity, etc., of, esp. by
  comparison with a standard;
- to estimate the relative amount, value, etc., of, by comparison with
  some standard;
- to judge or appraise by comparison with something or someone else;
- to bring into comparison or competition.

In not one of these definitions do we get an image of anything being created
in the measuring process; none give any inkling of the crucial contextuality of
quantum measurements, the context being a parameter ultimately set only
in terms of the agent. Measurement, in its common usage, is something
passive and static: it is comparison between *existents*. No wonder a slogan
like "unperformed measurements have no outcomes" (cf. Peres 1978) would

seem irreparably paradoxical. If a quantum measurement is not comparison, but something else, the only way out of the impasse is to understand what that something else is.

Correcting or modifying the word "measurement" is the prerequisite to a new ontology – in other words, prerequisite to a statement about the (hypothesized) character of the world that does not make direct reference to our actions and gambles within it. Therefore, as a start, let us rebuild quantum mechanics in terms more conducive to the QBism program. The best way to begin a more thoroughly delineation of quantum mechanics is to start with two quotes on personalist Bayesianism itself. The first is from Hampton *et al.* (1973):

> Bruno de Finetti believes there is no need to assume that the probability of some event has a uniquely determinable value. His philosophical view of probability is that it expresses the feeling of an individual and cannot have meaning except in relation to him.

And the second is from Lindley (1982):

> The Bayesian, subjectivist, or coherent, paradigm is egocentric. It is a tale of one person contemplating the world and not wishing to be stupid (technically, incoherent). He realizes that to do this his statements of uncertainty must be probabilistic.

These two quotes make it clear that personalist Bayesianism is a "single-user theory". Thus, QBism must inherit at least this much egocentrism in its view of quantum states ρ. The "Paulian idea" (Fuchs 2010b) – which is also essential to the QBist view – goes further still (cf. Figure 1). It says that the outcomes of quantum measurements are single-user as well! That is to say, when an agent writes down her degrees of belief for the outcomes of a quantum measurement, what she is writing down are her degrees of belief about her potential *personal* experiences arising in consequence of her actions upon the external world.

5 Basic Notions of Quantum Theory from a QBist Point of View

Before exploring this further, let us partially formalize in a quick outline the structure of quantum mechanics from the Bayesian point of view. At the moment we will retain the usual mathematical formulation of the theory, but we will begin the process of changing the verbal description of what the term "quantum measurement" means.

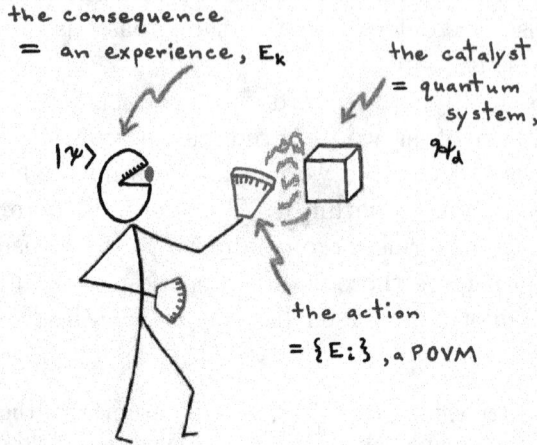

the consequence
= an experience, E_k

the catalyst
= quantum system, $\mathcal{H}_d$

$|\psi\rangle$

the action
= $\{E_i\}$, a POVM

Figure 1: The Paulian Idea (Fuchs 2010b) – in the form of a figure inspired by John Archibald Wheeler, whose vision of quantum mechanics has been greatly inspiring to us, and overtones of his thought can be found throughout our own. The figure of his that suggested the present one can be found in Patton and Wheeler (1975), Wheeler and Patton (1977), Wheeler (1979), Wheeler (1980), Wheeler (1982), Wheeler (1994).

In contemplating a quantum measurement (though the word is a misnomer), one makes a conceptual split in the world: one part is treated as an agent, and the other as a kind of reagent or catalyst (one that brings about change in the agent itself). In older terms, the former is an observer and the latter a quantum system of some finite dimension d. A quantum measurement consists first in the agent taking an *action* on the quantum system. The action is formally captured by some positive operator valued measure $\{E_i\}$(POVM, cf. footnote 2 below). The action leads generally to an incompletely predictable *consequence*, a particular personal experience E_i for the agent (Fuchs 2007, 2010a). The quantum state $|\psi\rangle$ makes no appearance but in the agent's head; for it only captures his degrees of belief concerning the consequences of his actions, and – in contrast to the quantum system itself – has no existence in the external world. Measurement devices are depicted as prosthetic hands to make it clear that they should be considered an integral part of the agent. (This contrasts with Bohr's view where the measurement device is always treated as a classically describable system external to the observer.) The sparks between the measurement-device hand and the quantum system represent the idea that the consequence of each quantum measurement is a unique creation within the previously existing universe (Fuchs 2010a, 2013). Wolfgang Pauli characterized this picture as a "wider form of the reality concept" than that of Einstein's, which he labeled "the ideal of the detached observer" (Pauli 1994, Laurikainen 1988, Gieser 2005).

1. Primitive notions: a) the agent, b) things external to the agent, or, more commonly, "systems," c) the agent's actions on the systems, and d) the consequences of those actions for her experience.

2. The formal structure of quantum mechanics is a theory of how the agent ought to organize her Bayesian probabilities for the consequences of all her potential actions on the things around her. Implicit in this is a theory of the structure of actions. This works as follows.

3. When the agent posits a system, she posits a Hilbert space $\mathcal{H}_d$ of dimension d as the arena for all her considerations.

4. Actions upon the system are captured by positive-operator valued measures $\{E_i\}$, briefly POVMs,[2] on $\mathcal{H}_d$. Potential consequences of the action are labeled by the individual elements E_i within the set. That is,

$$\text{ACTION} = \{E_i\} \quad \text{and} \quad \text{CONSEQUENCE} = E_k \ .$$

5. Quantum mechanics organizes the agent's beliefs by saying that she should strive to find a single density operator ρ such that her degrees of belief will always satisfy

$$\text{Prob}\Big(\text{CONSEQUENCE} \,\Big|\, \text{ACTION}\Big) \ = \ \text{Prob}\Big(E_k \,\Big|\, \{E_i\}\Big)$$
$$= \ \text{Trace } \rho E_k \ ,$$

   no matter what action $\{E_i\}$ is under consideration.

6. Unitary time evolution and more general quantum operations (completely positive maps) do not represent objective underlying dynamics, but rather address the agent's belief changes accompanying the flow of time, as well as belief changes consequent upon any actions taken.

7. When the agent posits *two* things external to herself, the arena for all her considerations becomes $\mathcal{H}_{d_1} \otimes \mathcal{H}_{d_2}$. Actions and consequences now become POVMs on $\mathcal{H}_{d_1} \otimes \mathcal{H}_{d_2}$.

8. The agent can nonetheless isolate the notion of an action on a single one of the things alone: These are POVMs of the form $\{E_i \otimes I\}$, and similarly with the systems reversed $\{I \otimes E_i\}$.

---

[2]See, e.g., Berberian (1966) for a precise definition. In contrast to a projection-valued measure (projector) with characteristic function $\{0,1\}$ and orthogonal eigenfunctions, the characteristic function of a POVM is the entire interval $[0,1]$, and the eigenfunctions are generally not orthogonal. POVMs generalize the idealized idea of quantum measurements as projections and lead to a more realistic picture.

9. Resolving the consequence of an action on *one* of the things may cause the agent to update her expectations for the consequences of any further actions she might take on the *other* thing. But for those latter consequences to come about, she must elicit them through an actual action on the second system.

With regard to the present discussion, the main points to note are items 4, 7, 8, and 9. Regarding our usage of the word "measurement," they say that one should think of it simply as an *action* upon the system of interest. Actions lead to consequences within the experience of the agent, and that is what a quantum measurement is. A quantum measurement finds nothing, but very much *makes* something.

It is a simple linguistic move, but it does crucial work for resetting the debate on quantum foundations. It might indeed have been the case that all this nonstandard formulation was for nought, turning out to be superfluous. That is, though we have spelled out very carefully in item 9 that, "for those latter consequences to come about, she must elicit them through an actual action on the second system", maybe there would have been nothing wrong in thinking of the latter (and by analogy the former) quantum measurement as finding a pre-existing value after all. But this, we have argued previously (Caves 2007, Fuchs 2013) would contradict item 8, i.e., that one can isolate a notion of an action on a single system alone.

Thus, in a QBist painting of quantum mechanics, quantum measurements are "generative" in a very real sense. But by that turn, the consequences of our actions on physical systems must be egocentric as well. Measurement outcomes come about for the agent herself. Quantum mechanics is a single-user theory through and through – first in the usual Bayesian sense with regard to personal beliefs, and second in that quantum measurement outcomes are wholly personal experiences.[3]

Of course, as a single-user theory, quantum mechanics is available to any agent to guide and better prepare her for her own encounters with the world. And although quantum mechanics has nothing to say about another agent's personal experiences, agents can communicate and use the informa-

---

[3]The usual belief otherwise – for instance in Pauli's own formulation (which is ultimately inconsistent with his taking measurement devices to be like prosthetic hands), that "the objective character of the description of nature given by quantum mechanics [is] adequately guaranteed by the circumstance that ... the results of observation, *which can be checked by anyone*, cannot be influenced by the observer, once he has chosen his experimental arrangement" (Pauli 1956, italics ours, to pinpoint the offending portion of the formulation) – we state for completeness, is the ultimate source of the Wigner's friend paradox. This will be expanded upon in a later work by the authors; for the moment see Fuchs (2013).

tion gained from each other to update their probability assignments. In the spirit of the Paulian idea, however, querying another agent means taking an action on him.

Whenever "I" encounter a quantum system, and take an action upon it, it catalyzes a consequence in my experience that my experience could not have foreseen. Similarly, by a Copernican-style principle, I should assume the same for "you": Whenever you encounter a quantum system, taking an action upon it, it catalyzes a consequence in your experience. By one category of thought we are agents, but by another category of thought we are physical systems. And when we take actions upon each other, the category distinctions are symmetrical. Like with the bistable perception of ambiguous images (e.g., the Rubin vase), the best the eye can do is flit back and forth between the two formulations.

6   The World View of QBism

The previous paragraphs should have made clear that viewing quantum mechanics as a single-user theory does not mean there is only one user. QBism does not lead to solipsism. Any charge of solipsism is further refuted by two points central to the Paulian idea (Fuchs 2002b). One is the conceptual split of the world into two parts – one an agent and the other an external quantum system – that gets the discussion of quantum measurement off the ground in the first place. If such a split were not needed for making sense of the question of actions (actions upon what? in what? with respect to what?), it would not have been made. Imagining a quantum measurement without an autonomous quantum system participating in the process would be as paradoxical as the Zen koan of the sound of a single hand clapping.

The second point is that once the agent chooses an action $\{E_i\}$, the particular consequence E_k of it is beyond his control. That is to say, the particular outcome of a quantum measurement is not a product of his desires, whims, or fancies – this is the very reason he uses the calculus of probabilities in the first place: they quantify his uncertainty (Lindley 2006), an uncertainty that, try as he might, he cannot get around. So, implicit in this whole picture – this whole Paulian idea – is an "external world ... made of *something*," just as Gardner calls for. It is only that quantum theory is a rather small theory: Its boundaries are set by being a handbook for agents immersed within that "world made of *something*".

But a small theory can still have grand import, and quantum mechanics most certainly does. This is because it tells us how a user of the theory sees his role in the world. Even if quantum mechanics – viewed as an addition to probability theory – is not a theory of the world itself, it is certainly conditioned by the particular character of this world. Its empirical content is exemplified by the Born rule, item 5 in the above list, which takes a specific form rather than an infinity of other possibilities. Even though quantum theory is now understood as a theory of acts, decisions, and consequences (Savage 1954), it tells us, in code, about the character of our particular world. Apparently, the world is made of a stuff that does not have "consequences" waiting around to fulfill our "actions" – it is a world in which the consequences are generated on the fly. When we on the inside prod that stuff on the outside, the world comes to something that neither side could have foretold.

Indeed, one starts to get a sense of a world picture that is part personal – *truly* personal – and part the joint product of all that interacts. It is almost as if one can hear in the very formulation of the Born rule one of William James' many lectures on chance and indeterminism. Here is one example (James 1956a):

> [Chance] is a purely negative and relative term, giving us no information about that of which it is predicated, except that it happens to be disconnected with something else – not controlled, secured, or necessitated by other things in advance of its own actual presence. ... What I say is that it tells us nothing about what a thing may be in itself to call it "chance." ... All you mean by calling it "chance" is that this is not guaranteed, that it may also fall out otherwise. For the system of other things has no positive hold on the chance-thing. Its origin is in a certain fashion negative: it escapes, and says, Hands off! coming, when it comes, as a free gift, or not at all.
>
> This negativeness, however, and this opacity of the chance-thing when thus considered *ab extra*, or from the point of view of previous things or distant things, do not preclude its having any amount of positiveness and luminosity from within, and at its own place and moment. All that its chance-character asserts about it is that there is something in it really of its own, something that is not the unconditional property of the whole. If the whole wants this property, the whole must wait till it can get it, if it be a matter of chance. That the universe may actually be a sort of joint-stock society of this sort, in which the sharers have both limited liabilities and limited powers, is of course a simple and conceivable notion.

And here is another (James 1956b):

> Why may not the world be a sort of republican banquet of this sort, where all the qualities of being respect one another's personal sacredness, yet sit at the common table of space and time?

To me this view seems deeply probable. Things cohere, but the act of cohesion itself implies but few conditions, and leaves the rest of their qualifications indeterminate. As the first three notes of a tune comport many endings, all melodious, but the tune is not named till a particular ending has actually come, – so the parts actually known of the universe may comport many ideally possible complements. But as the facts are not the complements, so the knowledge of the one is not the knowledge of the other in anything but the few necessary elements of which all must partake in order to be together at all. Why, if one act of knowledge could from one point take in the total perspective, with all mere possibilities abolished, should there ever have been anything more than that act? Why duplicate it by the tedious unrolling, inch by inch, of the foredone reality? No answer seems possible. On the other hand, if we stipulate only a partial community of partially independent powers, we see perfectly why no one part controls the whole view, but each detail must come and be actually given, before, in any special sense, it can be said to be determined at all. This is the moral view, the view that gives to other powers the same freedom it would have itself, – not the ridiculous "freedom to do right", which in my mouth can only mean the freedom to do as *I* think right, but the freedom to do as *they* think right, or wrong either.

This is a world of "objective indeterminism" indeed, but one with no place for "objective chance" in the sense of David Lewis (1986). From within any part, the future is undetermined. If one of those parts is an agent, then it is an agent in a situation of uncertainty. And where there is uncertainty, agents should use the calculus of Bayesian probability in order to make the best go at things.

But we have learned enough from Copernicus to know that egocentrism, whenever it can be shaken away from a *Weltanschauung*, it ought to be. Whenever "I" encounter a quantum system, and take an action upon it, it catalyzes a consequence in my experience that my experience could not have foreseen. Similarly, by a Copernican principle, I should assume the same for "you": Whenever you encounter a quantum system, taking an action upon it, it catalyzes a consequence in your experience. By one category of thought, we are agents, but by another category of thought we are physical systems. And when we take actions upon each other, the category distinctions are symmetrical.

In the common circles of the philosophy of science there is a strong popularity in the idea that agentialism can always be reduced to some complicated property arrived at from physicalism. But perhaps this republican-banquet vision of the world that so seems to fit a QBist understanding of quantum mechanics is telling us that the appropriate ontology we should seek would treat these dual categories as just that, dual aspects of a higher, more neutral realm.[4] That is, the concepts "action" and "unforeseen consequence in experience", both crucial for clarifying the very meaning of quantum measurement, might just be applicable after a fashion to arbitrary components of the world – i.e., venues in which probability talk has no place. Understanding or rejecting this idea is the long road ahead of us.

We leave an old teacher of ours with some closing words that touch on the challenge William James started us off with:

> *It is difficult to escape asking a challenging question. Is the entirety of existence, rather than being built on particles or fields of force or multidimensional geometry, built upon billions upon billions of elementary quantum phenomena, those elementary acts of "observer-participancy", those most ethereal of all the entities that have been forced upon us by the progress of science?*

John Archibald Wheeler 1982

References

Bell J.S. (1987): *Speakable and Unspeakable in Quantum Mechanics*, Cambridge University Press, Cambridge.

Bell J. (1990): Against measurement. *Physics World*, August 1990, 33–41.

Berberian S.K. (1966): *Notes on Spectral Theory.* Van Nostrand, New York, p. 5/6.

Bernardo J.M. and Smith A.F.M. (1994): *Bayesian Theory*, Wiley, Chichester.

Caves C.M., Fuchs C.A. and Schack R. (2007): Subjective probability and quantum certainty. *Studies in History and Philosophy of Modern Physics* **38**, 255–274.

de Finetti B. (1990): *Theory of Probability, 2 Volumes*, Wiley, New York. Originally published in 1974.

Diaconis R. and Zabell S.L. (1982): Updating personal probability. *Journal of the American Statistical Association* **77**, 822–830.

---

[4]For a few further suggestive things to read in this regard, we propose James (1940, 1996), Lamberth (1999), Taylor and Wozniak (1996), Wahl (1925). Neutral monism and dual-aspect monism have become influential frameworks of thinking in contemporary discussions in the philosophy of mind (cf. Velmans and Nagasawa 2012).

Fuchs C.A. (2002a): Quantum mechanics as quantum information (and only a little more). Manuscript available at `arXiv:quant-ph/0205039v1`. Abridged version in *Quantum Theory: Reconsideration of Foundations*, ed. by A. Khrennikov, Växjö University Press, Växjö, pp. 463–543.

Fuchs C.A. (2002b): The anti-Växjö interpretation of quantum mechanics. In *Quantum Theory: Reconsideration of Foundations*, ed. by A. Khrennikov, Växjö University Press, Växjö, pp. 99–116.

Fuchs C.A. (2007): Delirium quantum: Or, where I will take quantum mechanics if it will let me. In *Foundations of Probability and Physics – 4*, ed. by G. Adenier, C.A. Fuchs, and A. Yu. Khrennikov, American Institute of Physics, Melville, pp. 438–462.

Fuchs C.A. (2010a): QBism, the perimeter of quantum Bayesianism, accessible at `arXiv:1003.5209`.

Fuchs C.A. (2010b): *Coming of Age with Quantum Information: Notes on a Paulian Idea*, Cambridge University Press, Cambridge.

Fuchs C.A. (2013): *My Struggles with the Block Universe: Selected Correspondence, January 2001–May 2011*, ed. by B.C. Stacey. Preliminary version (2216 pages) available from the author.

Fuchs C.A. and Schack R. (2004): Unknown quantum states and operations, a Bayesian view. In *Quantum Estimation Theory*, ed. by M.G.A. Paris and J. Řeháček, Springer, Berlin, pp. 151–190.

Gardner M. (1983): Why I am not a solipsist. In *The Whys of a Philosophical Scrivener*, W. Morrow, New York, pp. 11–31.

Gieser S. (2005): *The Innermost Kernel: Depth Psychology and Quantum Physics*, Springer, Berlin.

Hampton J.M., Moore P.G. and Thomas H. (1973): Subjective probability and its measurement. *Journal of the Royal Statistical Society Series A* **136**(1), 21–42.

James W. (1922): *Pragmatism, a New Name for Some Old Ways of Thinking: Popular Lectures on Philosophy*, Longmans, Green and Co., New York.

James W. (1940): *Some Problems of Philosophy*, Longmans, Green and Co., London.

James W. (1956a): The dilemma of determinism. In *The Will to Believe and Other Essays in Popular Philosophy*, Dover, New York, pp. 145–183.

James W. (1956b): On some Hegelisms. In *The Will to Believe and Other Essays in Popular Philosophy*, Dover, New York, pp. 263–298.

James W. (1988): The many and the one 1903–1904. In *Manuscript Essays and Notes*, ed. by I.K. Skrupskelis, Harvard University Press, Cambridge.

James W. (1996): *Essays in Radical Empiricism*, University of Nebraska Press, Lincoln.

Jeffrey R. (2004): *Subjective Probability: The Real Thing*, Cambridge University Press, Cambridge.

Lamberth D.C. (1999): *William James and the Metaphysics of Experience*, Cambridge University Press, Cambridge.

Laurikainen K.V. (1988): *Beyond the Atom: The Philosophical Thought of Wolfgang Pauli*, Springer, Berlin.

Lewis D. (1986): A subjectivist's guide to objective chance. In *Studies in Inductive Logic and Probability, Vol. II*, Oxford University Press, Oxford, pp. 83–112.

Lindley D.V. (1982): Comment on A.P. Dawid's "The well-calibrated Bayesian". *Journal of the American Statistical Association* **77**, 604–613.

Lindley D.V. (2006): *Understanding Uncertainty*, Wiley-Interscience, Hoboken.

Patton C.M. and Wheeler J. A. (1975): Is physics legislated by cosmogony? In *Quantum Gravity: An Oxford Symposium*, ed. by C.J. Isham, R. Penrose, and D.W. Sciama, Clarendon Press, Oxford, pp. 538–605.

Pauli W. (1956): Relativitätstheorie und Wissenschaft. *Helvetica Physica Acta, Supp.* **IV**, 282–286. Reprinted as "The Theory of Relativity and Science", in W. Pauli: *Writings on Physics and Philosophy*, ed. by C.P. Enz and K. von Meyenn, Springer, Berlin 1994, pp. 107–111.

Peres A. (1978): Unperformed experiments have no results. *American Journal of Physics*, **46**, 745–747.

Savage L.J. (1954): *The Foundations of Statistics*, Wiley, New York.

Skyrms B. (1987): Dynamic coherence and probability kinematics. *Philosophy of Science* **54**, 1–20.

Taylor E. and Wozniak R.H., eds. (1996): *Pure Experience: The Response to William James*, Thoemmes Press, Bristol.

Velmans M. and Nagasawa Y., eds. (2012): Monist alternatives to physicalism. Special Issue of the *Journal of Consciousness Studies* **19**(9/10).

Vineberg S. (2011): Dutch book arguments. In *Stanford Encyclopedia of Philosophy*, ed. by E.N. Zalta, accessible at `plato.stanford.edu/entries/dutch-book/`.

Wahl J. (1925): *The Pluralist Philosophies of England and America*, Open Court, London.

Wheeler J.A. (1979): The quantum and the universe. In *Relativity, Quanta, and Cosmology in the Development of the Scientific Thought of Albert Einstein, Vol. II*, ed. by F. de Finis, Johnson Reprint, New York, pp. 807–825.

Wheeler J.A. (1980): Beyond the black hole. In *Some Strangeness in the Proportion: A Centennial Symposium to Celebrate the Achievements of Albert Einstein*, ed. by H. Woolf, Addison-Wesley, Reading, pp. 341–375.

Wheeler J.A. (1982): Bohr, Einstein, and the strange lesson of the quantum. In *Mind in Nature: Nobel Conference XVII, Gustavus Adolphus College*, ed. by R.Q. Elvee, Harper & Row, San Francisco, pp. 1–23.

Wheeler J.A. (1994): Time today. In *Physical Origins of Time Asymmetry*, ed. by J.J. Halliwell, J. Pérez-Mercader, and W.H. Zurek, Cambridge University Press, Cambridge, pp. 1–29.

Wheeler J.A. and Patton C.M. (1977): Is physics legislated by cosmogony? In *Encyclopedia of Ignorance: Everything You Ever Wanted to Know about the Unknown*, ed. by R. Duncan and M. Weston-Smith, Pergamon, Oxford, pp. 19–35.

Quantum Entanglement, Hidden Variables, and Acausal Correlations

Thomas Filk

Abstract

Entanglement seems to be the most distinguished feature of quantum theory, and it is the physical phenomenon closest to what C.G. Jung might have had in mind with his notion of "synchronicity". This article emphasizes the role of entanglement in the development of hidden variable interpretations of quantum mechanics during the years from 1927 to 1935. It is argued that psychological ideas might have played a much more dominant role in the history of quantum theory than is usually assumed.

> *Perhaps we do not conceive of matter, e.g. in the sense of life,*
> *"correctly" if we observe it as in quantum mechanics, namely*
> *altogether ignoring the inner state of the "observer". The famous*
> *"incompleteness" of quantum mechanics (Einstein) is in fact present*
> *somehow-somewhere, but certainly it cannot be eliminated by returning*
> *to classical field physics (this is but a neurotic misunderstanding of*
> *Einstein). Rather it has to do with holistic relations between "inside"*
> *and "outside" that are not contained in contemporary science.*

> Pauli to Fierz, August 10, 1954

1 Introduction

The consequences of entanglement – a particular type of acausal quantum correlations[1] – belong to the most surprising and counterintuitive effects of physics and our physical understanding of the world. Furthermore, entanglement endows quantum theory with a holism and a peculiar kind of nonlocality which is in complete contrast to the reductive nature of classical

---

[1]The term "acausal" here means that there is no continuous causal chain between the correlated parts of a system which can explain their correlations.

physics. In a certain sense one can consider entanglement as the most distinguished characteristic of quantum behavior. Many features of quantum theory can be related to classical local models, but in order to explain the quantum correlations arising from entanglement in a classical setting, nonlocal influences seem to be unavoidable. This is a consequence of the work of Bell (1966) on hidden variable approaches to quantum theory.

Entanglement may be that physical phenomenon which is closest to what C.G. Jung might have had in mind with respect to his concept of "synchronicity". Jung (1952) considered synchronistic phenomena as based on acausal, nonlocal, meaningful correlations. And for Pauli entanglement may have been a model for the relationship between mind and matter in the framework of the dual-aspect monism he proposed together with Jung (Atmanspacher 2012).

Ever since the beginning of quantum theory, scientists tried to formulate models which imbed quantum mechanics into a classical framework of thinking with hidden variables. One of the first attempts was Madelung's hydrodynamical description of quantum theory (Madelung 1926) and others followed soon after. For a detailed historical exposition see, e.g. Jammer (1974).

Numerous no-go theorems have been established which set severe constraints on such a classical formulation or extension of the quantum mechanical formalism (von Neumann 1932, Gleason 1957, Kochen 1967, Jauch 1969). Interestingly enough, most proofs of such theorems used the special kind of "uncertainty" inherent in quantum theory and (apart from the work of Bell) did not rely on entanglement. On the other hand, almost all arguments which have been brought forth against models with local hidden variables refer to entanglement.

In this article, I want to illuminate the concepts of entanglement and hidden variables from several perspectives: (i) the mathematical perspective (just a little bit), (ii) the physical perspective (slightly more), (iii) the philosophical perspective (only marginally), (iv) and the historical perspective (mostly). My historical account will be neither chronological nor complete, and I will mainly address some key events between 1927 and 1935, and this in reversed order. In addition I will argue that there is a (v) psychological perspective insofar as psychological factors were influential in the development of the concept of entanglement in particular and physics in general – to an extent which, at first sight, one would not expect in the natural sciences.

The article is organized as follows. In the next section I will briefly recall the famous article of Einstein, Podolsky, and Rosen (1935), and the reaction of some physicists, in particular Pauli, to this article. This will also give me the opportunity to briefly explain the notion of entanglement in a non-mathematical manner.

Section 3 deals with the proof by von Neumann (1932) of a theorem which states that, under very general conditions, quantum theory cannot be embedded into a classical theory with hidden variables. Already a few years later, the philosopher Grete Hermann pointed out that von Neumann's proof was based on an assumption which is physically not mandatory, but for various reasons her argument never made it into the discussion of physicists during that time.

In Section 4 I will go even further back in history into the year 1927. At the Solvay Conference of this year, Louis deBroglie presented a hidden variable model for quantum theory (of which a more elaborate version is known today as the Bohm-deBroglie model). DeBroglie might have hoped for Einstein's support at this conference, because his approach was quite close to Einstein's idea of a statistical explanation of quantum theory. But strangely enough, Einstein did not even comment on deBroglie's model, and I will speculate about the reasons why. Some conclusions finish the article.

2   The Article by Einstein, Podolsky, and Rosen

Let me briefly recall some of the highlights of the year 1935 in relation to the famous article by Einstein, Podolsky and Rosen (1935), briefly EPR: "Can quantum-mechanical description of physical reality be considered complete?". This article was received by the editors of the journal *Physical Review* on March 15th and appeared in the issue of May 15th. Exactly one month later, Pauli wrote a letter to Heisenberg, asking him to reply to the article by EPR. I will come back to this letter below.

Instead of a reply by Heisenberg, the *Physical Review* issue of October 15th contained a reply by Bohr (1935), under the same title as the original work by EPR. Shortly thereafter, an article by Schrödinger (1935) appeared in the November 29th issue of the German journal *Die Naturwissenschaften*, entitled "Die gegenwärtige Situation in der Quantenmechanik" (English: The current situation in quantum mechanics).

Even though Schrödinger's article was not a direct reply to that by EPR, it was a reaction to the type of situation which the EPR paper referred to. In his article, Schrödinger coined the German term "Verschränkung", English: entanglement. However, Schrödinger did not talk about the entanglement of two particles or two systems, but he spoke of an "entanglement of predictions" and an "entanglement of our knowledge about two objects". In one paragraph he even uses the German expression "sich verheddern" (to snarl up). For Schrödinger, the quantum state (also called the wavefunction) had no ontology in itself. He referred to it as a "catalogue of expectation".

As already indicated by the title of their article, the strategy of EPR was to prove that quantum mechanics is not complete. In accordance with Einstein's view that quantum theory is only a statistical theory, they wanted to convince the reader that physics contains "elements of reality" which are not taken into account by the mathematical formalism of quantum mechanics. EPR started with two definitions:

- *Physical Reality*: If, without disturbing a system in any way, we can predict the value of a physical quantity with certainty (i.e., with probability one), then there exists an element of physical reality which corresponds to this physical quantity.

- *Completeness*: Each element of physical reality must have its correspondence in a physical theory.

In order to illustrate the meaning of the definition for physical reality, let me consider a simple two-particle system. A billiard ball rests on a table and we shoot a second billiard ball against the first one. Without measurements (and unless we are an expert in billiard) we will not necessarily know the momenta of the two billiard balls after the collision. However, because of momentum conservation, we only have to measure one of the momenta, say p_1, in order to predict the exact value of the other one, p_2. EPR emphasized that we can make this prediction "with certainty" because we know that the two billiard balls *have* momenta with definite values ever since their collision. If one can assign definite values to properties of a system independent of whether a measurement is performed or not, EPR call these properties "elements of reality".

There are two-particle quantum states for which only the relative coordinate Q (the distance between the two particles) and the total momentum P have a well-defined meaning. For such states the positions of single particles, q_1 and q_2, and their momenta, p_1 and p_2, are not well-defined properties. In particular, they do not have specified values unless one of the two quantities is measured. A measurement of the position of the first particle, q_1, will also fix the position of the second particle, q_2, and a measurement of the momentum of the first particle, p_1, will fix the momentum p_2 of the second.

Now, EPR's argument was the following: A measurement of particle 2 cannot possibly have an influence on particle 1 (they can be light years apart from one another). Therefore, if the experimenter measures p_2, then the value of p_1 can be predicted for particle 1 "without disturbing it in any way". For

EPR the value of p_1 is an element of physical reality, i.e., it is fixed ever since the two-particle state as a whole has been prepared. Since quantum mechanics does not allow the assignment of definite p-values to the particles under these conditions (before one of the measurements is performed), it is, according to EPR, incomplete.

For Einstein this way of attacking quantum theory was, in a certain sense, a change of tactics. Already at several occasions before he had used similar arguments involving two particles and their entangled states. However, in earlier years Einstein tried to prove that quantum theory is plainly wrong. He attacked the uncertainty relations and used entangled states to demonstrate that the uncertainty relations can be violated.

His corresponding argumentation for the situation described above would have been the following: We can measure q_2 and, therefore, we know q_1 without disturbing particle 1. Next we can measure p_1 and know both, q_1 and p_1, and thus violate the uncertainty relation between p and q. Usually it was Bohr who replied to such arguments, and in this case he would have argued: There is no way to check whether q_1 has really the predicted value. If one measures q_1 before p_1, the momentum may have changed. If one measures q_1 after p_1, the value may no longer be the one which was predicted. The surprising new twist in the EPR article was that the attack was no longer directed against the uncertainty relations but rather against the completeness of the quantum mechanical description.

There had been many reactions to the EPR article even before its publication (see Jammer 1974). Here I will only refer to the reactions of Pauli and Bohr. On June 15th, 1935, Wolfgang Pauli wrote a letter to Heisenberg (von Meyenn 1985, pp. 402–405), starting:[2]

> Dear Heisenberg! ... Einstein has once again expressed himself publicly on quantum mechanics ... (with Podolsky and Rosen – no good company, by the way). As is well known, this is a catastrophe every time it happens. "And so he reasons pointedly: That cannot be which should not be" (Morgenstern). I'll grant him that if a student in the early semesters had made such objections to me, I would have regarded him as very intelligent and hopeful.

He then continues analyzing the EPR argument:

> He now has understood that much that two quantities which correspond to non-commuting operators cannot be measured at the same time and that one cannot assign numerical values to them simultaneously.

---

[2]The translation of this passage as well as most other translations of originally German texts are taken from Gilder (2008).

This refers to the old type of attacks of Einstein against quantum theory. But now Pauli turns to a sharp characterization of Einstein's new move:

> Now comes the "deep feeling" and he proceeds: "Because measurements of system 2 cannot disturb particle 1, there must be something called 'the physical reality', which is the state of particle 1 in itself, independent of which measurements have been performed at system 2."

After more detailed deliberations about EPR's arguments and some remarks about how to reply to them, Pauli addresses the issue of completeness (his italics):

> Elderly gentlemen like *Laue* and *Einstein* are haunted by the idea that quantum mechanics is *correct* but *incomplete*. They think that it can be *completed by statements which are not part of quantum mechanics, without changing the statements which are part of quantum mechanics.* (A theory with this property I would – in the logical sense – call *incomplete* ...) Maybe you could – in a reply to Einstein – clarify with authority that such a completion of quantum mechanics is impossible without changing its content.

Pauli now very brilliantly analyzes the two types of influences which quantum theory permits: the well known influence related to an interaction and the particular type of influence related to entanglement. As the term entanglement had not yet been coined, he defines the absence of such an influence (Pauli's italics):

> The total system is in a state where the partial systems 1 and 2 are *independent*. (Decomposition of the eigenfunction into a product.)
> *Definition*: This is the case if after the performance of a measurement of 2 of an arbitrary quantity F_2 with *known* result $F_2 = (F_2)_0$ (number) the expectation value of the quantities F_1 of 1 remains the same as if the measurement of 2 had not been performed.

Today we call such states separable, and when a state is not separable it is entangled. So, entanglement implies that the knowledge of the result of a measurement of one of the two subsystems changes the expectation value for certain measurements of the other subsystem.

Pauli's rather arrogant and presumptuous reaction was shared by many other (in particular young) physicists at the time. Carl Friedrich von Weizsäcker (1985) describes in his book *Aufbau der Physik* his perception of the EPR paper in much the same words as Pauli. A completely different response is reported by Leon Rosenfeld, who was the assistant of Bohr at Copenhagen in 1935. Rosenfeld (1967, p. 128) wrote:

> This onslaught came down on us as a bolt from the blue. Its effect
> on Bohr was remarkable ... as soon as Bohr had heard of Einstein's
> argument, everything else was abandoned: we had to clear up such a
> misunderstanding at once ... day after day, week after week, the whole
> argument was patiently scrutinized with the help of simpler and more
> transparent examples.

Obviously, the "philosopher" Bohr not only understood the subtlety of
the argument, but also its wider ranging consequences, in contrast to the
younger physicists who only looked at the mathematical formalism. Yet
Bohr's reply (Bohr 1935) in the *Physical Review* is not as crystal-clear as
Pauli's analysis, and its most cited phrase actually remains cryptic:

> We now see that the wording of the above-mentioned criterion of phys-
> ical reality proposed by Einstein, Podolsky, and Rosen contains an am-
> biguity as regards the meaning of the expression "without in any way
> disturbing a system". Of course there is in a case like that just con-
> sidered no question of a mechanical disturbance of the system during
> the last critical stage of the measuring procedure. But even at this
> stage there is essentially the question of an influence on the very condi-
> tions which define the possible types of predictions regarding the future
> behavior of the system.

Until today, physicists and philosophers of science debate on what exactly
the "influence" to which Bohr refers might be. As Bohr (in a similar way
as Schrödinger) emphasizes that the influence acts on "the possible types
of *predictions*", the Copenhagen interpretation is often characterized as a
"subjective" or "knowledge-related" interpretation.

3   Hidden Variables:
John von Neumann and Grete Hermann

In this section I want to concentrate on one particular aspect of hidden-
variable theories: the no-go-theorem by John von Neumann, the analysis of
this theorem by Grete Hermann, and the question of why the physicists at
the time did not take any notice of Grete Hermann's analysis.

John von Neumann published his famous book on the mathematical foundations of quantum theory in 1932. Here he laid the foundations for dealing
with infinite-dimensional vector spaces (so-called Hilbert spaces) and linear
operators on these spaces. For many scientists these concepts were completely new and only few could fully digested its content.

In chapter IV of his book, von Neumann gives a proof that quantum
mechanics cannot be completed by the introduction of hidden variables, and

– as a consequene – hidden-variable theories were considered an impossibility until the 1950s. In most cases, physicists referred to von Neumann's proof, even though only few may have really understood the argument.

When David Bohm (1952) presented his hidden-variable model of quantum theory and physicists had to accept that such extensions of quantum theory are possible, the reaction was quite astounding. Robert Oppenheimer commented: "If we cannot disprove Bohm, then we must agree to ignore him" (Gilder 2008). This can hardly be considered a scientific argument against a theory. Until today, the arguments against Bohmian mechanics mostly have nothing to do with science ("absurd", "nonsense", etc.). I will come back to this point at the end of the article.

There is one key assumption in von Neumann's proof: For the expectation value (German: "Erwartungswert") of physical quantities R and S he assumed the additivity relation:

$$\mathrm{Erw}(R + S) = \mathrm{Erw}(R) + \mathrm{Erw}(S). \tag{1}$$

This relation is trivial in the context of classical physics and follows mainly from the very definition of an expectation value. However, in quantum theory the sum $R + S$ of two non-commuting observables R and S is not defined operationally. It can be defined mathematically, but then this sum has (almost) nothing to do with the observables R and S. In particular, the possible results of measurements of $R + S$ are not related to the results of measurements of R and S if these two quantities do not commute (are not compatible).

Von Neumann was well aware that this assumption for the expectation value cannot be justified on physical grounds, but somehow most physicists did not take notice of this limitation of the applicability of the result. Whenever the question of hidden-variable theories was discussed, the reaction of most physicists was: "von Neumann has proven that this is impossible".

However, already in 1935 a young philosopher had analyzed von Neumann's proof and pointed exactly to the weak spot. Her name was Grete Hermann, and her article (Hermann 1935) "Die naturphilosophischen Grundlagen der Quantenmechanik" appeared in the *Abhandlungen der Fries'schen Schule*, a philosophy journal. Gilder (2008) writes: "this was hardly a place where the devotees of von Neumann's defective proof were likely to ever discover it". The reason why Grete Hermann had no problem to understand the proof was that she had received a PhD in mathematics (under Emmy Noether) before she studied philosophy. Here is what she wrote:

> A thorough examination of the proof of von Neumann reveals, however, that in his argumentation he makes an assumption which is equivalent

to the statement he wants to prove. ... von Neumann assumes for the expectation value Erw(R), which assigns to each physical quantity a number, that Erw($R+S$) = Erw(R)+ Erw(S). ... However, this relation is not evident for those quantum mechanical quantities for which an uncertainty relation holds ... because, in general, the sum of two of these quantities is not defined. ... It is only by means of mathematical operators ... that the formalism allows the introduction of a sum also for these quantities. ... Therefore, the proof is circular.

Presumably, the work of Grete Hermann would have been forgotten completely, if Heisenberg (1971) in his book "Physics and Beyond" had not dedicated a whole chapter to her. Referring to the years 1934/1935, this chapter begins (Heisenberg 1971, p. 117f):

> We were offered a special occasion for philosophical discussions one or two years later when the young philosopher Grete Hermann came to Leipzig for the express purpose of challenging the philosophical basis of atomic physics. ... Grete Hermann believed she could prove that the causal law – in the form Kant had given it – was unshakable. Now the new quantum mechanics seemed to be challenging the Kantian conception, and she had accordingly decided to fight the matter out with us.

And the chapter ends (Heisenberg 1971, p. 124):

> Science progresses not only because it helps to explain newly discovered facts, but also because it teaches us over and over again what the word "understanding" may mean. This reply, based partly on Bohr's teachings, seemed to satisfy Grete Hermann to some extent, and we had the feeling that we had all learned a good deal about the relationship between Kant's philosophy and modern science.

At first sight, these remarks by Heisenberg about Grete Hermann sound like compliments, and in a particular sense they are. But expressions like "the young philosopher" obtain a strange connotation when one takes into account that Grete Hermann not only held a PhD in mathematics but was also older than Heisenberg by about nine months.

With respect to hidden-variable theories, Grete Hermann's article is interesting irrespective of her dismissal of von Neumann's proof. During the first chapters of her article she argues, almost in the style of analytical philosophy, in favor of both deterministic causality and hidden variable theories:

- She emphasizes that hidden variables need not be classical observables (like position and momentum). Indeed, she considers such models as disproven by interference experiments: Precise values for position and momentum which are only unknown to us seem to contradict the interference fringes in double-slit experiments. In this context, she refers to an article of von Laue (1934) who also points out that non-classical hidden variables are not ruled out by such experiments.

- She cites an article by Schrödinger (1934) where he argues that measurements do not "disturb" a system by referring to a situation which later became known as "interaction-free measurements". From this argument it is obvious that the uncertainty relations for, say, position and momentum are not due to a "mechanical" disturbance during the act of measurement.[3]

- She points to an assumption in von Neumann's proof for which there is no physical justification (and which in her opinion is circular) and, thereby, reopens the possibility for hidden-variable theories despite von Neumann's proof.

- She discusses systems of identical particles and notices that these are entangled (without using this term, of course). Therefore, it is meaningless to speak of single particles in this context, and any theory of hidden variables has to refer to the system as a whole.

From the first eight chapters of her article the reader gets the impression that Grete Hermann is up to support hidden-variable theories, which would be in complete agreement with her Kantian philosophical convictions. Step by step she dismantles any argument which had been brought forth against hidden-variable theories and appears to pave the way for a model which can save the causality requirements of Kantian philosophy.

But then, suddenly, at the end of chapter 8 she changes her argumentation and the style of her writing: She uses complementarity, the correspondence principle, retrodiction etc. to argue in favor of the conventional interpretation of quantum mechanics. Quantum mechanics is, as she tries to convince the reader, already complete and capable of explaining every outcome of measurements in terms of causal chains. It is almost impossible to avoid the question: "What made Grete Hermann change her mind?"

The question remains unanswered until today and one can only speculate whether her host at the department (Werner Heisenberg) intervened with her original claims and persuaded her to argue along the lines of the Copenhagen interpretation. Whether this was really her own conviction we do not know.

---

[3]In the same article Schrödinger suggests to characterize the notion of a measurement by the term "Prokrustie". This refers to Greek mythology where the giant Procrustes forces his guests into his bed by stretching them when they were too small and crunching them when they were too tall.

And there are many other questions which remain unclear. Why did scientists ignore the fact that von Neumann's no-go-theorem was based on a physically unjustified assumption and, therefore, is inapplicable for quantum systems? It is very unlikely that nobody was aware of this. Heisenberg and von Weizsäcker were involved in many discussions with Grete Hermann, and presumably the whole group around Heisenberg at Leipzig knew about her work. And Einstein, in discussion with his assistants Bergmann and Bargmann around 1938, pointed to the additivity assumption in von Neumann's proof and asked: "Why should we believe in that?" (Jammer 1974).

Several physicists had their own proofs of no-go-theorems for hidden variables, for instance we know those by Pauli and Schrödinger. However, a careful analysis reveals (Bacciagaluppi and Crull 2009) that implicitly they made the same additivity assumption for the expectation value as von Neumann did. Heisenberg based his version on "the cut" between observer and observed, and his argument is quite interesting: Any hidden variable has to be assigned to the cut, because in the quantum regime the evolution of a state is described by the deterministic Schrödinger equation. In the classical regime the deterministic classical equations of motion hold. Therefore, the indeterminacy of quantum theory and the location of possible hidden variables had to be placed at the cut. However, the cut can be shifted arbitrarily, so there is no objective place for hidden variables.

4 Hidden Variables: deBroglie and Einstein

As early as at the Solvay Conference in 1927, Louis deBroglie had already presented a hidden-variable model for quantum theory. They were all there: Albert Einstein, Niels Bohr, Arnold Sommerfeld, Erwin Schrödinger, Wolfang Pauli, Werner Heisenberg, Max Born, Paul Dirac, Paul Ehrenfest, Max Planck, and many others. 1927 was five years before von Neumann presented his no-go-theorem in his book on the mathematical foundations of quantum theory. Until today the Solvay meeting is considered as one of the crossroads of quantum theory (Bacciagaluppi and Valentini 2009).

After deBroglie's presentation of his model, no one except Pauli, who dismissed the model on formal grounds, commented on deBroglie's suggestion. Decades later, in a Festschrift celebrating deBroglie's sixtieth birthday, Pauli wrote an article in which he explained his arguments against the model of deBroglie in more detail (Pauli 1955). After all, Pauli had to admit that he rejected deBroglie's model only because (i) it breaks a symmetry of quantum theory (the invariance under an exchange of position and momentum, which in a relativistic setting is broken anyhow) and (ii) it introduces variables which, by definition, cannot be observed.

We may speculate that deBroglie might have hoped that Einstein supported him at the conference. Everybody knew that Einstein was looking for a statistical interpretation of quantum theory, and deBroglie's model was a step in this direction. Why did Einstein not even make a single remark about deBroglie's suggestion?

One of the reasons may have been that Einstein had developed a hidden-variable model himself (Belousek 1996). On May 5th, 1927, five months before the Solvay Conference, Einstein read a paper at a meeting of the Prussian Academy of Sciences in Berlin, entitled: "Does Schrödinger's wave mechanics determine the motion of a system completely or only in the sense of statistics?"[4] Initially Einstein was very enthusiastic about his model. On May 19th, Heisenberg expressed in a letter a "burning interest" in this subject, writing that "it may be possible after all to know the orbits of particles more precisely than I would wish". Then, on May 21st, Einstein telephoned the editor of the journal to which the paper was submitted and withdrew it from publication: "Die Arbeit soll nicht erscheinen" ("The paper should not be published", Belousek 1996). Only a hand-written manuscript exists in the Einstein archives. What made Einstein change his mind?

Apparently, in analogy to general relativity, Einstein tried to define a metric which serves as a guiding field for a particle. (Already years earlier, the connection form derived from the metric field in general relativity had been called a "guidance field" by Weyl (1919).) He derived this metric from Schrödinger's wave function

$$g_{\alpha\beta}(x) \sim \frac{\partial^2}{\partial x^\alpha \partial x^\beta} \left[\ln\right] \psi(x) \,, \tag{2}$$

The right hand side of this equation determines the left hand side (but the actual relation is slightly more complicated). Moreover, while Einstein first tried the second derivatives of the wave function for this relation, it is likely that he later tried the second derivative of the logarithm of the wave function – the bracket [ln] indicates these two possibilities.

At this point Einstein might have realized the problem that $g_{\alpha\beta}$ does not vanish when x_α refers to one particle and x_β to another one. The motion of one system influences the motion of the other. Taking the logarithm in

---

[4]Originally in German (Einstein 1927): Bestimmt Schrödingers Wellenmechanik die Bewegung eines Systems vollständig oder nur im Sinne der Statistik? For more details about this paper and its recception see Belousek (1996).

Eq. (2), this is exactly the case when the two systems are entangled. Einstein concludes in his manuscript that his model "does not satisfy a general condition that must be placed upon a general law of motion of systems". Could it be that Einstein became aware already in 1927 of the possibility of a non-local influence of one particle onto the trajectory of another particle?

When deBroglie presented his model at the Solvay Conference five months later, Einstein might have realized that deBroglie's version also contains non-local influences and that his "general condition" which must be placed upon a general law of motion is not fulfilled in deBroglie's model either. Was this the reason for Einstein's silence after deBroglie's talk? Or was he still convinced to be able to prove that quantum theory is actually wrong (the old type of tactics)?

More open questions: Why did Einstein never again mention his attempt of formulating a hidden-variable model? Why did he, in a letter to Born in 1952, comment on Bohm's model: "it seems to cheep to me". Around the same time deBroglie also returned to his old hidden-variable ideas (very similar to Bohm's approach) and reported that Einstein "was happy to learn of this development and encouraged me to pursue my efforts in this direction".

5 Conclusions

Until today physicists are puzzled by the phenomena related to entanglement. This bewilderment may have been the reason why Einstein rejected his own hidden-variable model, why he did not comment on deBroglie's proposal in 1927, and why he considered Bohmian mechanics as "to cheep". And Pauli, who as a physicist largely constrained himself to the formal and mathematical aspects of entanglement, might have seen something much deeper in this phenomenon – deeper than the laws of physics, maybe even touching aspects of the psychophysical problem such as Jungian synchronicity, which he was extremely interested in.

I tried to illustrate that physics is not free of psychological influences. Discussions about different "interpretations" can be quite emotional and sometimes end up in almost fanatic disputes. What we accept as a "reasonable explanation" can be based on beliefs and convictions which have nothing to do with science. And such beliefs may decisively influence the historical development of the prevalent scientific view. The most common reaction to Bohm's model of quantum mechanics (or to similar "quasi-classical" explanations) is: "this would be a step back". Is it really?

6 Acknowledgment

I greatly acknowledge many stimulating discussions with the participants of the Pauli-Jung Workshop at Filzbach, Switzerland, September 2012. Furthermore, I am grateful to an anonymous referee for pointing out relevant errors as well as for interesting additional information.

References

Atmanspacher H. (2012): Dual-aspect monism à la Pauli and Jung. *Journal of Consciousness Studies* **19**(9-10), 96–120.

Bacciagaluppi G. and Crull E. (2009): Heisenberg (and Schrödinger and Pauli) on hidden variables. *Studies in History and Philosophy of Modern Physics* **40**, 374–382.

Bacciagaluppi G. and Valentini A. (2009): *Quantum Theory at the Crossroads: Reconsidering the 1927 Solvay Conference*, Cambridge University Press, Cambridge.

Bell (1966): On the problem of hidden variables in quantum mechanics. *Reviews of Modern Physics* **38**, 447–452.

Belousek D.W. (1996): Einstein's 1927 unpublished hidden-variable theory: Its background, context and significance. *Studies in History and Philosophy of Modern Physics* **27**, 437–461.

Bohm D.J. (1952): A suggested interpretation of the quantum theory in terms of "hidden" variables I and II. *Physical Review* **85**, 166–193.

Bohr N. (1935): Can quantum-mechanical description of physical reality be considered complete? *Physical Review* **48**, 696–702.

Einstein A., Podolsky B., and Rosen N. (1935): Can quantum-mechanical description of physical reality be considered complete? *Physical Review* **47** 777–780.

Einstein A. (1927): Bestimmt Schrödingers Wellenmechanik die Bewegung eines Systems vollständig oder nur im Sinne der Statistik? Manuscript in the Albert Einstein Archives, Archival Call Number 2-100, dated 5. 5. 1927.

Gilder L. (2008): *The Age of Entanglement*, Vintage Books, New York.

Gleason A. (1957): Measures on the closed subspaces of a Hilbert space: *Journal of Mathematics and Mechanics* **6**, 885–893.

Heisenberg W. (1971): *Physics and Beyond*, Harper Collins, New York. German original: *Der Teil und das Ganze*, Piper, München 1969.

Hermann G. (1935): Die naturphilosophischen Grundlagen der Quantenmechanik. In *Abhandlungen der Fries'schen Schule*, ed. by O. Meyerhof, F. Oppenheimer and M. Specht, Verlag Öffentliches Leben, Berlin, pp. 69–152.

Jammer M. (1974): *The Philosophy of Quantum Mechanics: The Interpretations of Quantum Mechanics in Historical Perspective*, Wiley, New York.

Jauch J.M. and Piron C. (1969): On the structure of quantum propositional systems. *Helvetica Physica Acta* **42**, 842–848.

Jung C.G. (1952): Synchronizität als ein Prinzip akausaler Zusammenhänge. In *Naturerklärung und Psyche*, ed. by C.G. Jung and W. Pauli, Rascher, Zürich, pp. 1–107.

Kochen S. and Specker E.P. (1967): The problem of hidden variables in quantum mechanics. *Journal of Mathematics and Mechanics* **17**, 59–87.

Madelung E. (1926): Quantentheorie in hydrodynamischer Form. *Zeitschrift für Physik* **40**, 322–326.

Pauli W. (1955): Bemerkungen zum Problem der verborgenen Parameter in der Quantenmechanik. In *Wolfgang Pauli – das Gewissen der Physik*, ed. by C. Enz and K. von Meyenn, Vieweg, Braunschweig 1988, pp. 251–257.

Rosenfeld L. (1967): Quoted in *Niels Bohr. His Life and Work as Seen by His Friends and Colleagues*, ed. by S. Rozental, North Holland, Amsterdam.

Schrödinger E. (1934): Über die Unanwendbarkeit der Geometrie im Kleinen. *Die Naturwissenschaften* **31**, 518–520.

Schrödinger E. (1935): Die gegenwärtige Situation in der Quantenmechanik. *Die Naturwissenschaften* **23**, 807–812, 823–828, 844–849.

von Laue M. (1934): Über Heisenbergs Ungenauigkeitsbeziehungen und ihre erkenntnistheoretische Bedeutung. *Die Naturwissenschaften* **22**, 439–441.

von Meyenn K., ed. (1985): *Wolfgang Pauli. Wissenschaftlicher Briefwechsel, Band II: 1930–1939*, Springer, Berlin.

von Neumann J. (1932): *Mathematische Grundlagen der Quantenmechanik*, Springer, Berlin.

von Weizsäcker C.F. (1985): *Aufbau der Physik*, Hanser, München.

Weyl H. (1919): Eine neue Erweiterung der Relativitätstheorie. *Annals of Physics* **59**, 101–133.

Dual Support for Pauli's Dual Aspects

William Seager

Abstract

Although Wolfgang Pauli's main interest in the mind began with Jungian self-exploration and therapy, he also made a number of suggestions about the mind-matter relation. His viewpoint appears to be a form of dual-aspect theory, but it is not entirely clear whether Pauli would have favored a neutral monism or a Spinozistic form in which mind and matter are co-fundamental features of reality. I begin with a discussion of this issue. But the main goal of this paper is to address a more general philosophical question: what is required to support a dual-aspect theory of nature? I think there are two central problems here which stem, unsurprisingly, respectively from matter and mind. About matter, dual-aspect theory requires a kind of completeness of the physical world. Some recent results in the foundations of quantum mechanics support the kind of completeness needed for a dual-aspect theory. With regard to mind, dual-aspect theory suggests that consciousness should be elemental, primitive or simple. I will review some phenomena which provide support for the claim that consciousness is intrinsically simple.

1   Mental Medicine

It is not uncommon for prominent physicists to put forth opinions on philosophical matters, but they typically reserve such indiscretions for late in their careers and try to avoid letting such marginal issues intrude on or influence their work in physics. In this regard, Wolfgang Pauli is rather unusual. He had a long standing philosophical interest in the nature of the mind that began when he was in his early thirties, albeit sparked by initially non-intellectual causes, and which continued up to his unfortunately early death in 1958. Somewhat ironically, Pauli regarded these philosophical interests as stemming from a "spiritual transformation" which overcame the influence of his "antimetaphysical descent" from his godfather Ernst Mach (see Enz 1973, p. 787).

It remains important to recognize that the proximate cause of Pauli's concern for the mind, especially his own mind, was a severe psychological crisis which befell Pauli around the age of thirty, rather than an academic interest in the philosophy of mind (of which, so far as I can tell and with respect to contemporary work, Pauli had no interest or knowledge).[1] As is well known, in 1931 Pauli sought treatment from Carl Gustav Jung for a poorly specified but debilitating mental condition. For some years Pauli had led a kind of double life: brilliant, highly focused, super rational, supremely mathematically gifted physicist on the one hand while on the other hand pursuing nightly exploits of bar hopping, womanizing and brawling in seedy areas of various cities (particularly Hamburg, where "seedy" takes on a whole new meaning). Finally, his growing self-loathing and a fear about this kind of life's final destination led him, on the advice of his father and after a surprisingly careful study of Jung's work, to approach the Swiss therapist.

Pauli met with Jung who was, as a healer, very concerned with Pauli's state of mind and, as a Jungian, extremely interested theoretically insofar as Pauli was "chock full of archaic material" (Miller 2010, p. 128). Indeed it was on the basis of this latter interest that Pauli and Jung's long friendship and philosophically fruitful association persisted. What is most interesting for our purposes is Jung's general diagnosis of Pauli's underlying problem, which was psychic disintegration. Jung wrote of Pauli's first visit (Miller 2010, p. 127):

> The reason why he consulted me was that he had completely disintegrated ... he had lost himself entirely. ... When [Pauli] came to consult me ... he was in such a state of panic that not only he but I myself felt the wind blowing over from the lunatic asylum.

It may be the Jung's theoretical interest trumped his therapeutic instincts for he refused to treat Pauli himself. He explained this in these words, referring to the "archaic material" (Miller 2010, p. 127):

> Now I am going to make an interesting experiment to get that material absolutely pure, without any influence from myself, and therefore I won't touch it.

Instead, Jung sent Pauli to a young and inexperienced analyst, Erna Rosenbaum, who Jung regarded as a "just a beginner" who he "was absolutely

---

[1]Pauli died in 1958 just as the philosophical resurgence of materialism in the form of the psycho-neural identity theory was getting up to full steam in the writings of Place (1956), Smart (1959) and the considerably more nuanced Feigl (1958). It seems that Pauli had little or no contact with or interest in this sort of academic analytic philosophy although it is interesting to speculate how he might have reacted to it.

sure ... would not tamper" (Zabriskie 2001) with this spectacular treasure trove from such an unexpected source.

Although somewhat miffed at first upon being shunted away from Jung himself,[2] Pauli entered into analysis with Rosenbaum and his condition improved substantially. He broke off analysis after a relatively short time but remained in frequent contact and collaboration with Jung until his death more than twenty-five years later. From that time on, the focus of their interaction was theoretical rather than therapeutic. Pauli later underscored this transition with the remark that (Pauli 1994a, p. 164)

> the further development of the ideas of the unconscious will not take place within the narrow framework of their therapeutic applications, but will be determined by their assimilation to the main stream of natural science.

The bulk of the writing by Pauli and Jung, both in their correspondence and numerous independent articles, was largely devoted to the structure and function of the mind as viewed from the Jungian perspective and especially on the relation of the conscious mind to the unconscious mind wherein resided the so-called archetypes critical, according to the Jungian picture, for a balanced life and genuine creativity.[3] But my focus here is on the metaphysics of mind endorsed by Pauli on which topic there is comparatively little source material and as yet scant scholarly investigation (however, see Atmanspacher (2012) and Atmanspacher and Primas (2006) for notable exceptions; see also Seager (2009)).

The main connection between Pauli's episode of analysis and his metaphysical outlook is the theme of disintegration. Psychic disintegration is overcome by the proper integration of the conscious and unconscious aspects of mind. Failure to achieve such integration leaves a subject emotionally and intellectually one-sided and subject to complaints of the sort that afflicted Pauli.[4] The "cure" is to achieve the appropriate degree of integration. It is more than tempting to see the problem of integration as lurking behind

---

[2] With his characteristic acidic humor, Pauli wrote to Rosenbaum that he had contacted Jung because of "certain neurotic phenomena which are connected with the fact that it is easier for me to achieve academic success than success with women. Since with Mr. Jung rather the contrary is the case, he appeared to me to be quite the appropriate man to treat me medically" but he coolly added that he wanted "nothing to be left untried" (Miller 2010, Chap. 8).

[3] Pauli's two best known contributions in this area are the lengthy essay "The Influence of Archetypal Ideas on the Scientific Theories of Kepler" (Pauli 1994b) and "Ideas of the Unconscious from the Standpoint of Natural Science and Epistemology" (Pauli 1994a).

[4] Jung diagnosed Pauli as "a highly educated person with an extraordinary development

Pauli's views on the nature of reality and the place of mind within it. The sort of view he sought is one where mind and matter are to be integrated not by the assimilation of the one into the other but rather as equally fundamental features of reality, each with its own, so to speak, appropriate place in reality as a whole. This theme is obviously also visible in Pauli's interpretation of quantum mechanics especially in the stress he puts on Bohr's notion of complementarity. Pauli explicitly links Jungian mental structure, quantum mechanics and the mind-body problem (Pauli 1994a, p. 164):

> but both formulations [i.e. archetypes & quantum mechanics] meet in their tendency to extend the old narrower idea of "causality" (determinism) to a more general form of "connections" in nature, a conclusion to which the psycho-physical problem also points.

Indeed, we shall see that by bringing in ideas from the interpretation of quantum mechanics, Pauli introduced an original embellishment of the venerable dual-aspect metaphysics of mind.

2   Metaphysical Options

In order to appreciate Pauli's view of mind, a brief review of alternative accounts will be useful.

It is fair to say that all the metaphysical options offered to solve the mind body problem over the last 400 years are fundamentally reactions to the picture formulated by René Descartes in his *Meditations* (Descartes 1984). Descartes' radical interactionist dualism is well known. He proposed that humans, alone amongst living creatures, were endowed with a non-physical mind, or soul, which was divinely coupled to the body via a single point of connection – the pineal gland. As one of the architects of the new scientific vision of the world, Descartes was well aware of the idea that the physical world should be seen as causally complete in itself. He was also aware that some basic physical quantities should be conserved and that physical transformations obey immutable laws of nature. It was thus immediately evident that any causal role for a non-physical mind in the physical world is deeply problematic.

Descartes' own solution is just about as elegant as is possible given a commitment to the causal efficacy of the mind. Descartes restricts the action of the mind to one point in the brain thus greatly limiting the scope of

---

of the intellect, which was, of course, the origin of his trouble; he was just too one-sidedly intellectual and scientific ... The reason why he consulted me was that he had completely disintegrated on account of this very one-sidedness" (Miller 2010, pp. 126f).

non-physical effects in the world. Furthermore, Descartes' physics mistakenly takes mass times motion or speed,[5] rather than velocity, as the physical quantity conserved through all interactions. Since motion is a scalar, the mind's power to change the direction of motion of the animal spirits in the brain ultimately responsible for (outgoing) bodily motion and (incoming or internally generated) sensory experience technically does not violate Descartes' conservation principle.

Nonetheless, the idea that the mind could, every so often, change the state of physical matter by direct influence was and remains difficult to comprehend, let alone accept. One of Descartes' favorite correspondents, Princess Elisabeth of the Palatinate, famously asked "how the soul of a human being (it being only a thinking substance) can determine the bodily spirits, in order to bring about voluntary action" (Descartes and Elisabeth 2007, p. 62).

In the face of this and other problems with interactionism, a plethora of alternative mind-body metaphysics were proposed. Any list would include these notable efforts: occasionalism, parallelism, materialism, idealism, epiphenomenalism and, most important here, dual-aspect theory. Each of these views is associated with important thinkers and each attempts to ameliorate Descartes' stark and uncompromising dualism in its own way.

Father Nicolas Malebranche accepted and expanded Princess Elisabeth's implicit criticism. His doctrine of occasionalism stems from the idea that absolutely nothing can cause anything except the divine power of God, who must intervene on the appropriate occasions to make sure the world evolves according to His plan. This puts dualistic interactionism in exactly the same metaphysical position as a purely physical world: all changes require intervention from without.[6] Few could take occasionalism seriously however.

If we insist that there are causal powers in things but also forbid any causal intercourse between disjoint metaphysical domains, we can embrace parallelism. Leibniz is the most famous philosopher associated with this doc-

---

[5]Rather bizarrely, Descartes refers to "size" rather than mass and his principles of conservation are peculiar and highly unintuitive. For example, according to Descartes, it is strictly impossible for a body of smaller size to move a larger body which is at rest. Descartes' laws were subjected to scathing criticism by Leibniz, among others, who correctly saw that the vector quantity of velocity should be used in the laws of motion. For an overview see Slowik (2009).

[6]It is worth pointing out that Malebranche was one of Hume's most important philosophical influences, and especially with respect to Hume's revolutionary work on causation. It is tempting to interpret Hume's skepticism about causation as following on Malebranche's argument that it is impossible to conceive of any necessary connections between things except those directly dependent upon God's will. For an extensive discussion of Malebranche's impact on Hume see Wright (1983).

trine and to him we owe the unforgettable metaphor of the two clocks, maintaining perfect synchrony yet causally isolated from each other.[7] I say "associated" because the opinion Leibniz considered is more reasonably viewed as a form of idealism but he remains the standard exponent of parallelism.

While perhaps a coherent view, the posit of two absolutely disjoint domains forces the doctrine into an unstable equilibrium. It is almost irresistible to let one domain slip away into oblivion as an unnecessary metaphysical luxury. This impulse leads either to materialism or idealism, depending upon which domain is to face execution.

Historically, until the twentieth century materialism was not a real contender. There were of course many materialist philosophers in the time of Descartes and the post-Descartes era, notably Thomas Hobbes and Julien de la Mettrie. Via its association with atheism, materialism was a potentially dangerous position to maintain and it remained highly marginal. Early materialists were also unclear about the distinction between mental substances and distinctively mental properties. Their primary concern was to deny the existence of mental substance (e.g. the soul) left materialism incomplete and, strange to say, insufficiently radical to truly eliminate the mental aspect of the world (see Seager 2007). The full flowering of materialism had to wait for the twentieth century.

By contrast, idealism's denial of the material world became almost philosophical orthodoxy. Once it was granted that all we are ever really, directly, aware of are mental entities (be they ideas, impressions or – much later – sense data) it was a small step to the economical elimination of the material realm. There were many forms of idealism, from the relatively straightforward version of Bishop Berkeley, through the nascent intricacies of Immanuel Kant's transcendental idealism to the impenetrably dense absolute idealism which came in both German and English models. But despite its philosophical credentials, I do not think idealism ever held much sway outside of dedicated metaphysical circles. The ever growing success and prestige of the physical sciences surely spoke irresistibly in favor of the reality of the material world. In the twentieth century, the steamroller of physical science overpowered philosophical scruples leading to the general triumph of materialism, belatedly embraced and anointed by philosophy itself.

---

[7]Leibniz used the clock analogy to illustrate all three of interactionism, occasionalism and parallelism. He was well aware of Huygens' discovery of resonance-based synchrony (interactionism), ridiculed the idea of having someone continuously adjusting the clocks to keep the same time (occasionalism and Newtonian interventionism) and celebrated the clockmaker so capable as to make two clocks that naturally kept perfect time and hence were automatically synchronized (see Leibniz 2006).

Finally, and anachronistically but most relevant to Pauli's views, is the multiple aspects view of Descartes' immediate successor, Baruch Spinoza. Spinoza held that there could, of necessity, be only one substance, in part because substance is that which is self-grounding whereas, in Descartes' picture, both matter and mind require the concordance of God to come into, and be sustained in, existence[8] and hence were not truly substances in themselves. This meant that, in some sense, Spinoza's "über"-substance had to be identified with God (leading to all sorts of personal trouble for Spinoza, up to and including excommunication).

Leaving aside theological issues, Spinoza held that nature was to be understood according to the traditional substance plus attribute metaphysical model, but he rigorously maintained that there could be but one substance. What Descartes had mistaken for the two distinct substances of mind and matter were merely attributes of the one, overarching substance which encompassed all of reality. The attributes were each capable of "expressing" their substance in its entirety from, as it were, their own point of view. In a famous passage, Spinoza (1985, Part 2, Prop. 7, scholium) writes that

> a circle existing in nature and the idea of the existing circle, which is also in God, are one and the same thing ... therefore, whether we conceive nature under the attribute of Extension, or under the attribute of Thought ... we shall find one and the same order, or one and the same connection of cause.

Given this view, it is evident that there is no problem of interactionism but there is equally no division of reality into distinct realms each of which seems to be, on its own, metaphysically optional (as attested by the mere possibility of materialism or idealism).

The substance plus attribute metaphysics raises a number of standard questions. One question involves the nature of substance. The only way to characterize substance is in terms of its attributes. It would be incoherent to say that a substance is without attributes or that it "transcends" them. What a substance is, is just a matter of the attributes that characterize it. For Descartes, there were two basic substances and their characteristics were either material or mental.

However, not all attributes of a substance need to be on an equal footing. Rather, attributes can be divided into fundamental and derived. Descartes

---

[8]This of course does not establish that there can only be *one* true substance. Spinoza (1985, Part 1) attempts to argue that there must be a substance which is infinite in all respects and that such a substance precludes there being any other (any putative second substance would, so to speak, be assimilated by the one true substance). The exact structure and ultimate success of his argument remains controversial.

and Spinoza were well aware of this distinction and its significance. Descartes defined the principal attribute of a substance to be the "property which constitutes its nature and essence and to which all the other properties are related" (Descartes 1983, Part 1, §53). For Descartes, each substance could have but one principal attribute.

Spinoza agreed that certain properties of substance were fundamental but allowed that a single substance could possess a plurality of such attributes. He defines an attribute as "what the intellect perceives of a substance, as constituting its essence" (Spinoza 1985, Part 1, D4) and says of the single existent substance which he called God that it consists "of an infinity of attributes, of which each one expresses an eternal and infinite essence" (Spinoza 1985, Part 1, D6).

It seems natural to regard Spinoza as holding that there is a single substance with a multiplicity of co-fundamental attributes, notable among which are those the human mind can grasp: thought and extension, or, in our terms, consciousness and matter. All other features of the world are to be regarded as derivative properties (what Spinoza called "modes") of substance, which are modifications of the attributes. For example, a circle is a modification of the attribute of extension; a finite mind is a modification of mentality. Not to enter the thickets of Spinoza scholarship, but taking this view is to implicitly accept the so-called objective view of the attributes, which has become the standard interpretation of Spinoza. The opposed subjective view holds that the attributes are merely appearances of some unitary attribute whose nature is flatly beyond human comprehension.[9]

But notice that the opposition between the subjective and objective view cuts across our other distinction, that between the fundamental and derivative properties of substance. One could deny that the attributes we are aware of are subjective appearances while still holding that they were not the fundamental attribute of substance. A view with essentially this characteristic arose at the end of the 19th century under the name of neutral monism.[10] This is the final entry in our catalog of metaphysical positions, but it is important to examine neutral monism because Pauli's views are in some ways suggestive of it and in other ways more suggestive of Spinozism.

In terms of the substance/attribute metaphysics, neutral monism holds that substance is fundamentally neither mental nor physical but rather has

---

[9]For a nice presentation of the seemingly decisive arguments against the subjective view see Shein (2009).

[10]The main proponents of neutral monism are Ernst Mach, William James and Bertrand Russell. For a comprehensive overview see Stubenberg (2008).

some other truly basic characteristic, difficult to specify and perhaps inaccessible to the human mind. Whatever it is, both mentality and materiality are derivative attributes ontologically dependent upon it.

Having roughly surveyed the metaphysical options available, it is now time to turn to Pauli's views.

3   Pauli's Dual-Aspect View of Mind

Although Pauli did not write very much about purely philosophical aspects of the mind-body problem, there is enough to glean something about his favored account. In terms of our catalog of options, we can find some which Pauli clearly rejected, others rejected by implication.

First, Pauli had no sympathy with classic Cartesian interactionism, accepting Elisabeth's point that causal relations holding across disjoint realms is paradoxical or even incoherent. Pauli (1994a, p. 154) wrote:

> ever since the 17th century [the psycho-physical interconnections] have been something of an embarrassment to the world picture of "classical" physics, in that it has been necessary to postulate ... a connection of a different, "parallelistic" kind, in addition to the ordinary causal connection.

Pauli further regarded this "parallelistic" connection as an *ad hoc* device, a rather empty concept which disguised rather than revealed the nature of the mind-body relation, for he goes on to write (p. 154):

> Is it *only* in the association of physical and psychical processes, and not in other situations as well, that a parallelistic relation exists? And does not a relation of parallelism mean that it is justifiable to demand that that which is associated, or "corresponds" (the corresponding) should also be embraced conceptually in a unity of essence?

The phrase "unity of essence" might suggest that Pauli falls within the purview of neutral monism if it is read as requiring a single attribute which provides this unity. On the other hand, if a single substance is sufficient then a more Spinozistic view is compatible with the quotation.

It is difficult to find any comments by Pauli which decisively favor either view, but here are some relevant remarks. Pauli (1994b, p. 260):

> it would be most satisfactory if physis and psyche could be conceived as complementary aspects of the same reality.

Or in a letter to Jung (Meier 2001, p. 159):

physis and psyche are probably two aspects of one and the same abstract
fact ... a mirror-image principle is a natural way to give an illustrative
representation of the psychophysical relationship.

It is reasonably clear that these remarks favor Spinozistic dual-aspect
monism over neutral monism. If Pauli had favored neutral monism it is
hard to believe he would not have mentioned the crucial fact that both
mentality and materiality are supposed to *reduce* to some third, more basic,
attribute of reality. The mirror-image metaphor is obviously closely akin
to Spinoza's example of the physical circle being correlative with the idea
of the circle. These attributes mirror each other and not some third more
fundamental attribute. I also take Pauli's reference to an "abstract fact" to
point more towards substance rather than the attributes (it should be noted
that Pauli was not bound to use the old philosophical distinctions or terms
in his thought).[11]

An extremely interesting facet of Pauli's version of dual-aspect thinking –
and one that should count as something of a philosophical innovation – is his
use of the quantum mechanical concept of complementarity (this has been
emphasized by Atmanspacher and Primas (2006)). In quantum mechanics,
complementary properties (1) cannot be jointly observed or measured with
arbitrarily high accuracy and (2) cannot even be assigned definite values in
one system at any one time. In the quote above Pauli (1994b, p. 260) intends
this technical meaning in his use of the term "complementary", although
in a generalized sense that goes beyond the confines of quantum theory in
physics.[12]

The application of the concept of complementarity to the mind-body
problem immediately raises intriguing and difficult questions. The first is
that it would seem to preclude any true parallelism between mind and matter
of the sort envisaged by Spinoza. Such a parallelism would assert that there
is a definite mirror-image relation (to use Pauli's own metaphor) between
the two domains or attributes. But such a relation would seem to imply a
definite "value" for both sets of properties. There does not seem to be any
general problem with assigning quite definite mental states with relevant
physical states (e.g. states of the brain). Much of cognitive neuroscience is

---

[11]Some more reasons for relating Pauli's view to Spinozistic dual-aspect theory have
been discussed by Seager (2009).

[12]As a follower of Bohr concerning the interpretation of quantum mechanics, Pauli would
have been very familiar with the idea that the concept of complementarity could be extended beyond the strict bounds of quantum physics and even beyond the bounds of science
itself.

involved with the search for such neural correlates of mental states and it is enjoying spectacular success.[13]

Perhaps a better approach would be to restrict the mental side of reality to consciousness in its subjective aspect (what is often called phenomenal consciousness or the "what it is like" to experience any particular mental state). Many have noted that reality has the peculiar feature that at least some of it can be regarded from either an objective, third person point of view (irresistibly identified with the "scientific" viewpoint) and a quite distinct and rather mysterious subjective, first person point of view (associated with our own personal states of consciousness as experienced "from within").

This division seems to me much more amenable to the application of the concept of complementarity. For if we take up an objective stance towards reality it seems quite impossible to even imagine how it can have a subjective aspect to it (this is the basis for the famous "hard problem" of consciousness (Chalmers 1996)). But when we take up the subjective stance, then the objective world seems to have only an arbitrary, merely causal relation to it. It does not seem possible to merge these two features of reality into one single overarching and unified picture of the world. I think something like this is what Pauli was trying to get at by applying the concept of complementarity to the mind-body problem.

One further point about this is worth noting. If – in some suitable sense of the term – subjective consciousness and objective physical reality stand in a relation of complementarity, then, whenever we try to think about the unity of nature it ought to be conceptually puzzling or paradoxical how the world can "integrate" them. From each particular point of view, the other should seem to be, as it were, inaccessible. This could help account for another aspect of the problem of consciousness which is called the "explanatory gap" (Levine 1983).

---

[13]For a recent remarkable example see Naci *et al.* (2013) who developed a method to "read" subject's brains enabling reliable yes-no communication via real-time magnetic resonance brain imaging. (Such work has important implications for recognizing "locked in" patients, fully conscious though paralyzed, who have been misdiagnosed as vegetative.) Of course, we remain far from anything like a catalogue of neural correlates of mental states. It is also worth mentioning that such correlates will in all likelihood be highly context sensitive and subject dependent (analogy: the physical correlates of "money"). One intriguing reason for such context dependence may be that the brain "reuses" neural circuitry for different functions as cognitive, sensory or emotional demands change from moment to moment (see Anderson 2010). Such "functional plasticity" does not undercut the existence of neural correlates of mental states, but obviously makes discovering and using them more difficult.

4 Explorations

4.1 Physical Completeness

I want now to consider some implications of Pauli's dual-aspect view. It would be wrong headed to look for any straightforward empirical verification of dual-aspect theory. By its nature such a view does not seem to make any straightforwardly testable predictions. Nor are there any dual-aspect theories which have been developed in sufficient detail for them to engage empirical investigation if this was possible. However, we can ask whether there are any general features of nature we might expect to observe if dual-aspect theory was true.

There are two sides to this question which arise simply because of the two aspects involved: mind and matter. On the physical side, what would we expect to find if dual-aspect theory is correct? Most dual-aspect theories regard both the mental and physical aspects as in some way reflecting or expressing all of reality. Although one can perhaps imagine a dual-aspect theory in which one aspect was associated with only some portions of reality, this would be an unstable theory. Such incompleteness would suggest instead that the incomplete aspect was somehow a product of the complete aspect. Modern physicalist accounts are of this sort: they see the material world as fully or exhaustively expressing nature while the mental aspect is an emergent feature depending on and partially expressing this physical nature.

It is true that we can easily envisage forms of dual-aspect theory that permit the interpenetration of each aspect into, as it were, the other. While it is partially a mere verbal dispute about the correct definition of "dual aspects", I think we should carefully distinguish property dualism (or pluralism) from dual-aspect theory. Pure dual-aspect theory sees the world as susceptible to description by two orthogonal conceptual systems reflecting a dual nature. Property dualism sees mental and physical properties as distributed over the world in various ways (e.g. it is generally believed that mental properties occur very rarely, applying to only a tiny fraction of the physical systems in the world). These properties have distinctive causal powers and may influence each others' distribution. By contrast, pure dual-aspect views see the mental and physical as different ways of characterizing reality as a whole. Moving away from the pure dual-aspect view should be regarded as collapsing into some form of property dualism.

Thus dual-aspect theorists should expect to see both the physical and the mental aspects "covering" the entire world; each aspect should exhaustively express the whole of nature from, so to speak, its own point of view. Unfortunately, it is extremely difficult to assess such a claim with respect to the aspect of mentality. We are each of us restricted to our own consciousness which all too obviously appears to express only a minuscule part of reality.[14]

We have a much better general access to the physical aspect of the world. The dual-aspect theory implies that the physical should exhaustively express reality which in turn suggests that our fundamental physical theories should, in principle, be such as to encompass all physical phenomena. Although our physics is currently incomplete and notoriously fragmented, there is some reason to think, or at least hope, that the goal of completeness is not unattainable. In fact, the structure of quantum theory encourages this hope.

To address this issue a little more precisely, we need a better idea of what it is for a physical theory to be a "total" theory. Totality can be defined in terms of three mutually definable concepts: completeness, closure and resolution. In terms of the physical, these are jointly defined as follows: Completeness is the doctrine that everything in the world is a physical entity or, in principle, has a non-trivial physical description and as such abides by closure and resolution. Closure entails that there are no "outside forces" – everything that happens, happens in accordance with fundamental physical laws so as to comply with resolution. Resolution requires that every process or object be resolvable into elementary constituents which are, by completeness, physical entities and whose abidance with physical laws governing these constituents leads to closure.[15]

These three concepts are meant to capture the sense in which the physical aspect is complete and expresses the entire world in its own (physical) terms. They forbid the existence of radically non-physical entities, such as Cartesian minds. Instead, they insist that everything that exists has a physical nature. Further, they demand that the dynamics of the world be similarly completely determined by physical systems and laws. Finally, they ensure that systems far from those which are dealt with in fundamental physics (trees, stars, oceans, etc.) do not escape the net of the physical. All of these things can be resolved into fundamental physical structures and processes which determine their complex natures.

_____

[14]Leibniz held this to be a kind of illusion so that our own states of consciousness actually do encode all of nature but in an obscure and confused way. Leibniz's own example of this was how when we hear the sound of a wave crashing on the shore we are not consciously aware of the components: the myriad sounds of each droplet hitting the seaside which make up the sound we do experience (Leibniz 1996, p. 54). Or it may be, as Spinoza held, that our minds by their nature have access to only a limited part of nature.

[15]For a more extensive discussion of these notions see Seager (2012, Chap. 7).

This last point should not be read as endorsing anything as strong as part-whole reductionism (though it is compatible with that doctrine). We know that physical resolution is much more complex and nuanced than simply summation of parts. Resolution is apparently non-local and physical systems can exhibit a kind of holism very different from atomistic reductionism. Nonetheless, the claim is that all physical systems have natures ultimately expressible in terms of the fundamental physical entities of the world.

Of course, we cannot claim that present theory even approaches this ideal state. But it does seem arguable that totality is the goal of physical theorizing. The signposts of totality should be the prospective empirical adequacy of our best theory and what I will call its prospective upward completeness. Upward completeness entails that there is no ontological domain which is not fully characterized by our physical theory (at least in principle) and which determines all physical states of the world. Failure of upward completeness would present us with a picture of the world in which there appear to be basic physical states which are not strictly determined by a purely physical underlying reality.

In this regard, some recent results in quantum mechanics are suggestive. They appear to show that the supposition of upward incompleteness conflicts with our best theories, under the assumption that our best theory is empirically adequate.[16] The result in question is due to Pusey, Barrett and Rudolph (2012) – henceforth PBR. It is explicitly aimed to refute an interpretation of quantum mechanics in which the state funcion (or wave function) is merely "epistemic", that is, encodes a state of knowledge about some physical system rather than representing the physical state of that system. Their result can however be re-purposed in aid of showing the prospective completeness of physical theory.

Very crudely, the assumption of upward incompleteness (see Fig. 1) entails that a state of the fundamental level, which (following PBR) I refer to as λ, is compatible with (at least) two quantum states, call them ϕ_1 and ϕ_2. Thus, when the world gets into the state λ there is some non-zero probability that the quantum state is ϕ_1 and some non-zero probability that it is ϕ_2. We can then define a Hilbert space in which $\mathbf{0} = \phi_1$ and $(\mathbf{0} + \mathbf{1})/\sqrt{2} = \phi_2$ (call this last state +). The PBR thought experiment then involves preparing

---

[16]This last assumption is necessary because there can of course be no guarantee that current theory is extensible into one which is empirically adequate. Any theory always risks falsification. On the other hand, the assumption of empirical adequacy does not in general preclude upward incompleteness.

World described by best physical theory

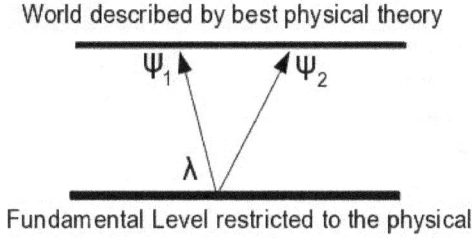

Figure 1: Upward incompleteness in the notation by Pusey *et al.* (2012).

two independent systems[17] into these states. This entails that there are four possible joint states of the two systems: $\mathbf{0} \otimes \mathbf{0}; \mathbf{0} \otimes +; + \otimes \mathbf{0}; + \otimes +$.

Now we bring together the two systems in order to perform a joint measurement. The crucial assumption here is that a measuring device's response is determined entirely by its physical state and the physical state of the system under observation. For our purposes, this requires the assumption that we can devise instruments which respond to λ (not intentionally of course, since we have no understanding of the nature of λ, but as a matter of fact we can produce instruments sensitive to the fundamental level represented by λ; this is a natural and plausible assumption). PBR then employ a clever measurement operator which projects onto four components, each one of which is guaranteed to *not* produce an output when the joint system is in one of the joint states given above. For example, the first component is simply $((\mathbf{0} \otimes \mathbf{1}) + (\mathbf{1} \otimes \mathbf{0}))/\sqrt{2}$. Thus, if the joint system was in the state $\mathbf{0} \otimes \mathbf{0}$ the probability of getting this first outcome would be zero. The other components are slightly more complicated but each leads to the same result. The upshot is that the experiment has zero probability of producing an output. Yet, by our assumption, the joint system sometimes gets into one of these states and our instrument would be able to "report" this fact. In that case quantum mechanics would be empirically inadequate, contrary to our assumption.

The point that matters to us is that on the assumption of empirical adequacy we can provide a kind of proof that no aspect of reality goes beyond what our best theory says about the world – at least, what it says about the *physical* aspect of the world. Thus empirical adequacy of quantum theory implies its completeness of coverage of the world. We can thus have some

---

[17]That this is possible is an important assumption of PBR, but it is one that is intuitively highly plausibly insofar as we appear to be free to create and manipulate experimental devices that are entirely remote from and disconnected from one another.

hope that our final physical theory will inherit this feature and thus verify this feature of the dual-aspect picture of reality.

It seems that the structure of quantum theory is rather special in this regard. Insofar as all phenomena are regarded as ultimately quantum mechanical in nature the completeness of quantum theory suggested by the PBR result provides support for our triad of desirable qualities: completeness, closure and resolution. Completeness is an obvious consequence if the PBR argument is correct and some descendant of quantum theory is empirically adequate. Closure would seem to follow since its denial would entail some "intrusions" of efficacious features of reality not taken into account in quantum theory, which would falsify our assumption of empirical adequacy. Finally, resolution follows from the guiding assumption that all phenomena are or are based entirely upon the fundamental physical features dealt with in our presumed empirically adequate quantum theory.

This result would mesh with a view of dual-aspect theory that saw the world as fully expressed by both aspects insofar as we would have shown that at least the physical aspect meets this condition. Given the difficulty of showing anything similar for the mental realm this may be the best we can hope for on this front. It at least supports to some extent Pauli's idea that the physical and the mental should mirror each other rather than influence each other. However, there may be some indirect support for the complementary mental aspect completeness arising from considerations about the nature of consciousness consonant with a dual-aspect outlook. The next section will address this.

It must be noted that a purely physicalist outlook in opposition to dual-aspect theory would also be happy to embrace the same conclusion of physical completeness, closure and resolution. I am not, however, trying to refute physicalism. My aim is the more limited one of showing that our current physical theorizing at least agrees with the kind of dual-aspect account of nature endorsed by Pauli. That is, it seems consistent with the direction of our physical theorizing that it should eventually attain completeness, closure and resolution at the physical level. This, in turn, is consistent with the idea that the physical aspect of nature expresses the totality of reality from, as it were, the physical point of view.

4.2 Mental Simplicity

I see no prospect of showing that the mental aspect of reality is similarly complete. A more modest goal might be attainable. If the mental aspect expresses all of reality than we would expect to find a spectrum of complexity on the mental side similar to the range of complexity exhibited by physical structures. We should, that is, expect to find in nature extremely simple forms of consciousness.[18]

There is a long and varied tradition which opposes this. As far back as Aristotle we find the claim that conscious states are in some way "aware of themselves" (see Caston 2002) and modern reflexive theories of consciousness from Brentano to the present day (see, e.g., Janzen 2008, Kriegel 2009) persist in this claim. No state of consciousness which involved such self-awareness could count as simple. The long phenomenological tradition finds consciousness to be a highly complex structure requiring a host of mental acts, again including some kind of intrinsic self-awareness of conscious states. See Gallagher and Zahavi (2008) for an introduction to the intricacies of the phenomenological analysis of consciousness.

Rosenthal's influential theory of consciousness[19] posits that consciousness results from a first-order mental state, α, generating a higher-order thought (HOT), $T(\alpha)$, about that state whose content is, roughly, "I am in α". This so-called HOT theory of consciousness obviously demands that any conscious creature have a system of concepts enabling thoughts explicitly about mental states. All these approaches preclude there being states of consciousness of a simplicity which comes anywhere near matching the pristine simplicity of the fundamental physical features of the world.

One or another of these theories of consciousness might be correct, but it is arguably the case that they all suffer from a kind of self-selection problem. It is, obviously, impossible to become introspectively aware of consciousness or to think about it in theoretical terms without having conscious states available to a complex, cognitively sophisticated mind. There is a danger of inferring from all the cases of consciousness available to us false conclusions about the essential structure of consciousness in itself.

Furthermore, we again face the problem that our access to consciousness is limited to our own case via introspection and, perhaps, a few not very different forms of consciousness which we can conjure up via imagination and empathy. But despite this epistemic limitation and theoretical prejudice it may be possible to grapple with the question of whether there are radically simple forms of consciousness.

---

[18]It may be worth noting that by "simple" here I mean no more than a commonsense notion of "lack of complexity". I am not making the claim consciousness is metaphysically simple, that is without parts or complexity of any kind.

[19]The theory was first enunciated by Rosenthal (1986) and has been under extensive development ever since by Rosenthal and many others. For an overview see Carruthers (2011).

Consider first an old issue: animal consciousness. Descartes' view was that animals are completely and totally unconscious. Although they exhibit physical reactions similar to those which, in us, lead to or from states of consciousness, they are purely mechanical beings devoid of inner life. Very few will follow Descartes nowadays. The evolutionary continuity between the animal world and ourselves is so abundantly clear that it seems absurd to deny animals minimally conscious states, such as that of sensory awareness, pain and pleasure, which are at least analogues of our own.

The orthodox and physicalist view of consciousness is that some animals are conscious, but that in general consciousness is not very common in the universe. Instead, consciousness is seen as an emergent feature stemming from the growth of biological complexity. On this view, at some point during the evolution of (animal) life consciousness sprang into existence. This would have happened in a way not that different from the way that, at some point, stars sprang into existence: once sufficient physical complexity arose, which would depend both on internal structures and the presence of a suitable environment, stars appeared in a universe as an ontological novelty.

Of course, the problem with this sanguine viewpoint is that the emergence of consciousness seems nothing at all like the emergence of other physical structures. Star formation is a predictable consequence of pre-existing conditions. Even the emergence of life seems, in principle, intelligible as a purely physical process, albeit one we still know relatively little about. Consciousness is quite different, hence the idea that there is a unique "hard problem" about explaining its presence in the world.

But even waiving this fundamental problem of consciousness, the orthodox view would be extremely implausible if it asserted that the first emergence of consciousness was in a form such as ours: fully articulated, introspectively accessible, cognitively complex and conceptually rich. It is much more plausible that the first consciousness arose at some point in the general evolution of animal (perhaps mere biological) complexity and was of a form of almost unimaginable simplicity. It is extremely hard to put any meaningful constraints on how simple an animal can be and still possess some kind of consciousness. It may well be that extremely simple forms of consciousness are to be found in extremely simple organisms.

It may even be that the problem of the emergence of consciousness is so intractable as to drive us towards the view that consciousness is a fundamental feature of the world. Panpsychists hold such a position and it is a tempting account of consciousness for dual-aspect theorists. Versions of panpsychism can be developed which can easily accommodate the mirroring of mind and matter which Pauli suggests. Spinoza's thinking may well be related to such a panpsychist view. There are some, very few so far as I can tell, traces of panpsychist thoughts in Pauli's writings. He makes some reasonably sympathetic remarks about the "hylopsychism" of Bernhard Rensch (Pauli 1994a, p. 155) and occasionally refers to the mental realism as analogous to a physical field – an analogy that can be understood in panpsychist terms.

A second line of argument undercuts the idea that all consciousness is reflexive, introspectively accessible and conceptualizable. There is a kind of experience, with which I think everyone is familiar, where everything but focus on the task at hand fades away. Described as involving feelings of serenity, a loss of feelings of self-consciousness, a sense of timelessness or a distorted sense of time, a sense of being and feeling so focused on the present that you lose track of time passing; this is often labeled as "flow" experience (Csíkszentmihályi 1996). These descriptions are from subjects' memory of such experience. Trying to introspect flow while it is happening will destroy it. I think it is possible to interpret flow experience as providing a window into a more primitive or basic kind of consciousness, which we struggle to attain because of the constant active overlay of cognitive, conceptual and self-conscious mechanisms of thought we humans have developed. It may be that animals are, more or less, in a constant state of flow or at least spend much of their time in this state. And it may be that the simplest form of consciousness is akin to flow insofar as it does not require or even permit introspective access or reflexive awareness.[20]

A third line of argumentation challenges the claim that consciousness requires conceptual abilities which depend upon language. While it is intuitively implausible that consciousness depends upon language, some accounts presuppose that conscious subjects must have access to sophisticated concepts referring to mental states, and it is hard to see how such concepts could be acquired or sustained in the absence of language. And one might worry that the claim that consciousness can exist without language is virtually unverifiable since only language users can report the presence of consciousness. In this regard, there is some fairly compelling evidence of consciousness without language that stems from patients' recall of epilepsy induced transient aphasia. In one case (Lecours and Joanette 1980), a man could recall experiences which occurred while he was totally unable to speak or understand language. There have also been studies which investigate whether conscious

---

[20]Here too we can take a panpsychist path and follow this idea to a hypothetical domain of ultimately simple forms of consciousness which pervade reality and are associated with the very simplest and most fundamental features of the physical aspect of the world.

thought without words is possible and they have reported some positive results (Hurlburt and Akhter 2008).

Finally, one can dispute claims about the nature of subjectivity which tend to make states of consciousness appear complex and sophisticated. The term "subjectivity" lacks a clear and agreed upon definition and it suffers from significant ambiguities. A weak reading of the term would refer only to the core feature of consciousness, the "what it is like"-ness or phenomenal character which individuates each conscious experience.

A stronger reading would demand that states of consciousness involve awareness of the subject of that experience. For example, in the HOT theory the higher-order thought which engenders consciousness and registers the content of the first-order state which becomes conscious via the HOT mechanism is of the general form "I am in mental state α". The "I" here is taken to be an essential part of any conscious experience. Many other accounts of consciousness include similar demands for some kind of "self-awareness". In short, if strong subjectivity is correct then all conscious experience should include a phenomenologically present sense of "mineness" or "belonging to me".

One might also include an ultra-strong reading involving an explicit or implicit awareness of a self in every conscious experience by which, it might be said, we recognize our ongoing identity over time.

Now, the weak reading is of course unobjectionable to a defender of simple consciousness. Nor would such defenders deny the existence of the strong form of subjectivity. It seems clear that human consciousness at least frequently involves a component of self-awareness and self-conscious introspective accessibility. The ultra-strong reading is highly controversial and was famously attacked by David Hume. Again, while it seems true that human beings can deploy a concept of a self which persists through change, there is little reason to think that all of our conscious experiences require the activation of such a concept. Many philosophers, from Hume on, have denied that there is any distinctive component of experience which answers to this concept. They regard it as something akin to a theoretical posit useful, perhaps even essential, to our self conception as beings with a determinate past and plans for a personal future but not a necessary component of every state of consciousness.

Defenders of simple consciousness need only deny that *every* conscious experience involves a self-awareness which correlates with the strong reading of "subjectivity". Yet again, we are hampered by the necessary restriction to our own, human, form of consciousness as we try to articulate this denial. Ordinary human consciousness is highly sophisticated and intimately connected to a host of complex concepts or which that of the self is an important element.

However, if we look to some pathologies of the mind, we may be able to discern a form of consciousness which lacks subjectivity (in the strong sense). A number of "deficits of consciousness" challenge strong subjectivity. I can only briefly consider a few of them here.

The first is a symptom of schizophrenia labeled "thought insertion". A familiar symptom of this illness is hallucinatory images and, more typically, auditions (voices) which are experienced as "inside the head". In these sorts of cases, patients distinguish between themselves and the source of the auditory hallucinations. In thought insertion, patients do not describe alien voices speaking to them but something much more strange. Gerrans (2001, p. 231) describes it thus:

> in thought insertion, the subject has thoughts that she thinks are the thoughts of other people, somehow occurring in her own mind. It is not that the subject thinks that other people are making her think certain thoughts as if by hypnosis or psychokinesis, but that other people think the thoughts using the subject's mind as a psychological medium.

Here is part of one patient's transcript (Hoerl 2001, p. 190):

> She said that sometimes it seemed to be her own thought "...but I don't get the feeling that it is." She said her "own thoughts might say the same thing...but the feeling isn't the same...the feeling is that it is somebody else's".

One must be careful interpreting such remarks, prompted as they are by undoubtedly very unusual mental states, but this does not preclude potential insights. Note that thought insertion does not involve a claim that someone is forcing the patient to think thoughts, but that someone else is using the patients mind to think their thoughts. This description violates strong subjectivity and it would be strange that patients could even come up with such descriptions if strong subjectivity was correct. Rather, it seems that it must be conceivable and within the bounds of possible experience for someone else to think their thoughts with my mind. Of course, weak subjectivity remains – it is like something for these patients to experience these thoughts – but they seem capable of interpreting their experience as lacking strong subjectivity. This suggests that the realm of consciousness extends beyond that permitted by strong subjectivity.[21]

---

[21]Denying strong subjectivity via appeal to thought insertion is of course controversial. For opposed views see Gallagher (2000) or Collivas (2000). However, it is important to

A very odd case concerns a man, known as DP, who, as the result of a mysterious episode that occurred while on a flight, complained of what he called "double vision". But when pressed and examined his symptoms were that he could perceive things without knowing *who* was doing the perceiving so that (Lane 2012, p. 257) :

> he was able to see everything normally, but that he did not immediately recognize that he was the one who perceives and that he needed a second step to become aware that he himself was the one who perceives the object.

The case of DP suggests that consciousness can occur in the absence of strong subjectivity. It seems that DP was able to "figure out" that it he who was perceiving on the basis of his current state of consciousness (which only involves weak subjectivity) and, presumably, his pre-existing knowledge about how the mind works which is encoded in our standard concepts of mental states and processes. But his consciousness did not include as part of its content the sense of mineness which should be an essential feature of any state of consciousness if the defenders of strong subjectivity were correct.

Cases such as these (and others discussed by Lane (2012) and Lane and Liang (2011)) suggest that there are forms of consciousness which lack any sense of self, do not involve self-awareness in any strong sense, and do not require that their subjects be aware of their own consciousness. This in turn lends support to the idea that consciousness can occur in simple forms.

Of course, we cannot approach more *radically* simple forms of consciousness via such indirect introspective reports. Very few non-human animals can be queried about the existence of their consciousness, let alone about its structure. Only purely theoretic arguments can support the existence of the sorts of radically simple states of consciousness which a dual-aspect account such as Pauli's should countenance. But at least the above discussion clears the path for these arguments and undercuts the claim that simple consciousness is impossible. Although many accounts of consciousness presume it to be highly sophisticated we have seen considerable evidence that consciousness can occur in very simple forms. This should give us some further encouragement in developing a dual-aspect metaphysics of the sort endorsed by Pauli.

---

distinguish the claim that thought insertion involves the sense that something is indeed happening in my mind (a judgment based on introspective access to ongoing consciousness) from a claim about the nature of the state of consciousness. I am claiming something about the latter, namely that in thought insertion there is lacking the sense of "mineness" which strong subjectivity would require as a component of any and all conscious experiences.

Acknowledgments

I would like to express my thanks to a referee whose close reading and comments have helped me correct some errors and improve the paper in several respects.

References

Anderson M.A. (2010): Neural reuse: A fundamental organizational principle of the brain. *Behavioral and Brain Sciences* **33**, 245–313.

Atmanspacher H. (2012): Dual-aspect monism à la Pauli and Jung. *Journal of Consciousness Studies* **19**(9-10), 96–120.

Atmanspacher H. and Primas H. (2006): Pauli's ideas on mind and matter in the context of contemporary science. *Journal of Consciousness Studies* **13**(3), 5–50.

Carruthers P. (2011): Higher-order theories of consciousness. In *Stanford Encyclopedia of Philosophy*, ed. by E.N. Zalta, accessible at `plato.stanford.edu/archives/fall2011/entries/consciousness-higher/`.

Caston V. (2002): Aristotle on consciousness. *Mind* **111**, 751–815.

Chalmers D. (1996): *The Conscious Mind: In Search of a Fundamental Theory*, Oxford University Press, Oxford.

Colivas A. (2000): Thought insertion and immunity to error through misidentification. *Philosophy, Psychiatry and Psychology* **9**(1), 27–34.

Csíkszentmihályi M. (1996): *Creativity: Flow and the Psychology of Discovery and Invention*. Harper, New York.

Descartes R. (1984): Meditations on first philosophy. In *The Philosophical Writings of Descartes, Vol. 1*, ed. by J. Cottingham, R. Stoothoff, D. Murdoch, Cambridge University Press, Cambridge, pp. 1–62.

Descartes R. (1983): *Principles of Philosophy*. Reidel, Dordrecht.

Descartes R. and Elisabeth P. (2007). Letter of 6 May 1643. In *The Correspondence between Princess Elisabeth of Bohemia and René Descartes*, ed. by L. Shapiro, University of Chicago Press, Chicago, pp. 61–62.

Enz C. (1973): W. Pauli's scientific work. In *The Physicist's Conception of Nature*, ed. by J. Mehra, Reidel, Dordrecht, pp. 766–799.

Feigl H. (1958): The "Mental" and the "Physical". In *Concepts, Theories and the Mind-Body Problem*, ed. by H. Feigl, M. Scriven, and G. Maxwell, University of Minnesota Press, Minneapolis, pp. 370–497.

Gallagher S. (2000): Self-reference and schizophrenia: A cognitive model of immunity to error through misidentification. In *Exploring the Self: Philosophical and Psychological Perspectives on Self-Experience*, ed. by D. Zahavi, Benjamin, Amsterdam, pp. 203–239.

Gallagher S. and Zahavi D. (2008): *The Phenomenological Mind: An Introduction to Philosophy of Mind and Cognitive Science*. Routledge, London.

Gerrans P. (2001): Authorship and ownership of thoughts. *Philosophy, Psychiatry and Psychology* **8**(2), 231–237.

Hoerl C. (2001): On thought insertion. *Philosophy, Psychiatry and Psychology* **8**(2-3), 189–200.

Hurlburt R. and Akhter S. (2008): Unsymbolized thinking. *Consciousness and Cognition* **17**, 1364–1374.

Janzen G. (2008): *The Reflexive Nature of Consciousness*. Benjamins, Amsterdam.

Kriegel U. (2009): *Subjective Consciousness: A Self-Representational Theory*. Oxford University Press, Oxford.

Lane T. (2012): Toward an explanatory framework for mental ownership. *Phenomenology and Cognitive Science* **11**(2), 251–286.

Lane T. and Liang C. (2011): Self-consciousness and immunity. *Journal of Philosophy*, **108**(2), 78–99.

Lecours A. and Joanette Y. (1980): Linguistic and other psychological aspects of paroxysmal aphasia. *Brain and Language* **10**, 1–23.

Leibniz G.W. (1996): *New Essays on Human Understanding*. Cambridge University Press, Cambridge. Although this work was not published by Leibniz it was an essentially complete work, composed around 1705.

Leibniz G.W. (2006): Letter to Basnage. In *Leibniz's "New System"*, ed. by R. Woolhouse and R. Francks, Oxford University Press, Oxford, pp. 62–64. See C.I. Gerhardt, ed.: *Die philosophischen Schriften von Gottfried Wilhelm Leibniz, Volume 4*, Weidmann, Berlin 1880, pp. 496ff.

Levine J. (1983): Materialism and qualia: The explanatory gap. *Pacific Philosophical Quarterly* **64**, 354–361.

Miller A. (2010): *137: Jung, Pauli, and the Pursuit of a Scientific Obsession*, Norton, New York.

Naci L., Cusack R., Jia V.Z., and Owen A.M. (2013): The brain's silent messenger: Using selective attention to decode human thought for brain-based communication. *Journal of Neuroscience* **33**, 9385–9393.

Pauli W. (1994a): Ideas of the unconscious from the standpoint of natural science and epistemology. In *Writings on Physics and Philosophy*, ed. by C. Enz and K. von Meyenn, Springer, Berlin, pp. 146–164.

Pauli W. (1994b): The influence of archetypal ideas on the scientific theories of Kepler. In *Writings on Physics and Philosophy*, ed. by C. Enz and K. von Meyenn, Springer, Berlin, pp. 219–280.

Meier C.A., ed. (2001): *Atom and Archetype: The Pauli/Jung Letters, 1932-1958*. Princeton University Press, Princeton.

Place U.T. (1956): Is consciousness a brain process? *British Journal of Psychology* **47**, 44–50.

Pusey M., Barrett J., and Rudolph T. (2012): On the reality of the quantum state. *Nature Physics* **8**(6), 476–479.

Rosenthal D. (1986): Two concepts of consciousness. *Philosophical Studies* **49**, 329–359.

Seager W. (2007): A brief history of the philosophical problem of consciousness. In *The Cambridge Handbook of Consciousness*, ed. by P. Zelazo, M. Moscovitch, and E. Thompson, Cambridge University Press, Cambridge, pp. 9–34.

Seager W. (2009): A new idea of reality: Pauli on the unity of mind and matter. In *Recasting Reality: Wolfgang Pauli's Philosophical Ideas and Contemporary Science*, ed. by H. Atmanspacher and H. Primas, Springer, Berlin, pp. 83–97.

Seager W. (2012): *Natural Fabrications: Science, Emergence and Consciousness*. Springer, Berlin.

Shein N. (2009): Spinoza's theory of attributes. In *Stanford Encyclopedia of Philosophy*, ed. by E.N. Zalta, accessible at `plato.stanford.edu/entries/spinoza-attributes/`.

Slowik E. (2009): Descartes' Physics. In *Stanford Encyclopedia of Philosophy*, ed. by E.N. Zalta, accessible at `plato.stanford.edu/entries/descartes-physics/`.

Smart J.J.C. (1959): Sensations and brain processes. *Philosophical Review* **68**, 141–156.

Spinoza B. (1985): Ethics. In *The Collected Works of Spinoza*, ed. by E. Curley, Princeton University Press, Princeton, pp. 401–617.

Stubenberg L. (2008): Neutral Monism. In *Stanford Encyclopedia of Philosophy*, ed. by E.N. Zalta, accessible at `plato.stanford.edu/entries/neutral-monism/`.

Wright J. (1983): *The Sceptical Realism of David Hume*, Manchester University Press, Manchester.

Zabriskie B. (2001): Jung and Pauli: A meeting of rare minds. In *Atom and Archetype: The Pauli/Jung Letters, 1932–1958*, ed. by C.A. Meier, Princeton University Press, Princeton, pp. xxvii–l.

Jung, Pauli, and the Symbolic Nature of Reality

Suzanne Gieser

Abstract

This article outlines Wolfgang Pauli's understanding and reading of Jung's definition of the concept of a "symbol" and follows his struggle from 1934 to 1956 to find a basic structuring principle beyond the dichotomy of psyche and matter that can explain how symbols both express psychological and material "truths". Following Kant, Pauli wanted to give the symbol the ontological status of a new "thing in itself", the first and fundamental unit of reality. Finally, Pauli's position will be compared with some recent claims about symbols, language and religion based on modern research.

1 Discovering the Autonomous Symbolic Activity of the Psyche

The physicist Wolfgang Pauli (1900–1958) came to the conclusion that "the most important and exceedingly difficult task of our time is to work on the construction of a new idea of reality" (Pauli to Fierz, August 12, 1948; von Meyenn 1993, p. 559).[1] For Pauli this included the notion that reality in itself is symbolic. The concept of a symbol became central to the discussions among the physicists who formulated quantum mechanics in the 1920s. Pauli became influenced by Jung's notion of the symbol, which differed from any other definition of this notion at the time. It is therefore important to look closely at Jung's definition of a "symbol" and the way this concept became important to Pauli.

As we know today Pauli sought the help of Jung in January 1932 due to some "neurotic symptoms" (Pauli to Rosenbaum, February 3, 1932, Enz 2002, p. 240). After a 20 minutes interview with Jung he was referred to

---

[1]Most of Pauli's letters concerning the topics of this contribution are written in German and have been translated by the author.

Erna Rosenbaum, a young woman doctor, just finishing her training as a psychotherapist at the time. He was in treatment with her until early October 1932, when Jung took over. During these eight months Pauli recorded 355 dreams and visions (von Meyenn, forthcoming). In August 1934 he wrote to his colleague Ralph Kronig that he, as a result of a one-sided development of consciousness, experienced a revolution from the inside, from the side of the unconscious, and so became acquainted with the "autonomous activity of the soul" and "its spontaneous growth products" that can be designated as symbols. In this early letter we can see that Pauli encompassed Jung's definition of the symbol as something "objective-psychic, which cannot and may not be explained as resulting from material causes". This specific definition of the symbol as something not deriving from material causes (such as biological drives) made him declare that Jung's psychology had the promise of becoming a true scientific psychology (Pauli to Kronig, August 3, 1934, von Meyenn 1985, p. 340).

In this statement we find Pauli's view on science condensed: It must be based on the observation of something autonomous and objective in the sense that it is not under the control of the subjective ego. It is important to observe a phenomenon in its own right, on the grounds that it produces some kind of effect, rather than to reduce it to underlying causes or substances. It was natural for Pauli to embrace this view because he had been partaking in the creation of quantum physics as a member of the Copenhagen school. As such, he subscribed to the famous "Copenhagen spirit" – the position that the building blocks of quantum physics are the physical phenomena, which are defined as the interaction between measuring instrument and measured object (not to be confused with "real material objects"). In the same way the symbolic products of the unconscious could be understood as psychological phenomena resulting from an interaction between consciousness and the unconscious psyche. This way of defining phenomena as products of an interaction between an autonomous reality and an observing instance can be found in Jung's definition of a symbol. But the concept contains many other aspects as well, always involving the demarcation of the interaction between the observer and the observed.

2 Jung's Definition of the Symbol

Firstly, the symbol can be defined as a renunciation of exhaustive knowledge, in the same way as the Copenhagen interpretation has been called a theory of renunciation. According to this first definition [1], a symbol is the best possible expression for something unknown (Jung 1921, par. 814). Here the symbol is understood as consisting of a visible expression that only reveals a certain limited aspect of something that cannot be fully conceptualized. This gives us the idea that our concepts and perceptions are limited and can only grasp bits and pieces of a complex reality. At its best this is a very humble view of what we can grasp with our rational scientific concepts and tools.

The second important aspect of a symbol is that it is something that takes hold of us and activates us through its numinous quality. Symbols are symbols because they have an effect on us, we react to them and feel the need to engage with them [2], which results in symbolic or ritual (rhythmic/recursive) activities. Through its numinous quality a symbol awakens our attention and keeps it focused, which in turn stimulates the development of consciousness. By holding our attention and also triggering us to dwell on the symbolic object, to manipulate it and work on it, the symbol becomes an energy transformer [3] (Jung (1928, par. 47) calls it a libido analogue). This indicates that humans have a fair amount of excess libido that expresses itself in a symbolic way; in fact Jung sees symbols as the *typical* way in which psychic energy (libido) expresses itself [4]. This excess libido constitutes the essential precondition to developing a culture (Jung 1928, par. 91). As such, the symbol is also the *mother of science* [5] "initiating" a sustained playful interest in the object that allows man to make all sorts of discoveries which would otherwise have escaped him (Jung 1928, par. 90).

The symbol can be alive or dead [6], it is alive as long as it has the above mentioned function of being numinous and pregnant with meaning. For instance, Christian symbols can be numinous for some in our contemporary society, but not for others. The symbol is alive when it generates specific behaviors and feelings. Symbols can also be dead in the sense that we infer that a certain artefact must have had symbolic value at a certain time, but today it is only a historical relic or remnant. An example of this are the carvings found in the Blombos caves that are dated back 77'000 years ago, and are today considered as the cradle of human symbolic thinking (Henshilwood and D'Errico 2011). But Jung also uses the concept of a "living symbol" in an even more specific sense, as the product of intense, active inner and personal work, which produces a living and unique relationship to an emerging symbol that has a creative function in a person's life. These are described as products of a yearning and highly developed mind belonging to a category different from the already existing symbols in a culture, and sometimes the source of new seminal symbols for a certain epoch (Jung 1921, par. 823). In this sense Jung's concept of the symbol includes the whole process from awakening interest in an object, its exploration and production of knowledge (or culture) to its becoming a fully explored object that has lost its interest, like a toy thrown away (Jung 1921, par. 816.)

Symbols can be both individual and collective, i.e. carriers of meaning for a whole culture or age, as for instance expressed in a specific paradigm, world-view, or belief system. In this sense culture and science are symbolic systems which are temporary and often limited expressions of recurring archetypal patterns. For instance the figure of the Virgin Mary can be seen as a specific version of the mother archetype, in this case having lost its dark dimension and only retaining the aspect of the good and caring mother. With this view of the symbol as an archetypal image, both revealing and alluding to a more complex underlying reality, Jung tries to encompass *both* the limited expression of the symbol (its phenomenological aspect) *and* its connection to the archetypal matrix (non-causal, ordering principles of nature) that gives it its numinous quality.

Here we also find a conception of a hierarchy of symbols. The more inclusive ones, those which express more complexity – as for instance those referring to the "unity of opposites" – are in some way closer to the ungraspable reality (Jung 1935, par. 373; Jung 1954a, par. 156).

3   Pauli's Reading of Jung

In Pauli's own copy of *Psychological Types* he made 13 marginal notes in the definitions part of the book under the entry symbol, which is about seven pages long.[2] The passages marked are the following:

1. The symbol always presupposes that its chosen expression is the best possible description, or formula, of a relatively unknown fact which is nonetheless recognized or postulated as existing.

2. The symbolic expression is the best possible formulation of a relatively unknown thing.

3. The interpretation of the cross can only be seen as symbolic when it puts the cross beyond all imaginable explanations, regarding it as an expression of an unknown and as yet incomprehensible fact of a mystical or transcendent psychological character which simply finds its most striking and appropriate representation in the cross.

4. The symbol is alive only as long as it is pregnant with meaning.

---

[2]This book is in "La Salle Pauli" at CERN, where Pauli's scientific and private library is kept. It contains seventeen works by Jung, most of them with notes in the margins.

5. The way in which St. Paul and the early mystical speculators handle the symbol of the cross shows that for them it was a living symbol which represented the inexpressible in an unsurpassable form.

6. Whether a thing is a symbol or not depends chiefly upon the *attitude* of the consciousness considering it – for instance, a mind that regards the given fact not merely as such but also as an expression of the yet unknown.

7. There undoubtedly are products whose symbolic character not merely depends upon the attitude of the considering consciousness, but manifests itself spontaneously in a symbolic effect upon the observer.

8. Concerning Jung's concept of the *symbolic attitude* Pauli emphasizes that this is the outcome of a definite view of the world which *assigns meaning* to events, whether great or small, and attaches to this meaning a greater value than to bare facts.[3]

9. The living symbol shapes and formulates an essential unconscious factor, and the more widespread this factor is, the more general is the effect of the symbol.

10. Jung describes the way symptoms and symbols are connected to each other: When man suppresses a vital part of himself, this results in psychic imbalance and produces symptoms of psychic impairment due to a tension of opposites. Working through this tension by giving heed to the products of the unconscious, like dreams and fantasies, gives rise to the birth of a new uniting symbol: "Since life cannot tolerate a standstill, a damming up of vital energy results, which would lead to an insupportable condition did not the tension of the opposites produce a new uniting function that transcends them" (Jung 1921, par. 824).

11. There emerges a new content from the activity of the unconscious, constellated by thesis and antithesis in equal measure standing in a *compensatory* relation to both. Jung then introduces the term "transcendent function" which refers to the totality of this psychological process, proceeding from one-sidedness to a position where one endures the opposites and finds a new unifying position.

---

[3] Here I would like to insert a comment by Marcus Appleby: The idea of a so-called "bare fact" – a fact devoid of psychic accretions – can *itself* be considered a symbol. Some bare facts receive a much greater value than others (e.g. for some scientists facts about spacetime), and it would not be difficult, if one put one's mind to it, to tear the idea of a bare fact to pieces. There are no bare facts: the notion is pure myth. Facts always come dressed, with symbolic and other significance.

12. The stability of the ego and the superiority of the new mediatory product towards thesis and antithesis are equally important, they are correlates conditioning one another (Jung 1921, par. 826).

13. The raw material, shaped by thesis and antithesis is the living symbol.

4 Why do Symbols Take on Specific Forms?

We do not know at what time Pauli made these marginal notes, but one guess is that his careful reading or re-reading of the text on symbols occurred around 1947–1948, during the period when he had just returned to Europe after his "exile" in America during the war. At this time he worked on several ideas linking the world of physics with the world of Jung's archetypal psychology, with a strong emphasis on the role of the symbol. We can also see this increase of interest in his correspondence at the time. He was intensely studying the hypothesis that the development of Johannes Kepler's scientific ideas was influenced by archetypal Christian ideas of the trinity.

In 1948 he also drafted an unpublished manuscript entitled "Modern Examples of Background Physics", where he explores how in his dreams and spontaneous fantasies symbols from physics are "misused". They obviously deviate from the well-defined way they are used in physics, and rather express a transferred qualitative and figurative sense. Pauli thought that the "misuse" of the terminology of physics in dreams must be a kind of free association in analogies which can probably be seen as a preliminary stage of conceptual thinking (Pauli to Jung, June 22, 1935; July 4, 1935; October 2, 1935; Meier 2001).

Pauli asks himself why his dreams and fantasies disregard his knowledge of physics and choose physical terminology to point to something else, to a deeper psychological process, using a language that belongs to his rational and conscious world.[4] This question had bothered him since 1935, when he started to have dreams using language and symbols from physics.

---

[4]Pauli might have been interested in a recent pilot study on lucid dreamers, who are trained in having deliberate conversations with their dream characters. The study refers to previous work which showed that dream characters can be creative and ingenious, but that they seem to struggle with more logical tasks, such as doing arithmetic. The pilot study explores this issue deeper and finds that dream characters are not especially good at mathematics. When asked to solve mathematical tasks only about a third of their answers were correct and their arithmetic abilities do not surpass those of primary school children. Surprisingly, they were more successful with multiplication and division than with addition and subtraction (Stumbrys *et al.* 2011).

Before this time, the symbolism in his dreams, as we can see in the dream material published in *Psychology and Alchemy*, is of the more figurative and narrative kind that uses everyday symbols and symbols which can be linked to mythological motifs. Examples are many sheep pasturing, veiled women figures, women guiding him, wild beasts in the Jungle, children, dwarfs, traveling with boats, trains and airplanes, etc. (von Meyenn, forthcoming; cf. Jung 1944). Suddenly, in 1935, he starts dreaming of spectral lines, frequencies, rotation, radioactivity, fine structure and the like. One thing that bothered him was that neither Jung, nor later most other Jungians, could help him interpret the symbols stemming from physics.

A radical conclusion he made already in 1935 was that the language of physics is much more precise and so can also describe psychological processes in the finest detail, more so than other languages, based on everyday imagery, mythology or art. Pauli compiled a tentative *translator's glossary* for how, at a symbolic level, physical terms express psychological processes. An example is the image of the splitting of spectral lines in a magnetic field which Pauli likens to the process of differentiation in psychological development. Another example is the radioactive nucleus, which can be compared to the indestructible center of the personality (the self in Jung's terms). The radioactive nucleus refers to a gradual transformation of the center, but also to an effect radiating outwards, in the same sense that an individual who has become more centered, more in touch with the inner self, also has a transformative effect on his surroundings.

Pauli would not agree with the statement that a symbol "denotes a sign that has no natural connection or resemblance to its referent" (Henshilwood and D'Errico 2011, p. 89). Like Jung, Pauli believed that the shape of the symbol is linked to its deeper organizational (i.e. archetypal) levels that refer to some kind of psychophysical processes that are operative in nature "wilfully insisting on its own form and effect" (Jung 1922, par. 116). Recurring themes in Pauli's dreams were rhythms, periodicity, stripes, measuring proportions, arranging colors, scales, spectral lines, and mirroring (von Meyenn, forthcoming).

5 Symbols, Concept Formation in Science, and the Neutral Language of Nature

During the Second World War Pauli had started to study Kepler's original writings at the *Institute of Advanced Study* at Princeton, under the guidance of, amongst others, art historian Erwin Panofsky. This would eventually result in the publication of his essay *The Influence of Archetypal Ideas on the Scientific Theories of Kepler* first published in 1952. The Kepler essay (Pauli 1952) begins with the question of the relationship between sensory impressions and conceptualization. Pauli discusses this in a letter to Markus Fierz of January 1948, directly referring to Jung's definition of the symbol in *Psychological Types* as a "postulated but as yet unknown objective fact".

A very important aspect of Pauli's early interest in Jung's concept of the symbol is his experience of the autonomy of the inner world – that it behaves as if it has its own agenda driving a process of inner growth. It is not logic that relates sensory perceptions to concepts, but rather a kind of *pictorial viewing* whose origin cannot be reduced to the sensory perceptions but has to do with the creative autonomy of the mind, an instinct of the imagination, producing similar images in different individuals, independently of each other. In order to gain insight into the nature of scientific conceptualization, one must take into account its preliminary stage on the preconscious level, i.e. fantasies and archaic images that lie just under the surface of rational thought. Pauli called this area of study the "psychology of scientific conceptualization" (Pauli to Jung, December 12, 1950, Meier 2001; Pauli to Fierz, January 7, 1948, von Meyenn 1993, pp. 495–497).

According to Pauli, the real link between sensory impressions and concept formation is most likely the archetypal image, i.e. the archetypal symbol.[5] So we get the picture of the sensory perceptions being received and creatively processed by imagination – an imagination structured by the same archetypal forces that are active behind the formation of matter. This archetypal

---

[5]Here it is important to be aware of the distinction between archetypes as such and archetypal images, which are synonymous with archetypal symbols. As symbols can be both private, collective (typical for a certain culture) and archetypal (universally recurring through different cultures and times), this distinction is important. Archetypes themselves are not images or conceptions but structural elements which function like the axial system in a crystal, which preforms the crystal structure in the mother liquid without having a material existence of its own. The archetype is described as an empty, formal element, or as an a priori *possibility* of representational form. What is inherited in man is not the representation or the image but potentials for formal structures that correspond to the formally determined instincts (Jung, 1954a, par. 155).
This can be compared to the statement (Ellis in Henshilwood and D'Errico 2011, p. 177): "What is inherited, then, are basic cognitive abilities rather than specific cognitive modules, plus the basic sensory and emotional systems that guide the use of cognition. Any effective cognitive modules that result develop from interactions of these systems with the social and physical environment, with the salience of reactions guided by the inherited emotional systems. Overall this process is of Darwinian rather than Lamarckian nature, because it does not propose genetically determined modules with specific cognitive content, but rather genetically determined emotional systems that guide cognitive development."

imagining is a combination of psychological meaning and a *factual exposure of the archetypal basis of the physical concepts*, Pauli states (Appendix 3, Meier 2001).

This was the beginning of Pauli's hypothesis of a neutral language, a further development of the heritage from his Godfather Ernst Mach. Mach spoke about experience as *psychophysically neutral*, because our experience is always based on a combination of psychological factors and physical input from the senses. For Pauli the hypothesis of a neutral language means that there is common "structural" language describing universal processes occurring in nature, independent of whether this nature is outside our mind or inside our mind (it is psychophysically neutral).

In contrast to the structuralist point of view of Lévi-Strauss and others, where the "universal grammar" is supposed to consist of a couple of known and logical rules or relations typically expressed as opposites (up-down, left-right, inside-outside, dark-light), Pauli's view is that the neutral language is not defined in the first place. It can only be discovered by detecting similar patterns in several parallel disciplines. It cannot be defined statically, but is rather something more processual and dynamic.

The inclusion of the observer is an important precondition of this view of the neutral language. However, the point is not that our perceptions are filtered through the categories of our mind or senses, but that all our observations constitute an interaction between observer and observed. What makes this complicated is that the products of the mind (like dreams) are also something observed by an observer. Apparently Jung validated Pauli's hypothesis of a neutral language and commented that developing such a language would be the ultimate goal for a unified science of physics and psychology (Pauli to Fierz, August 12, 1948, von Meyenn 1993, pp. 558–561).

If I understand Pauli's view of the neutral language correctly, it is first of all not a "language" in the common sense of a "socially acquired system of semantic and syntactic processing", and not a conscious process of ratiocination (Henshilwood and D'Errico 2011).[6] It is neutral in the sense that

---

[6]It is noteworthy that a recent publication on the origins of symbols (Henshilwood and D'Errico 2011, 3ff) stresses that we have to differentiate between communication, language, speech, symbol-use, non-human and human language. Here symbols are defined as mental (conceptual) representations of phenomena, concrete or abstract, something to be distinguished from mere signs, which are defined as basic perceptual signals, that do not require mental representations. It is maintained that symbol-use is a necessary condition for language (defined as "socially-acquired system of semantic and syntactic processing"), but in itself something more basic. The argument is based on the fact that

it cannot be reduced to exclusively coming from the outside (acquired) or from the conscious mind (rationally processed). It is something that presents itself spontaneously and phenomenally.

Pauli would disagree with the definition that the meanings of the symbol "are construed and depend on collectively shared belief" (Henshilwood and D'Errico 2011, p. 76). Meaning is more than something merely construed and shared together with others. It is also an original emotional experience of numinosity. The language of our dreams is the model example of a neutral language. However, it is not the dream imagery itself that represents the neutral language, but the underlying structuring process *and* its emotional affect that manifests itself in the dream.

Pauli speculates, and sometimes even states that he "firmly believes", that this kind of "processing" has an objective meaning (collectively valid) and can be discovered by detecting similar patterns in distinct domains of experience. The neutral language has to do both with the objectivity of mathematics and with emotion. The same factor that expresses itself as order in matter can manifest itself as an apprehension of meaning (emotion) in the internal world (Pauli to von Franz, October 30, 1953, "Die Klavierstunde", von Meyenn 1999, pp. 329–340; Pauli to Fierz, October 17, 1954, von Meyenn 1999, pp. 801–805; Pauli to Fierz, December 10, 1955, von Meyenn 2001, pp. 434–438).

The hypothesis of a "neutral language" means that certain structures and processes observed in outer nature are also autonomously active within the psyche of man. Pauli believes that these inner products are expressions of general objective organizing principles underlying our psyche. To explain this he postulates a third level of existence where psyche and matter are not separate, but ruled by the same organizing principles. Here Pauli is interested in Jung's concept of the archetypes, but it is essential to him that archetypes cannot be defined as purely psychic or cognitive organizing principles. A neutral language should be able to formulate principles that can be observed both in psyche and in matter: "a symbol is on the one hand a product of human effort, on the other a sign of objective order in cosmos, of which man is only a part". It is "two-sided in the sense of the understanding of the cognitive process" and "it has a relationship with the 'observed' and with 'concepts'" (Pauli to Fierz, August 12, 1948, von Meyenn 1993, pp. 558–561). This philosophical position has been labeled a psychophysically neutral

---

animals have communicative abilities that are not "languages", that symbols can function non-linguistically (e.g. computer code) and that certain apes have shown the ability to use symbols as communication.

ontic monism combined with epistemic dualism (Atmanspacher and Primas 2009, p. 4).

6   The Double Nature of Symbols: Veiling and Revealing

Pauli was convinced that mathematical language is truly symbolic because it unites information about man and about nature. It is both a product of human effort and a sign of objective order in the cosmos. It has a relationship both with the observed and with the conceptual. For a mathematically gifted person, mathematics is a creative tool which is also created by man and seems to possess a life of its own, both producing and showing new facets of reality (Pauli to Fierz, August 12, 1948, von Meyenn 1993, pp. 558–561). It moreover possesses the quality of numinosity and beauty that so often characterizes living symbols.[7] In quantum mechanics the symbolic character of concepts is especially obvious since the wavefunction works as a reconciling symbol by uniting our conceptions of continuity and discontinuity, i.e. wave and particle picture (Pauli 1950, p. 40).

Pauli uses the concept of a "symbol" in a way different from Niels Bohr. Pauli and Bohr were both most strongly associated with the Copenhagen interpretation of quantum physics. To Bohr it is not the wavefunction that is a symbol, but the concepts of wave and particle. They are symbolic descriptions because they are "clear" visualizations of a certain aspect of the nature of matter, and precisely therefore they are limited. Bohr's concept of symbolic reality is directly related to the inadequacy of our visual concepts. Wave and particle are symbols because they give an incomplete picture of reality, they are symbols which only describe one aspect of reality on a phenomenological level (Bohr 1929b, Folse 1985).

Pauli, influenced by Jung *and* Bohr, has a double-aspect view of the symbol. The symbol is visible in its limited, phenomenological and/or rational expression, but it also opens up to a deeper archetypal level of inclusiveness and complexity, e.g. the unity of opposites. This double-aspect definition of the symbol may cause confusion as Pauli sometimes uses the concept "symbol" to denote the one-sidedness of a certain symbolic image and therefore to emphasize that knowledge is always connected to making a choice of seeing

---

[7]It should be noted that Jung (1950, par. 141) states that a living symbol goes beyond our sense of the aesthetic being both "sublime, pregnant with meaning, yet chilling the blood with its strangeness, ... grotesque; it bursts asunder our human standards of value and aesthetic form."

one aspect of reality and sacrificing another. There is always the risk of being deluded by believing that a symbolic concept corresponds to properties of concrete reality.

Pauli mentions this when he stresses the importance of becoming conscious of how easily we attach primitive or egotistical expectations to symbols, like security, success, power or eternal life, instead of understanding symbols as "work in progress" pointing towards individuation, inner value and expanded consciousness. He compares this with the degenerate belief of some alchemists that they could manufacture gold with the help of the philosopher's stone (Pauli to Jung, May 24, 1937, Meier 2001).

Another example, which Pauli discussed years later, concerns how symbols appear in science to cover up for lack of knowledge, like the concept of chance and selection in biology. In a letter to Pascual Jordan, he writes (Pauli to Jordan, March 31, 1954, von Meyenn 1999, pp. 550):

> The concept of selection seems very *formal* to me – it tells us nothing of the developments, i.e. hereditary changes that are possible to the organism. But I am no Lamarckian (I trust the instinct of the Soviet Russians to always hit the wrong note) and "selection" is for me mainly a symbol of the *absence of a direct causal connection* between *hereditary* changes of an organism and its environment (I include other species into the concept of environment).

In the same vein he considered the concept of random mutation in the neo-Darwinian theory of evolution as a concept that was used to fill the gaps of our inadequate knowledge about the evolutionary process, a critique that modern epigenetics seems to confirm (Gieser 2005, p. 300; Atmanspacher and Primas 2009, p. 316ff).

In this sense concepts can become symbols that mislead us, "filling in" gaps in a construction of reality based on our biases. This way of seeing the symbol and its archetypal foundation comes close to what modern research highlights when it links cognition to visual thinking and to the recognition of geometric patterns. This makes us able to predict and fill in information when presented with partial information and "noise" (redundancy). According to cognitive scientists these "conceptual schemas" arise from our common genetic heritage and shared developmental experiences before the learning of language (Feldman 2006).

Daniel Kahneman (2011), psychologist and winner of the 2002 Nobel Prize in Economic Sciences, highlights this feature of our mind very clearly. His research has shown that the so-called system 1 part of our thinking (unconscious, automatic, non-logical) is built upon our accumulated learned experiences organized in stereotypes of emotional and associative coherent "stories".[8] This also explains the feeling of confidence, which is generated by the coherence of a good story. Kahneman states that this results in a construction of the world in our minds that is much simpler and much more coherent than reality itself. We pick simple stories and tend to ignore what does not fit (pattern recognition). It is interesting to contrast this view with the scientific ideal that simpler theories are better theories, closer to reality.

Moreover, there are common, collective biases based on how our fears and hopes control us (risk aversion and over-confidence for instance). In Kahneman's view of reality there is only one way to escape biases, and that is through statistical truth: the odds. He thinks that statistical truth is the only anchor to guarantee a reality check. Pauli, on the other hand, was very sceptical about seeking truth in statistics. He (and Jung) searched for a deeper reality in the pattern-making factors themselves, both in psyche and matter.

The visible, limited, phenomenological appearance of the symbolic image can be understood as emerging from projection. Projecting an archetypal image (symbol) onto the canvas of physical reality amounts, according to Jung, to confusing psychological reality and psychological needs with the world of outer objects. Here it becomes very tricky to distinguish different levels of reality, as both Jung and Pauli believed that there is an objective reality to be discovered beyond our projections. One way to try to understand this is to compare symbols to windows – sometimes they reflect our own image and sometimes they allow us to see what is beyond the glass, all depending on the angle.

But this simile is deceptive. It rather cements the illusion of a dualistic either-or of two completely different realities (either psyche or matter), and does not take into account that what we see is a co-creation of the angle we take, the glass and the object behind the glass. Here the psychologist and the physicist have to do similar work: the work of sorting out and becoming conscious of what is projected and what belongs to the object, a process of uncovering and approximation that is continuously ongoing and may be impossible to bring to a close (Jung to Pauli, May 4, 1953, Meier 2001; Jung 1952b, par. 1511).

---

[8] An internal story is understood as a jumping to conclusions on the basis of little evidence.

7   The Symbol as the "Ding an sich"?

Pauli argued that the concept of the symbol as described above should become the new fundamental unit of reality. It should occupy an epistemological position which Kant assigned to the "thing in itself" (Pauli to Fierz, August 12, 1948, von Meyenn 1993, pp. 558–561). This "thing in itself" is unknowable according to Kant, as all observation is always filtered through the categories of our cognitive faculties. In a model where the concept of the unknowable "thing in itself" is replaced by the concept of the reality of the symbol, our knowledge still depends on our cognitive faculties, but these are in turn also organized by the same principles that organize matter – i.e., the archetypes.

Archetypes as objective ordering factors cannot be apprehended directly; it can only be inferred that they are operative on a phenomenological level. These categories of order are not only rational categories like space, time and causality (as in Kant), they are also typical centering processes (center versus periphery), the interplay and unity of opposites (the paradox), differentiation (autonomy, refinement of details and function), interconnectedness (affinity, belonging, caring), rhythm and repose, mirroring (symmetries), repetition and deviation, duplication (multiplication), sublimation, sacrifice, condensation, purification and many more. In Jung's description of archetypes it is obvious that they are not so much recurring images or even motifs, rather they are systemic networks of recurring themes of interactions (e.g. the father-son interaction including "killing of the son or father and their subsequent resurrection").

Regarding the concept of the symbol as the essential starting point of knowledge means that it is not the independent "elementary parts" of the world that are fundamental reality, but the inseparable interconnectedness of observer and observed.

8   God, Nature, and the Emergence of Consciousness and Ethics

In a letter to Jung in May 1952 Pauli defines the neutral language as a "psycho-physical standard language whose function is symbolically to describe an invisible, potential form of reality that is only indirectly inferrable through its effects". This letter was written after an evening conversation with Jung about his recently published book *Answer to Job*. During the evening, Pauli realized the central role of the concept of "incarnation" in

Jung's thinking and how he used it as a scientific working hypothesis. Pauli was sympathetic to it and came to endorse it (Pauli to Jung, May 17, 1952, Meier 2001).

In his book on Job, Jung understands the biblical story as representing an evolution of consciousness through history, especially a consciousness of good and evil. The drama of Job, and later of God becoming man through Christ, is described as the unconscious (or even nature itself) ambivalently striving towards higher consciousness (Jung 1952a, par. 740):

> The unconscious wants to flow into consciousness in order to reach the light, but at the same time it continually thwarts itself, because it would rather remain unconscious. That is to say, God wants to become man, but not quite.

The metaphysical process addressed in this quote is, described in the language of Jung's psychology, the individuation process. It can run its course unconsciously, but it is man's duty to carry through this process consciously. For Jung the symbol of incarnation is about the essential question of taking responsibility for our choices and actions (Jung 1952a, par. 746):

> The only thing that really matters now is whether man can climb up to a higher moral level, to higher plane of consciousness, in order to equal to the superhuman powers which the fallen angels have played into his hands.

Seeing the ideologies of the 19th century as destructive products of the rational mind, Jung did not believe that rationality and logic would bring sustainable solutions. Jung instead puts his faith into symbols produced spontaneously by the unconscious and amplified by the conscious mind (Jung 1952a, par. 755):

> But if the individuation process is made conscious, consciousness must confront the unconscious and a balance between the opposites must be found. As this is not possible through logic, one is dependent on *symbols* which make the irrational union of opposites possible.

After his conversation with Jung in May 1952 it became clear to Pauli how Jung links the concept of incarnation to ethics, based on the identification of the self with one's fellow men on deeper psychological levels ("what one does to others, one also does to oneself"). He asks if Jung's point of view can be called *incarnatio continua*, a continuous incarnation in the sense that each step towards higher consciousness also is a materialization of a potential reality.

Driven by his search for a neutral language, it becomes important to Pauli to find the parallel to this principle in physics. In the field of physics this would correspond to the fact that empirical reality is always a "realization" of a potential in connection with a specific moment of interaction or measurement (Pauli to Jung, May 17, 1952, Meier 2001). In a letter to the German philosopher and theologian Günther Jacoby, Pauli writes on August 4, 1954 (von Meyenn 1999, pp. 735–738):

> there is also a *middle way*, which is conceived of as a dynamic equilibrium. One should then try to find concepts that can denote an abstract, non-visual reality, which can manifest itself inside as well as outside, depending on the conscious attitude of the "observer", which should be understood as relative to all statements about the reality of the non-immanent. ... To me it seems that a first, rather little step has been made in our occidental natural science towards such a middle position, through quantum mechanics and the departure from ordinary causality (in the narrow sense) and in the inclusion of the observer into a symbolic reality.

9 Pauli's Correspondence with Aldous Huxley

Pauli was eager to find a conversation partner outside the circle of Jungians, and in mid April 1956 he was encouraged to send his essay on Kepler to Aldous Huxley. Pauli had read Huxley's books *Grey Eminence* and *The Devils of Loudun*, which both deal with 17th century France, and with religion, politics and mysticism.[9] From 1948 to 1953 he discussed Huxley's work in his correspondence.

He had followed the development of Huxley's writing and saw in Huxley's vedic mysticism the strength and weaknesses of a sudden conversion. Huxley speaks of the "divine ground" and the "the one reality" that can be accessed by certain spiritual practices, a constituent of the mind that is not visible in the ordinary circumstances of everyday life, but shows itself when the mind

---

[9]We know that Pauli had read many of Huxley's books over the years and, in particular, that he had already read *Grey Eminence* in 1949. We also know that he had read the work on evolution by Huxley's brother Julian.

There are ten books by Aldous Huxley in Pauli's private library, some with marginal notes (MN). The books are *Crome Yellow* [1933] (2 MN), *Two or Three Graces und Other Stories* [1934] (12 MN), *Beyond the Mexique Bay* [1934], *Grey Emminence. A Study in Religion und Politics* [1941], *Time Must Have a Stop* [1944], *The Perennial Philosophy* [1945] (16 MN), *Science, Liberty, and Peace* [1946], *Ape and Essence* [1948] (1 MN), *The Devils of Loudun* [1952], and *Heaven and Hell* [1955/56]. Also contained is the book *Evolution: The Modern Synthesis* [1942] by Julian Huxley.

is subjected to "certain rather drastic treatment". In many of his books, Huxley turns to man's deep need for transcendence, a need that can also take destructive paths ("downward self-transcendence"), such as drug abuse or immersion in a crowd – so-called "herd-intoxication" (Huxley 1952).

Huxley compares how we can find out about the nature of matter by making physical experiments with the fact that we can discover "the intimate nature of mind and its potentialities" by psychological and moral experiments. If we would realize these potentialities of the mind, "we must fulfill certain conditions and obey certain rules, which experience has shown empirically to be valid" (Huxley 1945, p. 3).

These words must have appealed to Pauli. On the other hand he thought that Huxley too quickly left the level of the manifold material world in order to step into the realm of the divine ground of the mystics. Pauli notices that Huxley, when he deals with the 17th century in *Grey Eminence*, ignores the alchemists, who for Pauli constitute such an important link between the mystical and scientific approach to reality. The double-aspect symbols of the alchemists point both to the material and to the psychological and spiritual world – as for instance in the symbol of Mercurius, being both a chemical element and a transformative spiritual agent.

Pauli complains that Huxley goes too quickly to the "last reality" and forgets about the "second last reality", where there are symbols of different quality. Pauli labors with symbols organized in a hierarchical order at different levels. They form a continuum from concrete, singular symbols (i.e., the good mother, particle and wave) to more inclusive levels (embracing opposites, the great mother including good and destructive aspects of the mother archetype, wavefunction) and from there to symbols expressing wholeness (i.e., unus mundus, divine ground, etc.). What we need, says Pauli, is to find links between symbols at different levels (Pauli to Goldschmidt, March 2, 1949, in Goldschmidt 1990, p. 51).

Pauli returns to Huxley again and again, as he sees in him someone who has seriously dealt with the important issues of the relationship between the spiritual and the material, and between the dark and light forces. Pauli sees two ways to deal with this relationship in human culture. The first is exemplified by the Taoist view where opposites are equally present and always in balance in nature, with no interaction with human consciousness. This is a kind of *static* worldview, which had always appealed to Pauli. The other view, typical for the occidental world, is an evolutionary model with a starting point in one substance and a linear or cyclic evolutionary process in which the world develops towards a final or higher condition.

In some of these models human consciousness plays an important and interactive part in the evolutionary process. But it must be explained why this process got started at all, especially when the "first substance" is described as perfect or good. Pauli thinks that Huxley's position in this respect is too much on the "spiritual" side, for instance in passages like the following from his *Perennial Philosophy* (Huxley 1945, p. 209):

> In the Hebrew-Christian tradition the Fall is subsequent to creation and is due exclusively to the egocentric use of a free will, which ought to have remained centred in the divine Ground and not in the separate selfhood. The myth of Genesis embodies a very important psychological truth, but falls short of being an entirely satisfactory symbol, because it fails to mention, much less to account for, the fact of evil and suffering in the non-human world. To be adequate to our experience the myth would have to be modified in two ways. In the first place, it would have to make clear that creation, the incomprehensible passage from the unmanifested One into the manifest multiplicity of nature, from eternity into time, is not merely the prelude and necessary condition of the Fall; to some extent it is the Fall. And in the second place, it would have to indicate that something analogous to free will may exist below the human level.
>
> That the passage from the unity of spiritual to the manifoldness of temporal being is an essential part of the Fall is clearly stated in the Buddhist and Hindu renderings of the Perennial Philosophy. Pain and evil are inseparable from individual existence in a world of time; and, for human beings, there is an intensification of this inevitable pain and evil when the desire is turned towards the self and the many, rather than towards the divine Ground.

Huxley goes on and speculates that even sub-human existences may be endowed with something which resembles the power of choice, i.e., some kind of consciousness. He defines the "divine ground" as spirit, the eternal now as consciousness and the ultimate being (Brahman) as knowledge. And he states that it is man's task to achieve knowledge of this ultimate being in his present life.

Pauli underlined these passages in his copy of the book and wrote in the margin: "Why this Fall took place at all? Answer: knowledge by creatures? Why 'the ground' wants to be known?" If all knowledge was already there in a perfect state at the beginning, why was there a creation? Why a temporal world full of suffering when the only purpose is to be enlightened about the original perfect state? This can only be interpreted in the sense that creation, man and consciousness was pointless, a mistake, to begin with. Therefore Pauli thinks Huxley's view is too one-sidedly Buddhist-Platonic and does not include the paradox of complementary opposites. The mixing of the material and the spiritual has to have a purpose – or, to put it another way, both the rational and the irrational are needed for a complete understanding of existence.

When Pauli sent Huxley his essay on Kepler, he explained that he wanted to show the influence of archetypal visionary experiences on the scientific ideas of Kepler and also throw light on the relation of the spiritual and empirical elements in the origin of the natural sciences. Pauli states that both types of elements seem always to be present and interact with each other. Neither of them is entirely reducible to the other.

Pauli sees in Kepler a particular example of the relation between science and mysticism, something which Huxley deals with in his books. According to Pauli Kepler was driven into a new form of science, (astronomy) by the mystical qualities in his thinking. Pauli believes, just as he understands Huxley to believe, that mysticism, whatever form it will take in the modern scientific age, should have a place and expression in any well-balanced culture. Only mysticism can make room for certain specific human experiences. Pauli might have thought about the passage in Huxley's *Grey Eminence*, where he states (Huxley 1941, p. 98):

> The mystics are channels through which a little knowledge of reality filters down into our human universe of ignorance and illusion. A totally unmystical world would be totally blind and insane.

Pauli looks for ways to address the topic of alchemy with Huxley and finds that he alludes to the problem of the relation between spirit and matter in *Heaven and Hell* (Huxley 1956). Pauli points to the controversy between Kepler and Robert Fludd in his essay and says that the alchemists, represented by Fludd, had some intuitive knowledge of the connections between matter and spirit that has been lost in the age of modern science. Pauli tries to introduce his idea of the neutral language by describing how the "hermetic philosophy" tries to express a general connection between material (chemical) processes and visionary experiences with the help of a concretistic monistic unifying language, which was based on the assumption that in all matter there lives spirit.

If this vague and archaic philosophy could be combined with our modern chemical and physiological knowledge, Pauli thinks that a new way of uniting the opposites of spirit and matter at a higher level might be found in the future (Pauli to Huxley, April 1956, von Meyenn 2001, p. 553):

> Perhaps it will be possible then to characterize a wholeness of the state both of mind and of matter (particularly in living organisms) by a new monistic language which, however, will refer to an abstract reality, which is only indirectly observable like the atoms and the unconscious.

And he ends with this beautiful passage (Pauli to Huxley, April 1956, von Meyenn 2001, p. 553, originally written in English by Pauli):

> Personally I do not share the creed of any church, nor the particular christian belief of Kepler, and I have until now resisted to any temptation of adopting for myself any particular metaphysical language (may it be Christian or Indian or the language of an individual philosopher). In an age, in which the old fashioned form of rationalism has lost its convincing power some time ago and in which the ethical position of physics is getting so problematical, I am living therefore in a spiritual house without a roof. Just for this reason your words did not find any obstacle to pour into it and they often give me the beautiful sensation of falling rain.

Huxley's answer to Pauli shows that he does not believe in exploring the intermediate level between spirit and matter expressed in hermetic philosophy. Quite the opposite, he believes that the dictum "as above, so below" is misleading and led those who accepted it into all kinds of false analogies. Huxley mentions Paracelsus, who advocated the use of antimony in medicine on the grounds that antimony was an effective agent for the purification of gold, and therefore must be an effective agent for the purification of the body. Because of this false analogy, based on the idea of a one-to-one correspondence between microcosm and macrocosm, sick people in Europe had to suffer for a century and a half from the administration of a dangerous and debilitating poison. Huxley maintains that the connection between inner and outer reality is at a level far deeper than that on which any kind of symbol can be perceived.

Huxley refers to a state of "obscure knowledge", in which there is an immediate experience of All in One and One in All that the mystics talk about. This resembles a kind of total omniscience, with no clear knowledge of any particular aspect of the world, which has nothing to do with concrete problems in astronomy, biology or physics. That in Kepler's case a religious symbol contributed to the development of a correct scientific hypothesis was purely accidental, according to Huxley. Other religious symbols, before and after Kepler, led to the formulation of incorrect scientific hypotheses and, moreover, Kepler's hypothesis was only partially correct.

Huxley sees no merit with the symbolic approach, rather he regards it as misleading, and accuses Jung's followers of assuming a medieval mindset with their obsession with symbols. The world of the scholastics and the alchemists, where everything "means" something, is a world which is antipathetic to Huxley. He finds it impossible to breathe in such a world, and says (Huxley to Pauli, June 10, 1956, von Meyenn 2001, p. 583): "It is like a closely shuttered room, crowded with people, smelling of humanity, with no outlets into fresh air and open spaces."

Huxley declares that the most wonderful thing about the mind is that it is more than the personal ego, more than the collective unconscious "stocked with archetypal images". It is also a series of not-selves, culminating in the supreme Not-Self, the Atman Brahman, the Void, the Suchness, which is at the same time the Self of every sentient being. And he gives the example of Zen Buddhists, who have no use for symbols and aim at getting out of their own light "in such a way that they may be filled with an obscure knowledge" of the non-particular that is in particulars, the not-thought that lies in thought.

The goal is to turn away, not only from the "nonsense of a world accepted at its face value", but also from kinds of religion and philosophy in which everything is a parable of something else. This is all too human. We must avoid taking innate symbols too seriously, and should not attribute to them values and virtues which they do not possess. Huxley distinguishes true mysticism from occultism: while the latter perceives everything transient as a parable, true mysticism sees the transient as an intersection of "a ray of the Godhead with an event in time". In true mysticism the ultimate insight is not verbal, and is not at the level of any symbolism. Instead, it is "obscure knowledge" of unity in multiplicity. This kind of certainty and enlightenment cannot be put in the form of expressible content.

Pauli read Huxley's letter two months later, after finishing his duties at ETH for the summer term. He found it important to reply that the differences in viewpoint between him and Huxley are only apparent, and could be clarified in conversation, beginning with defining their terminology. He agrees with Huxley's distinction between occultism and mysticism, in that occultism uses (alongside alchemy) a "concretistic" language, in which "everything means something". He also agrees that images are never an ultimate reality and always very human. He also hastens to add that he distances himself from "the followers of Jung" as he finds their therapeutic approach too narrow. The problem that really interests Pauli is the relation between mysticism and science (Pauli to Huxley, August 10, 1956, von Meyenn 2001, pp. 632–633):

> I can assure you that for me, as a modern scientist the difference is less obvious than for the layman. Both mystics and scientists have the same aim to become aware of the unity of knowledge, of man and the universe and to forget our own small ego.

Again he emphasizes that symbols exist at different levels of abstraction. Mathematical models of reality in physics are symbols, expressions like "obscure knowledge", "ray of the Godhead", which Huxley uses, are also symbols. What Huxley rejects are only symbols on a too "concretistic" level, but for Pauli these are steps toward more abstract symbols. He concludes: "And who believes that our present form of science is the last word in this scale? Certainly not I."

Pauli therefore thinks that Huxley judges Kepler's success by using symbolic images as "accidental" too quickly. Pauli is convinced, maybe through his own experience with dreams, visions, symbols and their relationship to his scientific work, that symbols have a relation both to mysticism and to science. "Does not that remain true, however incomplete and imperfect Kepler's symbols have been in comparison with the modern scientific picture of the universe and in comparison with the 'obscure knowledge' and the 'ray of the Godhead' of the great mystics of West and East?" he asks. He ends his letter with the words (Pauli to Huxley, August 10, 1956, von Meyenn 2001, p. 633):

> With the expression of my suspicion, that in our science are also traces of these mystical elements (even when it appears not "obscure" but "clear" to the layman, the layman sees "rational clarity" rather than "obscure knowledge" in science) and with the hope to see you at some occasion, (I do not know yet when I shall go to the States again), I remain, yours sincerely ...

Pauli went back to the United States, to Berkeley, during the spring of 1958. He left Zürich at January 17 and returned at June 1. During this stay he presented his cooperation with Heisenberg on the *Unified Field Theory* that at first had given him such high hopes in November 1957. He describes these in a letter to Aniéla Jaffé as him and Heisenberg being in the grip of the same archetype: *quaternity and reflection* (mirroring). In this, Pauli wanted to see a confirmation that the ancient symbols, which Jung had explored in his psychology, were now clearly reflected in physics and mathematics. He went so far as to say that the theory of Heisenberg and himself constituted a *realization of the Self*.

In this letter we also find the last recorded dream by Pauli, a dream that he interpreted as a promise for a fertile cooperation with Heisenberg (Pauli to Jaffé, January 5, 1958, von Meyenn 2004, p. 808):

> In our matrimonial bedroom I discover two children, one boy and one girl, both blond. They resemble each other a lot as if they just shortly before still were one and the same. They both tell me: "We have been here for three days. We like it here, nobody has just noticed us yet." Exalted I call my wife. She can't be far off, the children will soon have

her wrapped around their fingers (in reality my wife is very yielding towards children) and they will from now on always stay here.

In February 1958 he presented his and Heisenberg's work to some of his colleagues in Berkeley and was met with scepticism, to say the least. From his euphoric heights Pauli returned to earth with a bump, and by April 1958 he had finally decided to withdraw from the joint project with Heisenberg (Pauli to Heisenberg, April 7, 1958, von Meyenn 2004, pp. 1124–1126).

We do not know if Pauli met Huxley during his stay in Berkeley. There are no indications of such a meeting in the remains of Pauli's correspondence. But maybe what happened to him was an example of what Huxley had expressed as "mislead by false analogies"? After returning to Europe, Pauli only lived another seven months. He died on December 15, 1958.

10   Ritual and Reason: Contemporary Research on the Emergence of Symbolic Thought

In a late letter to Jung Pauli described his vision of future scientific knowledge as a house where the opposition of ritual and reason is resolved. With this he means that the value of religious rituals, i.e. the willingness to be transformed by participating in a process, is no longer in opposition to rational reason based on the ideal of the detached, unaffected observer. The idea of sacrifice in relation to the pursuit of knowledge had just begun to enter into quantum physics, and Pauli expected this tendency to grow in importance in the future (Pauli to Jung, October 23, 1956, Meier 2001).

To shed some further light on the questions Pauli was struggling with it could be interesting to compare them with a recent anthology exploring the emergence of symbolic thinking. In *Homo Symbolicus: The Dawn of Language, Imagination and Spirituality* (Henshilwood and D'Errico 2011) several aspects of and questions about the emergence of symbolic thought and culture are addressed. First of all it is worthwhile to notice that there is no general agreement on a definition of symbolism among academic researchers. Secondly there is agreement that symbolic activity consists of many levels and components and must be approached in an interdisciplinary way.[10]

The theories presented and discussed by Henshilwood and D'Errico (2011) show a variety of theoretical frameworks. Most of them are reductive and

---

[10]Some of the disciplines crucially involved in tracing the emergence of symbolic thinking are archaeology, social and religious anthropology, linguistics, psychology, cognitive science and biology.

therefore not satisfactory when compared to the ambitions of Pauli and Jung. Nevertheless, the general interdisciplinary approach to explore the topic of symbolism makes it an interesting contribution. All scholars in the volume agree that symbols and symbolic activity are hierarchically structured from simpler to more complex forms. An example is the unconscious use of symbols for decorative purposes (something that humans share with animals) through several steps before reaching full-blown time/space-factored symbolism including beliefs, myth and stories (Pettitt in Henshilwood and D'Errico 2011).

Some of the key functions of symbolic systems (including verbal language) identified today can be compared to Jung's more intuitive concepts. When it comes to understanding the psychological aspect of the development of symbolic thinking, researchers focus on the development of a specific bifurcation of the human mind: the capacity to experience self-awareness, being both a "self" and "a doer", i.e. simultaneously being an observer and an agent of the same action. It is easy to compare this point of view with Jung's distinction between the ego (agent) and the Self (observer) from which the ego emerges (Jung 1951).[11]

In modern research one hypothesis explaining this bifurcation is the way in which mothers interact with infants. The human infant's incapacity to cling to the mother and therefore "become one" with her is replaced by the mother monitoring the child through eye-contact and sounds. This has been connected to specific movements of hands and feet of infants and the expansion of attention towards joint attention and self-observation (Savage-Rumbaugh and Fields in Henshilwood and D'Errico 2011).

Another hypothesis emphasizes that we have two different conceptual systems that are both used to make sense of human behavior. One system deals with the properties and motions of physical bodies, and the other, called "theory of mind" (ToM), tries to explain and predict behavior based on mental states. The application of ToM in developmental psychology and the theory of mentalization explore the human capacity of reflective functioning, i.e. the ability to establish and use mental representations of our own and

---

[11]This reminds us of Niels Bohr's famous parable of the man with the cane in a dark room who tries to find his bearings with the aid of a cane. If he holds the cane tightly, it feels as if the point of the cane were an extension of the self; it becomes an extended arm. But if he holds the cane loosely, then he is more likely to receive an impression of the cane as an object. The cane can be both subject and object depending on how we relate to it. Bohr (1929a) used this parable to illuminate the epistemology of the measurement problem in quantum physics, i.e. the difficulty of demarcating the measuring instrument from the object to be measured.

other people's emotional states. It is a relational theory based on social biofeedback and mirroring, especially focused on the development of the inner world of fantasy and pretend play (Fonagy 2002).[12]

The two systems, one dealing with physical bodies and the other with intentional minds, must be held together to make sense of human behavior, but it is stated that their tenuous relationship leads to a sort of intuitive dualism. We tacitly assume some minded part of an individual that is separable from the body, so the idea that some part of a person can continue existing and even acting after death becomes natural in the light of intuitive dualism. This could explain why the imagination that some kind of human spirit survives bodily death and continues to act in the world belongs to the most widespread and oldest religious concepts. The step from there to infer other non-visible "minded" agents, like gods, or anthropomorphic beliefs in animated objects, is not too far.

A worldview allowing living interaction with minded entities requires the ability to deliberately and reflectively ponder the contents of other's thoughts. These meta-representational abilities are specifically human insofar as we can consider what mental state or intention lies behind an utterance or gesture, including the possibility that the intention was to change the mental state, not just a behavior. For these reasons it has been argued that the meta-representational ability makes human language possible. It is this capacity that changes signaling (that we share with mammals) into linguistic communication and symbolism more generally. Meta-representation is more flexible than a signaling system and triggers both behavioral routines and private mental, epistemic states (see Barrett in Henshilwood and D'Errico 2011).

The concept of meta-representations which trigger both behavior and epistemic states seems not so distant from Jung's notion of the psychoid archetype – a true *complexio oppositorum*, playing with the image of a spectrum, ranging from instinct (infrared) to archetype (ultraviolet), on which

---

[12]It is interesting that this psychological theory emphasizes "marked mirroring" as an important factor for the development of resilience and sound psychological development of the sense of self in children. In marked mirroring the parent mirrors the child's affective state, but not exactly. By exaggerating some features of the affect (facial and gestural displays) and temper others (affective pitch) sensitive caregivers differentiate as-if (or pretend) communications from realistic ones. This helps the child to assemble a symbolic representational system of affective states and assists in developing affect regulation (and selective attention), which make up the basis of secure attachment (Fonagy 2002). The emphasis on this asymmetry of mirroring for favorable development resonates with Pauli's fascination with the topic of mirroring, symmetry and asymmetry as fundamental factors for our physical universe and for psychology (Pauli to Jung, August 5, 1957, Meier 2001).

consciousness "slides". A spiritual experience can be as "compulsive" as an instinctual urge. When consciousness approaches the infrared part, it is controlled by instinct (behavioral schemes); close to the ultraviolet part it is dominated by the spirit (epistemic state). In this picture, instinct may be seen as a latent archetype manifesting itself on a longer wavelength, and the archetype itself may be seen instinct raised to a higher intensity. Reflexive consciousness can only be reached towards the spiritual end of the spectrum. We can only reflect on the instinct when it is transformed into imagery. Conscious processing and assimilation can only take place by means of an integration of the instinct as an image which both signifies it and at the same time brings it to life (Jung 1954b, par. 414).

In this context it is interesting that Pauli in his search for a neutral language wanted to change the term archetype into the term *automorphism* taken from mathematics. Loosely speaking, an automorphism maps a system onto itself, revealing the inner symmetries and the wealth of relationships within the system. This concept is ideally tailored for neutral qualities: it can describe processes in both psyche and matter. In alchemy, Pauli found the notion of *multiplicatio* (generative power) to describe a similar principle (Pauli to Jung, October 23, 1956, Meier 2001).

In this light it is noteworthy that the principle of recursion (self-similarity, nested hierarchies of patterns) plays a central role in current thinking about the emergence of symbolic thinking. Symbolic activities, especially in the form of music and dance, contain recursive structures that are essential for their generative nature. The emergence of music and dance is considered to have encouraged the same intellectual abilities and motor skills that are needed for language, and that they even played a more specific role in language development than vision.

Music and dance, with their strong emotional power, are also considered to have had a deep effect on evolutionary psychological development, playing a major role in social bonding and cohesion. Hierarchy and recursion are naturally developed through music, song and dancing. They are associated with play, which is important for higher-level integrative processes. A key step in language development could be the development of neural connections allowing recursion, perhaps related to the use of tools and first realized in relation to imaginative play (Barrett in Henshilwood and D'Errico 2011). It is crucial that non-rational, emotional and motivational factors drive these developments.

Emphasizing the emergence of recursive patterns for symbolic thinking, for music and rhythmic activity, play and imaginary worlds, we come close to Jung's view of the function of the archetypal symbol for raising consciousness. Recursion is a process that we find both in physical nature (fractals), in biological nature, and in psychological nature. It could therefore be a paradigm example of a basic principle of a "neutral language".

11   Summary

What then is the neutral language that Pauli proposes? First of all, it is a concept that describes processes which we can observe in different domains of our experienced world, notably in matter as well as in the psyche. A neutral language is not reducible to one specific domain such as matter, brain, psyche or spirit. Pauli could not accept a worldview assigning unequal status to material and spiritual aspects of reality. Regardless of how much he liked Huxley's fervor in approaching these topics, he could not accept his choice of spirit (divine ground) over matter.

The concept of an "automorphism" is an example of a term of a neutral language. It describes processes that we can observe in psyche and in matter. Mirroring could be another such term. A neutral language is an "objective" language describing a psychophysical or holistic nature – with the important distinction that "objective" respects the epistemological lesson that all knowledge is an interaction between observer and observed. All acts of knowledge imply a choice (and a sacrifice) of observing reality in a certain way. This can be readily extended to the existential dimension of choice (and sacrifice) in everyday life, addressing issues of morality and ethics.

Beyond the world of visible phenomena there is a postulated structural level of general processes governed by lawful regularities. However, these laws must be conceived beyond the conventional understanding of causal or statistical laws in science. With the idea of a neutral language, Pauli attempted to formulate a new kind of laws in the sense of general correlations and symmetries manifesting themselves in both psyche and matter (Pauli 1954). The neutral language tries to grasp these lawful aspects of our psychophysical, symbolic reality.

Acknowledgment

The author is grateful for helpful discussions with Marcus Appleby.

References

Atmanspacher H., Primas H., eds. (2009): *Recasting Reality: Wolfgang Pauli's Philosophical Ideas and Contemporary Science*, Springer, Berlin.

Bohr N. (1929a): The quantum of action and the description of nature. In *Niels Bohr Collected Works, Vol. 6*, ed. by J. Kalckar, North-Holland, Amsterdam 1985, pp. 201–208.

Bohr N. (1929b): Atomic theory and description of nature. Introductory survey. In *Niels Bohr Collected Works, Vol. 6*, North-Holland, Amsterdam 1985, pp. 255–257.

Enz C.P. (2002): *No Time to Be Brief. A Scientific Biography of Wolfgang Pauli*, Oxford University Press, Oxford.

Feldman J.A. (2006): *From Molecule to Metaphor. A Neural Theory of Language*, MIT Press, Cambridge.

Folse H. (1985): *The Philosophy of Niels Bohr: The Framework of Complementarity*, Elsevier, Amsterdam.

Fonagy P., ed. (2002): *Affect Regulation, Mentalization, and the Development of the Self*, Other Press, New York.

Gieser S. (2005): *The Innermost Kernel. Depth Psychology and Quantum Physics*, Springer, Berlin.

Goldschmidt H.L. (1990): Begegnung mit Wolfgang Pauli. In *Nochmals Dialogik*, ETH Stiftung Dialogik, Zürich, pp. 57–63.

Henshilwood C.S., d'Errico F., eds. (2011): *Homo Symbolicus: The Dawn of Language, Imagination and Spirituality*, John Benjamins, Amsterdam.

Huxley A. (1941): *Grey Eminence. A Study in Religion and Politics*, Chatto and Windus, London.

Huxley A. (1945): *The Perennial Philosophy*, Harper, New York.

Huxley A. (1952): *The Devils of Loudun*, Chatto and Windus, London.

Huxley A. (1956): *Heaven and Hell*, Chatto and Windus, London.

Jung C.G. (1921): Psychological types. In *Collected Works, Vol. 6*, Princeton University Press, Princeton 1971.

Jung C.G. (1922): On the relation of analytical psychology to poetry. In *The Spirit in Man, Art, and Literature, Collected Works Vol. 15*, Princeton University Press, Princeton 1966.

Jung C.G (1928): On psychic energy. In *The Structure and Dynamics of the Psyche, Collected Works, Vol. 8*, Princeton University Press, Princeton 1969.

Jung C.G. (1928): Analytical psychology and "Weltanschauung". In *The Structure and Dynamics of the Psyche, Collected Works, Vol. 8*, Princeton University Press, Princeton 1969.

Jung C.G. (1935): The relations between the ego and the unconscious. In *Two Essays on Analytical Psychology, Collected Works, Vol. 7*, Princeton University Press, Princeton 1966.

Jung C.G. (1944): Individual dream symbolism in relation to alchemy. In *Psychology and Alchemy, Collected Works, Vol. 12*, Princeton University Press, Princeton 1968.

Jung C.G. (1948): A psychological approach to the trinity. In *Psychology and Religion, Collected Works, Vol. 11*, Princeton University Press, Princeton 1969.

Jung C.G. (1950): Psychology and literature. In *The Spirit in Man, Art, and Literature, Collected Works, Vol. 15*, Princeton University Press, Princeton 1966.

Jung C.G. (1951): Aion. Researches into the phenomenology of the self. In *Aion, Collected Works, Vol. 9/II*, Princeton University Press, Princeton 1968.

Jung C.G. (1952a): Answer to Job. In *Psychology and Religion, Collected Works, Vol. 11*, Princeton University Press, Princeton 1969.

Jung C.G. (1952b): Religion and psychology. A reply to Martin Buber. In *The Symbolic Life, Collected Works, Vol. 18*, Princeton University Press, Princeton 1975.

Jung C.G. (1954a): Psychological aspects of the mother archetype. In *The Archetypes and the Collective Unconscious, Collected Works 9/I*, Princeton University Press, Princeton 1969.

Jung C.G. (1954b): On the nature of the psyche. In *The Structure and Dynamics of the Psyche, Collected Works 8*, Princeton University Press, Princeton 1969.

Kahneman D. (2011): *Thinking, Fast and Slow*, Farrar, Straus and Giroux, New York.

Meier C.A., ed. (2001): *Atom and Archetype: The Pauli/Jung letters 1932–1958*, Routledge, London.

Pauli W. (1950): The philosophical significance of the idea of complementarity. In *Writings on Physics and Philosophy*, ed. by C. Enz and K. von Meyenn, Springer, Berlin 1994, pp. 35–42.

Pauli W. (1952): The influence of archetypal ideas on the scientific theories of Kepler. In *Writings on Physics and Philosophy*, ed. by C. Enz and K. von Meyenn, Springer, Berlin 1994, pp. 219–279.

Pauli W. (1954): Ideas of the unconscious from the standpoint of natural science and epistemology. In *Writings on Physics and Philosophy*, ed. by C. Enz and K. von Meyenn, Springer, Berlin 1994, pp. 149–164.

Stumbrys T., Erlacher D., Schmidt S. (2011): Lucid dream mathematics: An explorative online study of arithmetic abilities of dream characters. *International Journal of Dream Research* **4**(1), 35–40.

von Meyenn K., ed. (forthcoming): *Wolfgang Pauli - Die Traumaufzeichnungen*, Springer, Berlin.

von Meyenn K., ed. (1985): *Wolfgang Pauli. Wissenschaftlicher Briefwechsel, Band II: 1930–1939*, Berlin, Springer.

von Meyenn K., ed. (1993): *Wolfgang Pauli. Wissenschaftlicher Briefwechsel, Band III: 1940–1949*, Springer, Berlin.

von Meyenn K., ed. (1999): *Wolfgang Pauli. Wissenschaftlicher Briefwechsel, Band IV, Teil II: 1953–1954*, Springer, Berlin.

von Meyenn K., ed. (2001): *Wolfgang Pauli. Wissenschaftlicher Briefwechsel, Band IV, Teil III: 1955–1956*, Springer, Berlin.

von Meyenn K., ed. (2004): *Wolfgang Pauli. Wissenschaftlicher Briefwechsel, Band IV, Teil IV: 1957–1958*, Springer, Berlin.

Notes on Psychophysical Phenomena

Harald Atmanspacher

Abstract

In the mid 20th century, the physicist Wolfgang Pauli and the pschologist Carl Gustav Jung proposed a conceptual framework, not more than speculative at the time, which may help us to clarify psychophysical phenomena *beyond* what our knowledge about the mental and the physical in separation are capable of achieving. Their conjecture of a dual-aspect monism, with a complementary relationship between mental and material aspects of an underlying, psychophysically neutral reality, is subtler and more sophisticated than many other attempts to discuss the problem of how mind and matter are related to one another.

1 Dual-Aspect Monism According to Pauli and Jung

1.1 The Overall Picture

The Pauli-Jung version of *dual-aspect monism*[1] merges an ontic monism, reflected by a psychophysically neutral backgroud reality, with an epistemic dualism of the mental and the physical as perspectival aspects of the underlying ontic reality. Jung coined the notion of the *unus mundus*, the one world, for this domain. In dual-aspect monism, the aspects are not *a priori* given, but depend on epistemic issues and contexts. Distinctions of aspects are generated by "epistemic splits" of the distinction-free, unseparated underlying realm, and in principle there can be as many aspects as there are contexts.[2]

---

[1] For a more comprehensive description see Atmanspacher (2012).

[2] In somewhat more abstract terms, distinctions can be conceived as symmetry breakings. Symmetries in this parlance are invariances under transformations. For instance the curvature of a circle is invariant under rotations by any arbitrary angle. A circle thus exhibits complete rotational symmetry. Symmetry breakings are a powerful mathematical tool in large parts of theoretical physics, but we can only speculate which symmetries must be ascribed to the psychophysically neutral *unus mundus*.

According to the Pauli-Jung conjecture, mind and matter appear as *complementary* aspects: they are mutually incompatible but both together necessary to describe mind-matter systems exhaustively. A straightforward reason for this is the fundamentally non-Boolean nature of the underlying reality. As is well known in mathematics, representations of non-Boolean systems are generally incompatible, and complementarity can be formally characterized as a maximal form of incompatibility.

There are important respects in which this framework differs from *neutral monism* á la Mach, James, or Russell. In neutral monism, the mental and the physical are *reducible* to the underlying domain, whereas they are *irreducible* in dual-aspect monism. The reason for this diffference is that neutral monism conceives the underlying domain to consist of psychophysically neutral elements whose combinations determine whether the compound products appear mental or physical. In dual-aspect monism, the underlying domain does ultimately not consist of separate elements at all.[3] It is radically holistic, and the mental and physical aspects emerge by a *decomposition of the whole rather than a composition of elements*.

In the Pauli-Jung conjecture, the psychophysically neutral domain is apprehensible only indirectly, by its manifestations in the aspects. Their dual-aspect monism is a metaphysical position including both epistemic and ontic elements. Although large parts of the 20th century witnessed an often pejorative connotation with metaphysics, insights into the nature of reality are in general impossible without metaphysical assumptions and regulative principles. The metaphysical nature of the Pauli-Jung conjecture *implies* a lack of concrete illustrative examples which is not due to missing imagination but represents an importan feature of their approach.

This alludes to the situation in quantum theory, repeatedly expressed by one of its main architects, Niels Bohr (1934):

> we are concerned with the recognition of physical laws which lie outside the domain of our ordinary experience and which present difficulties to our accustomed forms of perception.

Accordingly, so-called "intuitively appealing thinking" may mislead us by inhibiting rather than advancing our ways to insight.[4] Along the same lines,

---

[3]This is crucial because it avoids the so-called "combination problem" in various accounts of panpsychism; see Seager (2010) for a detailed discussion and a proposed solution.

[4]This implies a plea against misplaced concreteness and simplification. As cognitive scientists found not long ago, learning processes can be substantially improved if abstract principles are learned first and concrete examples for them thereafter (Kaminski *et al.* 2008). This result counters a carefully nurtured long-time dogma in education.

Heisenberg (1971) remembers a conversation with Bohr at Göttingen in 1922. He asked Bohr:

> If the inner structure of the atoms is inaccessible to an illustrative [anschauliche] description, as you say, if we basically have no language to speak about this structure, will we ever be able to understand the atoms? Bohr hesitated for a moment, then he replied: Yes we will. But at the same time we will have to learn what the word "understanding" means.

As we will see below, it may not be entirely accidental that the issue of meaning arises here – pretty astonishing for a typical physics discussion but absolutely pivotal for Jung's concept of synchronistic events and the symbolic expression of their meaning.

1.2 Synchronicities as Psychophysical Correlations

Conceiving the mind-matter distinction in terms of an epistemic split of a psychophysically neutral reality implies psychophysical correlations between mind and matter as a direct and generic consequence. Pauli and Jung discussed psychophysical correlations extensively in their correspondence between June 1949 and February 1951 (Meier 1992, pp. 40–73) when Jung drafted his article on "synchronicity" for the book that he published jointly with Pauli (Jung and Pauli 1952). In condensed form, two (or more) seemingly accidental, but not necessarily simultaneous events are called synchronistic if the following three conditions are satisfied.

1. Each pair of synchronistic events includes an internally conceived and an externally perceived component.

2. Any presumption of a direct causal relationship between the events is absurd or even inconceivable.

3. The events correspond with one another by a common meaning, often expressed symbolically.

The first criterion makes clear that synchronistic phenomena are intractable when dealing with mind or matter alone. The second criterion expresses that synchronistic correlations cannot be explained by (efficient) causation in the narrow sense of a conventional cause-and-effect-relation as usually looked for in science. And the third criterion suggests the concept of meaning (rather than causation) as a constructive way to characterize psychophysical correlations.

Since synchronistic phenomena are not necessarily "synchronous" (in the sense of "simultaneous"), synchronicity is a somewhat misleading term. For this reason Pauli preferred to speak of "meaningful correspondences" under the influence of an archetypal "acausal ordering". He considered both Jung's synchronicity and the old teleological idea of finality (in the general sense of a process oriented toward a goal) as particular instances of such an acausal ordering. Meaningful coincidences cannot be set up fully intentionally or controlled reproducibly. On the other hand, "blind" chance (referring to stochastically accidental events) might be considered as the limiting case of meaning*less* correspondence.

For a psychologist like Jung, the issue of meaning is of primary significance anyway. For a long time, Jung insisted that the concept of synchronicity should be reserved for cases of distinctly numinous character, when the experience of meaning takes on existential dimensions. With this understanding synchronistic correlations would be extremely rare, thus contradicting their supposedly generic nature. In later years, Jung opened up toward the possibility that synchronicity might be a notion that should be conceived as ubiquitous as indicated above. Meier (1975) has later amplified this idea in an article about psychosomatics from a Jungian perspective.

1.3   From Quantum Physics to (Depth) Psychology

According to Pauli and Jung, the role which measurement plays as a link between epistemic and ontic realities in physics is mirrored by the act in which subjects become consciously aware of "local mental objects", as it were, arising from unconscious contents in psychology.[5] In this sense, they postulated the possibility of transitions between the mental and/or the material mediated by the psychophysically neutral *unus mundus*. This idea is most clearly elaborated in Jung's supplement to his *On the Nature of the Psyche* (Jung 1969).[6] Let me first quote from a letter by Pauli which Jung cites in footnote 130 in this supplement (Jung 1969, par. 439):[7]

_____

[5]We use the term "local mental objects" to emphasize the analogy with local material objects, meaning that neither of them are non-local or non-Boolean. More concretely, local mental objects should be understood as distinct mental representations or categories endowed with a Boolean (yes-no) structure: a mental state is either in a category or it is not. Using the formal apparatus of the theory of complex systems, such categories can be defined, e.g., as attractors in an appropriately defined phase space (van Gelder 1998, Fell 2004).

[6]The German version of this essay was first published as "Der Geist der Psychologie" in 1946, and later revised and expanded (essentially by the mentioned supplement) as "Theoretische Überlegungen zum Wesen des Psychischen" in 1954.

[7]This letter is contained neither in the published Pauli-Jung correspondence (Meier 1992) nor in Pauli's correspondence edition by von Meyenn. Since Jung presents the

... the epistemological situation regarding the concepts of "conscious-ness" and the "unconscious" seems to offer a close analogy to the situa-tion of "complementarity" in physics, sketched below. On the one hand, the unconscious can only be made accessible in an indirect way by its (ordering) influence on conscious contents, on the other hand every "ob-servation of the unconscious", i.e. every attempt to make unconscious contents conscious, has a *prima facie* uncontrollable reaction back onto these unconscious contents themselves (as is well known, this precludes that the unconscious can be "exhaustively" brought to consciousness). The physicist will *per analogiam* conclude that precisely this uncon-trollable backlash of the observing subject onto the unconscious limits the objective character of its reality and, at the same time, provides it with some subjectivity. Although, moreover, the *position* of the "cut" between consciousness and the unconscious is (to a certain degree) up to the free choice of the "psychological experimenter", the *existence* of this "cut" remains an inevitable necessity. Thus, the "observed system" would, from the viewpoint of psychology, not only consist of physical objects, but rather comprise the unconscious as well, whereas the role of the "observing device" would be ascribed to consciousness. The development of "microphysics" has unmistakably led to a remarkable convergence of its description of nature with that of the new psychology: While the former, due to the fundamental situation known as "comple-mentarity", faces the impossiblity to eliminate actions of observers by determinable corrections and must therefore in principle relinquish the objective registration of all physical phenomena, the latter could basi-cally complement the merely subjective psychology of consciousness by postulating the existence of an unconscious of largely objective reality.

This excerpt describes Pauli's position concerning objective and subjec-tive aspects of the mental, a distinction that he adopted from Jung quite early. Already in a letter to Kronig of August 3, 1934 (letter 380 in von Meyenn 1985, pp. 340–341), he talks about the "autonomous activity of the soul" as "something objectively psychical that cannot and should not be explained by material causes." Hence, the "objective reality" at the end of the quote refers to the psychophysically neutral background reality, while the "subjective" relates to its contextual, epistemic, manifestation in the psyche.

As a consequence of Pauli-Jung style dual-aspect monism, mind-matter relations, or psychophysical relations, can be understood due to their com-mon origin in the underlying domain of reality. Although there is no di-rect causal pathway between the mental and the physical, Pauli and Jung conjectured indirect kinds of influence via their underlying domain. These

quotation with the remark that Pauli "was gracious enough to look over the manuscript of my supplement", the letter is likely of 1954.

influences are possible because the relation between ontic (psychophysically neutral) and epistemic (mental and material) domains is conceived as *bidirectional* (see also Sec. 3.2 below).

If, for instance, unconscious contents become conscious, this very transition changes the unconscious left behind. Analogously, physical measurement entails a transition from an unobserved to an observed state, and this very measurement changes the state of the system left behind. This picture, already outlined in Pauli's letter to Fierz of October 3, 1951 (von Meyenn 1996, p. 377), represents a genuine interdependence between ontic and epistemic domains. It can entail mind-matter correlations in addition to those *unidirectional* correlations that are due to mere epistemic manifestations of the ontic realm.

The Pauli quote above emphasizes parallels between basic conceptual structures of quantum theory and psychology. One of the key common features in these two scientific areas is arguably the fact that an observation does not only register an outcome, as in classical thinking, but also changes the state of the observed system in a basically uncontrollable manner. This holds for physical quantum systems as well as for mental systems and, as simple as it sounds, it has far-reaching consequences which psychology and cognitive science are just about to realize (cf. Aerts *et al.* 1993, Atmanspacher *et al.* 2002, Khrennikov 2010, Busemeyer and Bruza 2012).

A most evident effect of this backreaction on mental states is the almost ubiquitous appearance of order effects in surveys and questionnaires. This has recently been addressed in detail (Atmanspacher and Römer 2012) on the basis of non-commutative structures of mental observables. Since the mathematics of such structures is at the heart of quantum theory as well, this parallel is not a mere analogy – it points to a constitutive joint principle underlying the mental and the physical: "almost too good to be true", as one recent commentator expressed it (Tresan 2013).

2   Relative Onticity

As appealing and compact as the sketch outlined in the preceding section may appear, it is not subtle enough. For instance, the boundary between the mental and physical aspects on the one hand and their underlying domain on the other is unsharp: there is always a grey zone between conscious and unconscious states, and no physical state is ever exactly disentangled from the rest of the material world.

Rather than speaking of a grey zone, one might conceive of a whole spectrum of boundaries, each one indicating the transition to a more comprehensive level of wholeness until (ultimately) the distinction-free *unus mundus* is approached. A viable idea in this context might be archetypal levels with increasing degrees of generality: the *unus mundus* at bottom, the mental and physical on top, and intermediate levels in between. Depending on the status of the individuation process of the individual concerned, Jung's *transcendent function* regulates the exchange among these levels.

This entails that a tight distinction of one fundamentally ontic and two derived epistemic domains is too simplistic. However, an idea originally proposed by Quine (1969), developed by Putnam (1981, 1987) and later utilized by Atmanspacher and Kronz (1999) comes to help here: *ontological relativity* or, in another parlance, *relative onticity*.[8]

The key motif behind this notion is to allow ontological significance for any level, from elementary particles to icecubes, bricks, and tables – and all the same for elements of the mental. One and the same descriptive framework can be construed as either ontic or epistemic, depending on which other framework it is related to: bricks and tables will be regarded as ontic by an architect, but they will be considered highly epistemic from the perspective of a solid-state physicist. Schizophrenia, depression, and dissociative disorders will be considered as basic ontic features in psychiatry, yet a detailed psychological or philosophy-of-mind analysis will try to find its own ontic terms with which these impairments can be described as epistemic manifestations.

Quine proposed that a "most appropriate" ontology should be preferred for the interpretation of a theory, thus demanding "ontological commitment". This leaves us with the challenge of how "most appropriate" should be defined, and how corresponding descriptive frameworks are to be identified. Here is where the notion of *relevance* becomes significant. For particular degrees of complexity, the "most appropriate" framework is that which provides those features that are relevant for the question to be studied. And the referents of this descriptive framework are those which Quine wants us to be ontologically committed to.

This can be applied to the Pauli-Jung conjecture in an interesting way: An archetype which may be regarded as ontic relative to the perspective of the mind-matter distinction, can be seen epistemic relative to the *unus mundus*. This twist is additionally interesting because it also relativizes Jung's (overly) stern Kantian stance that archetypes *per se* as formal order-

---

[8]Similar ideas have been developed independently by van Fraassen (1980) in terms of "relevance relations" and Garfinkel (1981) in terms of "explanatory relativity", though with less, or less explicit, emphasis on issues of ontology.

ing factors in the collective unconscious must be *strictly inaccessible* epistem-
ically, and thus empirically (cf. Kime 2013 for more discussion). A relativized
notion of ontology allows us to see clearer why and how a more sophisticated
blend of epistemic and ontic realms in dual-aspect monism can acquire systematic and explanatory status.

Taken seriously, this framework of thinking entails a farewell to the
centuries-old conviction of an absolute fundamental ontology (usually that
of basic physics). This move is in strong opposition to many mainstream
positions in the philosophy of science until today. But in times in which
fundamentalism – in science and elsewhere – appears increasingly tenuous,
Quine's philosophical idea of an ontological relativity offers a viable alternative for more adequate and more balanced worldviews. And, using the
scientifically tailored concept of relative onticity, this is not merely a conceptual idea but can in fact be used for an informed discussion of concrete
issues in the sciences.

Coupled with an ontological commitment to context-dependent "most
relevant" features in a given situation, the relativization of onticity does
not mean dropping ontology altogether in favor of a postmodern salmagundi
of floating beliefs. The "tyranny of relativism" (as some have called it)
can be avoided by distinguishing more appropriate descriptions from less
appropriate ones. The resulting picture is more subtle and more flexible
than an overly bold reductive fundamentalism, and yet it is more restrictive
and specific than a patchwork of arbitrarily connected opinions.

3   More than Physics "plus" Psychology

3.1   A Semi-Fictitious Historical Excursion

Imagine a scientist specializing in the science of electricity in the early 19th
century. At this point in time, Faraday just started investigations that ultimately led him to the concept of electric and magnetic fields, which Maxwell
picked up and developed into a unified theory of electromagnetism, culminating in the set of four basic equations which Maxwell published under the
title *On Physical Lines of Force*.[9]

At the beginning of the 19th century, however, electricity and magnetism
were regarded as basically unrelated phenomena. Now consider our imag-

---

[9] *On Physical Lines of Force* is a four-part article that appeared in the *Philosophical
Magazine* in 1861 and 1862. The four parts are devoted to "the theory of molecular
vortices applied to magnetic phenomena" (I), "... to electric currents" (II), "... to statical
electricity" (III), and "... to the action of magnetism on polarized light" (IV).

ined scientist experimenting with electric currents on his laboratory desk. Incidentally, a compass, unwittingly left by a visitor the other day, is sitting on a side-table not far from the desk. The scientist starts his experiments and connects the wires on the table with a battery (invented by Volta just a few years back).

He looks around in the room to look for some additional equipment, and suddenly rivets on the compass. The compass needle trembles, and points into an entirely wrong direction – not north, not south, but something completely different! What happened? An outright spooky apparition it seems, inexplicable by anything he ever learned. Which impudent specter tries to fool him with such a kind of nuisance? Did the compass get inhabited by naughty spirits, moving the needle at their pleasure?

Indeed, the body of knowledge in physics at the time of this fictitious story does not offer any compelling explanation of the distorted behavior of the compass. Of course, this changed half a century later, when it became well known that electric currents generate a magnetic field, and that this field naturally moves the compass needle, such that it deviates from the orientation of the magnetic field of the earth.

Maxwell's electrodynamics succeeded in describing both electric and magnetic phenomena in the same compact framework, specifying the relations by which the two are linked together. Without this framework, magnetic phenomena in the presence of electricity and electric phenomena in the presence of magnetism were regarded as inexplicable magic, miracles, or misconduct – depending on who reported them and for what purpose.

What can this little story teach us? It expresses a historical analogy of the contemporary situation concerning the psychophysical problem of how mind and matter are related. Exactly as the moving compass needle (due to electric current), a moving hand (due to mental decision) represents a paradigm example of an anomaly not understood by current science. Needless to say, there are more stunning psychophysical anomalies such as out-of-body experiences, premonitions, etc. – more about them later.

At present, we do not have a theoretical framework for psychophysical phenomena, just as the early 19th century did not have electrodynamics. The analogy tells us also that it is misleading to try and study psychophysical phenomena as if they were either mental or physical, exactly as electromagentic phenomena are neither solely electric nor magnetic. They are not, and it is likely that they need to be recast in a way even more radical than Maxwell's breakthrough has been.

3.2   Structural and Induced Psychophysical Correlations

The development of Pauli's and Jung's views about psychophysically neutral archetypes and their role in manifesting psychophysical correlations (e.g., "synchronicities") suggests a distinction between two basically different kinds of mind-matter correlations for which we propose the notions of "structural" and "induced" correlations.

Structural correlations refer to the role of archetypes as ordering factors with an exclusively *unidirectional* influence on the material and the mental (Pauli's letter to Fierz of 1948, von Meyenn 1993, pp. 496–497). They arise due to epistemic splits of the *unus mundus*, and manifest themselves as correlations between mental and material aspects. These correlations are a straightforward consequence of the basic structure of the Pauli-Jung conjecture, and they are expected to be ubiquitous, persistent and empirically reproducible.

Induced correlations refer to the backreaction that changes of consciousness induce in the unconscious and, indirectly, in the physical world as well.[10] (Likewise, measurements of physical systems induce backreactions which can lead to changes of mental states.) In this way, the picture is extended to a *bidirectional* relation (Pauli's letter to Jung of 1954, Jung 1969, par. 439). In contrast to structural, persistent correlations, induced correlations depend on all kinds of contexts (e.g., personal situation, environment). They occur occasionally, and are evasive and not (easily) reproducible.

What Pauli wrote to Fierz on June 3, 1952 (von Meyenn 1996, pp. 634–635), yields an almost seamless fit with this distinction:

> ... synchronistic phenomena ... elude being captured in natural "laws" since they are not reproducible, i.e., unique, and are blurred by the statistics of large numbers. By contrast, "acausalities" in physics are precisely described by statistical laws (of large numbers). ... I would personally prefer to begin with always reproducible acausal dispositions (incl. quantum physics) and try to understand psychophysical correlations as a special case of this general species of correlations.

Pauli's proposal to begin with "always reproducible acausal dispositions" relates perfectly well to the structural mind-matter correlations due to epistemic splits of the *unus mundus*. What he referred to as special cases of

---

[10]Jungian psychology describes this in more detail: When a subject becomes aware of some problematic unconscious content, the corresponding unconscious complex may be (partially) dissolved. This affects the archetypal core that is constellated in the complex, which in turn is supposed to manifest itself in the physical world.

psychophysical correlations can then be mapped to the induced correlations superimposing those structural, "general species of correlations".

Pauli speculated that synchronicities exhibit a kind of lawful regularity beyond both deterministic and statistical laws, based on the notion of *meaning* and, thus, outside the natural sciences of his time (and also, more or less, of today): "a third type of laws of nature consisting of corrections to chance fluctuations due to meaningful or purposeful coincidences of causally unconnected events" (von Meyenn 1999, p. 336). It remains to be explored how the key issues of meaning and purpose can be implemented in an expanded worldview not only comprising, but also exceeding both psychology and physics.[11] A comprehensive substantial account of psychophysical phenomena needs to address them beyond the distinction of the mental and the physical. This excludes considering them as a simplistic ("additive") composition of these two domains.

While structural correlations define a baseline of ordinary, robust, reproducible psychophysical correlations (such as mind-brain correlations or psychosomatic correlations), induced correlations (positive or negative) may be responsible for alterations and deviations (above or below) this baseline. Induced positive correlations, above the baseline, are experienced as unconventional "coincidence" phenomena – similar to "salience" phenomena (cf. Kapur 2003, van Os 2009). Numinous synchronistic events in the sense Jung proposed originally clearly belong to this class. Induced negative correlations, below the baseline, are experienced as unconventional "dissociation" phenomena. In Sec. 3.4 below we will relate these features to the phenomenology of exceptional human experiences.

It is important to keep in mind that in both induced and structural correlations there is no direct causal relation from the mental to the physical or *vice versa* (i.e. no direct "efficient causation"). The problem of a direct "causal interaction" between categorically distinct regimes is thus avoided. Of course, this does not mean that the correlations themselves are causeless. The ultimate causes for structural correlations are the epistemic split of the *unus mundus* and the ordering influence of psychophysically neutral archetypes. The causes for induced correlations are interventions in the conscious mental domain or the local material domain, whose backeffects on archetypal activity must be expected to manifest themselves in the complementary domain, respectively.

---

[11] See also the contribution by Main in this volume.

3.3   Formal and Experienced Meaning

In the characterization of synchronistic events given above, the common meaning of mental and material events figures prominently. However, meaning is a notoriously difficult notion, used differently in different areas. Formally speaking, meaning is a two-place relation between a sign and what it designates, or a representation and what it represents. Meaning in this formal sense is simply a reference relation, in accordance with the philosophical usage of the term intentionality since Brentano (1874).

What Jung had in mind when he emphasized meaning is different, however. He did aim at meaning as an element of experience, not as a formal relationship. This can be rephrased in Metzinger's (2003) representational account of the mental, where intentionality – a reference relation between a representation and its referent – is itself encoded as a (meta-)representation. In Metzinger's parlance this (meta-)representation is a "phenomenal model of the intentionality relation" (PMIR).[12]

Mental representations have intentional content and they have phenomenal content. While the intentional content explicates their reference, as mentioned above, their phenomenal content refers to "what it is like to" instantiate a representation, in other words: to experience it. So the phenomenal content of a PMIR refers to "what it is like to" experience a particular meaning. Jung's usage of meaning refers to the phenomenal content of PMIRs: the subjectively experienced meaning of a synchronistic event.

It should be stressed that the meaning of synchronistic events, although being subjectively ascribed (by the experiencing subject), is not completely arbitrary. It depends on a subject's life situation as a whole, likely including conditions that are not consciously available to the subject. According to Jung, synchronistic events arise due to constellated archetypal activity. This activity limits the range of possibly attributable meanings by "objective", metaphysical constraints; see Sec. 3.4 for more details.

In typical situations of "ordinary" structural psychophysical correlations, the *formal intentionality* due to plain reference is hardly experienced explicitly – subjects are not actually aware of its phenomenal quality. This is different for induced psychophysical correlations: their deviation from the ordinary baseline stimulates that *experienced intentionality* is incurred, re-

---

[12]It should be noted that Metzinger's general position is far from dual-aspect monism – it is basically an attempt to naturalize mental processes such that they are understood as a result of physical brain activity. Nevertheless, his (epistemic) categories of self model and world model are in one-to-one correspondence with the (epistemic) mental and material aspects of the Pauli-Jung conjecture.

ferring to the phenomenal content of the appropriate PMIR. In this case, the corresponding meaning is distinctly and phenomenally inflicted upon the experiencing subject.

It is plausible to assume that the extent to which contextually induced correlations deviate from the baseline of persistent structural correlations complies with the degree of intensity to which the corresponding PMIR is phenomenally experienced. Small deviations indicate quasi-persistent, almost reproducible correlations, while large deviations signify what Jung insisted on for truly synchronistic events: the "numinous" dimension of the experience.

In his concept of synchronicity, Jung typically emphasized induced psychophysical correlations in the sense of meaningful coincidences, i.e. positive correlations above the ordinary baseline. The more comprehensive approach presented here also includes baseline correlations and negative correlations below the baseline, appearing in dissociation events rather than coincidence events. Jungian synchronicities may be regarded as special cases of induced positive psychophysical correlations with large deviations above the baseline.

Occasionally, Jung also characterized out-of-body experiences as synchronicities (Jung 1952, pars. 949–955). This expands his understanding of synchronistic events from positive deviations from baseline correlations to deviations in general, including negative ones – which will be addressed in more detail in the following subsection.

3.4 Exceptional Human Experiences

The rich material of exceptional psychophysiological correlations comprehensively reviewed by Kelly (2007) suggests various concrete types of psychophysical correlations deviating from the correlation baseline. Moreover, a recent statistical analysis of a large body of documented cases of extraordinary human experiences, also called exceptional experiences (Fach 2011, Belz and Fach 2012) provides significant evidence that the Pauli-Jung conjecture matches with existing empirical material surprisingly well. For more details see Atmanspacher and Fach (2013).

Particularly relevant with respect to the discussion of psychophysical correlations are exceptional experiences which refer to the way in which mental and physical states are merged or separated, connected or disconnected, above or below ordinary baseline correlations. In coincidence phenomena, ordinarily disconnected elements of self and world, inside and outside, appear connected; in dissociation phenomena, ordinarily connected elements of self and world appear disconnected.

1. *Coincidence phenomena* refer to experiences of positive psychophysical correlations above the persistent ordinary baseline. Typically, these correlations are experienced as acausal meaningful links between mental and material events, e.g. meaningful coincidences such as Jungian "synchronicities". Spatiotemporal restrictions may appear as inefficacious, as in several kinds of "extrasensory perception".

2. *Dissociation phenomena* refer to experiences of negative psychophysical correlations below the persistent ordinary baseline. For instance, subjects are not in full control of their bodies, or experience autonomous behavior not deliberately set into action. Out-of-body experiences, sleep paralysis and various forms of automatized behavior are among the most frequent phenomena in this class.

In order to assess whether and how these classes are empirically relevant, they have been compared with empirical data from the counseling department of the Institute for Frontier Areas of Psychology (IGPP) at Freiburg (Germany) since 1996. For details of the documentation system and the statistical analyses see Bauer *et al.* (2012). It is important to note that the patterns obtained by statistical factor analyses reflect the subjective views of the clients about their experiences – not their veridicality. The collected data yield an exclusively phenomenological classification scheme, not a system for clinical diagnosis. [13]

It turned out that coincidence and dissociation phenomena represent key patterns in the documented material from IGPP clients.[14] An additional study, together with the Psychiatric University Hospital Zurich, based on subjects from ordinary population (rather than advice-seeking clients) was recently published (Fach *et al.* 2013). As expected, the average intensity of their reported experiences is rated significantly lower than for IGPP clients. However, the patterns extracted from the ordinary population sample as well as their relative frequencies are in good agreement with the IGPP sample.

Exceptional experiences are typically difficult to communicate in conventional language. This often leads to paradoxical formulations (Bagger 2007)

---

[13]Such systems are the "Diagnostic and Statistical Manual of Mental Disorders" (DSM) of the American Psychiatric Association or the "International Classification of Diseases" (ICD) of the World Health Organization. While both DSM and ICD are continually developed based on more or less heuristic criteria, the classification scheme used by Belz and Fach (2012) can be systematically derived from the basic structure of a dual-aspect picture.

[14]The full spectrum of exceptional experiences reported by those clients contains internal and external phenomena in addition to coincidence and dissociation phenomena. See also Fach, this volume.

or metaphorical descriptions in which (Boolean) categories are used to cir-
cumscribe the experience. One way to do this amounts to projections onto
physical or mental phenomena, e.g. experiences of joy, bliss and lucidity are
then referred to as experiences of "inner light". Repeating a point made earlier, this should be understood as a genuinely *psychophysical phenomenon*
– neither a physical (electromagnetic) field within the body, nor a mental
image of light.

As discussed before, it is crucial for such experiences to be experiences of
meaning. Insofar as explicit (or explicated) meaning is a two-place relation
between a representation and what it represents,[15] psychophysical phenomena might be conceived as meaningful *relations* between the physical and the
psychological. Dual-aspect monism suggests that this relation of meaningfulness arises due to the epistemic split of the psychophysically neutral *unus
mundus*: without this split there would be no mental and physical referents
which could be related by meaning.

Alternative to an explicitly relational view, it might also be possible to
understand the experience of meaning *implicitly*, not as a relation between
distinguishable entities. Such experiences transcend the realm of Boolean
categories and could be examples for the refinement indicated by relative
onticity in Sec. 2. Elements of the psychophysically neutral reality could
be apprehensible without a mind-matter distinction, thus relaxing Jung's
neo-Kantian conviction that elements of the unconscious are immutably inaccessible in themselves.

More systematically speaking, archetypal activity could be the carrier
of that implicit meaning which can be explicated in terms of meaningful
psychophysical phenomena. This adds a further kind of "meaningfulness"
to formal and experienced meaning as addressed above. As Aziz (1990)
indicated and Main (2004, Chap. 2) demonstrated, Jung referred extensively
to such a kind of "objective" meaning in his synchronicity essay (Jung 1952).
Dual-aspect monism provides places for all these kinds of meaning, from
the purely formal notion of intentionality to the metaphysical dimension of
archetypal activity itself.

In this spirit, a key difference between the experience of archetypal activity and psychophysical phenomena would be the difference between implicit
and explicated meaning. Maybe Pauli's understanding of the "reality of the
symbol" (in Jung's sense) comes close to the notion of implicit, not yet ex-

---

[15]On Metzinger's account, this relation would be (meta-)represented by a PMIR. This
is logically consistent as long as PMIRs are neither ascribed as belonging to the physical
nor to the mental – possibly a problematic point in Metzinger's approach.

plicated meaning (letter of Pauli to Fierz of August 12, 1948, von Meyenn 1993, p. 559):

> When the layman says "reality", he usually thinks that he is talking about something evident and well-known; by contrast it seems to me that it is the most important and exceedingly difficult task of our time to work out a new idea of reality. ... What I have in mind concerning such a new idea of reality is – in provisional terms – the idea of the reality of the symbol.

4   Conclusions

The conceptual framework of dual-aspect monism according to Pauli and Jung stipulates that phenomena based on psychophysical correlations are misconstrued if they are described physically (plus some mental context) or mentally (plus some physical context). It is suggested that genuinely psychophysical phenomena are more properly regarded as *relations* between the physical and the mental rather than *entities* in the physical or mental realm. This challenging idea elucidates why meaning is so essential for psychophysical phenomena – either as an explicitly relational concept or an implicitly holistic experience.

In a recent commentary, Tresan (2013) expressed the intuition that the theory of complex systems, which has been widely applied to the description of synchronicities and their archetypal origin, still relies "on dependency (neither strictly reductive nor random, but nonetheless still causal, albeit diluted)". By contrast, the more radical vision of the Pauli-Jung conjecture hits the core of psychophysical phenomena: a holism in which *wholes do not consist of parts* to begin with. Elements of the theory of complex dynamical systems, such as networks of attractors (archetypes) and their basins of attraction (complexes) can be useful descriptive tools within epistemic contexts (cf. Cambray 2009). However, Pauli's and Jung's daring ideas in their full scope may persuade us to believe that the repertoire of complex dynamical systems is not deep enough.

Unlike numerous neuroscientists and philosophers of mind seem to assume – including essential elements of Metzinger's position – the Pauli-Jung conjecture implies that brain science alone will be unable to unveil the mysteries of psychophysical phenomena, neither in the "decade of the brain" nor in decades to come. What is needed is a new idea of reality, implying novel and refined metaphysical structures. If we can make progress on this route, it will provide us, and our culture, with a satisfactory and beneficial worldview – a key element of Jungian psychology besides its therapeutic values.

References

Aerts D., Durt T., Grib A., Van Bogaert B., and Zapatrin A. (1993): Quantum structures in macroscopical reality. *International Journal of Theoretical Physics* **32**, 489-498.

Atmanspacher, H. (2012): Dual-aspect monism à la Pauli and Jung. *Jornal of Consciousness Studies* **19**(9/10), 96–120.

Atmanspacher, H., and Fach, W. (2013): A structural phenomenological typology of mind-matter correlations. *Journal of Analytical Psychology* **58**, 219–244.

Atmanspacher, H., and Kronz, F. (1999): Relative onticity. In *On Quanta, Mind and Matter*, ed. by H. Atmanspacher, A. Amann, and U. Müller-Herold, Kluwer, Dordrecht, pp. 273–294.

Atmanspacher, H., and Römer, H. (2012): Order effects in sequential measurements of non-commuting psychological observables. *Journal of Mathematical Psychology* **56**, 274–280 (2012).

Atmanspacher, H., Römer, H., and Walach, H. (2002). Weak quantum theory: Complementarity and entanglement in physics and beyond. *Foundations of Physics* **32**, 379-406.

Aziz R. (1990): *C.G. Jung's Psychology of Religion and Synchronicity*, University of New York Press, Albany.

Bagger, M. (2007): *The Uses of Paradox*. Columbia University Press, New York.

Bauer, E., Belz, M., Fach, W., Fangmeier, R., Schupp-Ihle, C. and Wiedemer, A. (2012): Counseling at the IGPP – An overview. In *Perspectives of Clinical Parapsychology*, ed. by W.H. Kramer, E. Bauer, and G.H. Hövelmann, Stichting Het Johan Borgman Fonds, Bunnik, pp. 149–167.

Belz, M. and Fach, W. (2012): Theoretical reflections on counseling and therapy for individuals reporting ExE [exceptional experiences]. In *Perspectives of Clinical Parapsychology*, ed. by W.H. Kramer, E. Bauer, and G.H. Hövelmann, Stichting Het Johan Borgman Fonds, Bunnik, pp. 168–189.

Bohr, N. (1934): *Atomic Theory and the Description of Nature*, Cambridge University Press, Cambridge, p.5.

Brentano, F. (1874): *Psychologie vom empirischen Standpunkt*, Duncker & Humblot, Leipzig.

Busemeyer, J.R., and Bruza, P.D. (2012): *Quantum Models of Cognition and Decision*, Cambridge University Press, Cambridge.

Cambray, J. (2009): *Synchronicity. Nature and Psyche in an Interconnected Universe*, Texas A&M University Press, College Station.

Fach, W. (2011): Phenomenological aspects of complementarity and entanglement in exceptional human experiences. *Axiomathes* **21**, 233–247.

Fach, W., Atmanspacher, H., Landolt, K., Wyss, T., and Rössler, W. (2013): A comparative study of exceptional experiences of clients seeking advice and of subjects in an ordinary population. *Frontiers in Psychology* **4**:65, 1–10.

Fell, J. (2004): Identifying neural correlates of consciousness: The state space approach. *Consciousness and Cognition* **13**, 709–729.

Garfinkel, A. (1981): *Forms of Explanation*, Yale University Press, New Haven.

Heisenberg, W. (1971): *Physics and Beyond. Encounters and Conversations.* Harper and Row, New York. German original: *Der Teil und das Ganze*, Piper, München 1969, p.64.

Jung C.G. (1952): Synchronicity: An acausal connecting principle. In *The Structure and Dynamics of the Psyche, Collected Works Vol. 8*, Princeton University Press, Princeton 1969.

Jung C.G. (1954): On the nature of the psyche. In *The Structure and Dynamics of the Psyche, Collected Works Vol. 8*, Princeton University Press, Princeton 1969.

Jung, C.G., and Pauli, W. (1952): *Naturerklärung und Psyche*, Rascher, Zürich. English: *The Interpretation of Nature and the Psyche*, Routledge and Kegan Paul, London 1955.

Kaminski J.A., Sloutsky, and Heckler A.F. (2008): The advantage of abstract examples in learning math. *Science* **320**, 454–455.

Kapur, S. (2003): Psychosis as a state of aberrant salience: A framework linking biology, phenomenology, and pharmacology in schizophrenia. *American Journal of Psychiatry* **160**, 13–23.

Kelly, E.W. (2007): Psychophysiological influence. In *Irreducible Mind*, ed. by E.F. Kelly, E.W. Kelly, *et al.*, Rowman and Littlefield, Lanham, pp. 117–239.

Khrennikov, A. (2010): *Ubiquitous Quantum Structure*, Springer, Berlin.

Kime, P. (2013): Regulating the psyche: The essential contribution of Kant. *International Journal of Jungian Studies* **5**(1), 44–63.

Main R. (2004): *The Rupture of Time: Synchronicity and Jung's Critique of Modern Western Culture*, Brunner-Routledge, New York.

Meier, C.A. (1975): *Psychosomatik in Jungscher Sicht*. In *Experiment und Symbol*, ed. by C.A. Meier, Walter, Olten, pp. 138–156.

Meier, C.A., ed. (1992): *Wolfgang Pauli und C.G. Jung. Ein Briefwechsel 1932–1958*, Springer, Berlin. English: *Atom and Archetype. The Pauli/Jung Letters 1932–1958*, Princeton University Press, Princeton 2001.

Metzinger, T. (2003): *Being No One*, MIT Press, Cambridge.

Putnam, H. (1981): *Reason, Truth, and History*, Cambridge University Press, Cambridge.

Putnam, H. (1987): *The Many Faces of Realism*, Open Court, La Salle.

Quine, W.V.O. (1969): Ontological relativity, in *Ontological Relativity and Other Essays*, Columbia University Press, New York, pp. 26–68.

Seager, W. (2010): Panpsychism, aggregation and combinatorial infusion. *Mind and Matter* **8**, 167–184.

Tresan, D. (2013): A commentary on 'A structural phenomenological typology of mind-matter correlations' by H. Atmanspacher and W. Fach. *Journal of Analytical Psychology* 58, 245–253.

van Fraassen, B. (1980): *The Scientific Image*, Clarendon, Oxford.

van Gelder, T. (1998): The dynamical hypothesis in cognitive science. *Behavioral and Brain Sciences* **21**, 615–661.

van Os, J. (2009): A salience dysregulation syndrome. *British Journal of Psychiatry* **194**, 101–103.

von Meyenn, K. (ed.) (1985): *Wolfgang Pauli. Wissenschaftlicher Briefwechsel, Band III: 1930–1939*, Springer, Berlin.

von Meyenn, K. (ed.) (1993): *Wolfgang Pauli. Wissenschaftlicher Briefwechsel, Band III: 1940–1949*, Springer, Berlin.

von Meyenn, K. (ed.) (1996): *Wolfgang Pauli. Wissenschaftlicher Briefwechsel, Band IV/1: 1950–1952*, Springer, Berlin.

von Meyenn, K. (ed.) (1999): *Wolfgang Pauli. Wissenschaftlicher Briefwechsel, Band IV/2: 1953–1954*, Springer, Berlin.

Are Synchronicities Really Dragon Kings?

George Hogenson

Abstract

Discussions of synchronicity tend to focus either on the meaningful content of the experience, or on speculation about possible mechanisms underlying the phenomena. The present paper suggests that the symbolic or meaningful content of some synchronistic phenomena are themselves governed by identifiable dynamics associated with the emergence of symbol systems generally. Specifically, these dynamics are associated with complex dynamical systems theory and give rise to phenomena governed by power laws such as Zipf's law. It is suggested that synchronicities, which display distinctly symbolic features, behave in ways that conform to power-law distributions in which highly coupled systems form rare outlier aggregations referred to as "dragon kings." This terminology is explained and related to the experience of synchronistic phenomena.

1   The Scarab Beetle

Jung's paradigmatic example of a synchronistic event, the famous example of the scarab beetle, is well known to anyone interested in Jung's theory of synchronicity. While Jung provided a host of other examples of phenomena that he wanted to include within the framework of synchronistic events, this case remains particularly salient and instructive. In 2008, at the *Journal of Analytical Psychology* conference in Italy, following the work of both Main (2004, 2007) and Bishop (2000), I argued that the entire situation Jung describes in this case required much more attention to the role of the scarab beetle in Jung's own iconography and symbolic lexicon. Main, in particular, has commented on the various instances where scarabs are remarked on by Jung, or otherwise appear in materials that interested him, particularly in the alchemical tradition. With the publication of the *Red Book* we can push the significance of the scarab even further in its relationship to Jung's symbolic world, given the appearance of several scarabs both in the text and in his paintings.

In all, I suggested at that time that the occasion of the scarab beetle dream was in all likelihood a major transference/counter-transference event which may well have been of greater significance to Jung than it was to the dreaming patient. While Jung characterized the event as if the scarab dream was uniquely his patient's experience of a moment of rebirth, I believe we must really take the situation more as a transferential response on the part of the patient to Jung's own symbolic preoccupations. The sudden movement in the treatment may then be seen as much a counter-transferential response on Jung's part as a breakthrough on the part of the patient.

I will therefore argue in what follows that the scarab synchronicity is an instance of complex dynamics involving intense symmetric coupling within the patient-analyst system. This dynamics gave rise to a system-wide amplification of the symbolic interactions associated with the dream, resulting in a global reorganization, or transformation, of the system as a whole. This is a form of emergent phenomenon that has been called a "dragon king." I will attempt to make good on this proposition in what follows.

On the other hand, in my 2008 paper I also proposed that, because of the complexity of this case, the symbolism of the scarab and the response to the dream, as well as the fortuitous appearance of the chaffer beetle, created a moment of "symbolic density".[1] One objective in this paper is to flesh out the concept of symbolic density and show how it rests on certain features of symbolic processes in general that add to our understanding of synchronicity while at the same time removing some of its more perplexing qualities.

Let me also say that I do not want to represent this work as my last word on synchronicity, given the variety of forms it takes in Jung's writings on the subject. My focus will be on those instances of synchronicity that are most explicitly associated with the occurrence or generation of symbolic meaning in the present moment, leaving aside precognitive and other phenomena Jung broadly associated with synchronicity.

2 Symbolic Density

Let me begin with a more general point of view on Jung regarding the importance of his early work on the word association test and on the linguistic

---

[1] I first presented this concept at the Congress of the *International Association for Analytical Psychology* at Barcelona in 2004, within a larger context of discussing what has come to be called the emergence model of the archetype. In that presentation I outlined a body of research on the nature of symbolic systems that cast a light on the structure of archetypal symbolism including those rare events we call synchronicities.

patterns of *dementia praecox*. Jung's work in these areas is, in my experience, too easily overlooked in preference for the later materials on archetypes, typology, and the alchemical writings. It is important for our purposes in these two areas that the word association test revealed important foundational elements of psychological functioning in the phenomenology of associative networks, particularly the affective content of these networks.

As Spitzer (1992) argues in his historical review of the word association experiment, Jung significantly enlarged the scope of the test and, more importantly for psychiatry, used it to argue for the deep coherence of psychotic discourse, thereby influencing Bleuler's model of schizophrenia. Jung's understanding of schizophrenic discourse emphasized the presence of relatively coherent associative paths connecting the manifest elements of the discourse, but large portions of that connective network were obscured. This point of view on psychotic discourse has been investigated more recently by Rebeiro (1994) among others, who mapped the subterranean – or unconscious – networks of association that exist in the often seemingly disjointed ramblings of severely psychotic patients.

The essential element in this early work of Jung is the centrality of associative relationships among elements that we can reasonably call symbols, or at least symbolically significant markers in the individual's psychic world. For an example, I recommend reading Jung's masterful unpacking of the symbol of the linden tree in his study of *dementia praecox*. What I now want to do is begin to tie Jung's work on symbolic networks to a larger body of research on the nature of the symbolic, beginning with the American pragmatist and founder of semiotics, Charles Sanders Peirce (1839–1914). Peirce divided sign systems into three essential levels, the icon, the index and the symbol. Deacon (1997), in his essential study *The Symbolic Species*, draws directly on Peirce's work as part of his discussion of the emergent nature of truly symbolic systems.

We do not need linger on icons, as they are relatively direct representations of the object of concern, such as a portrait of Louis XIV or a photograph of Jung. Where matters become interesting is in the move to the indexical level of signification. At this level, a sign begins to aggregate instances into larger sets such as monarchs or famous psychologists. There remains, however, a known, and essentially unquestioned referential relationship between the sign stimuli and the object or action in question.

Deacon identifies a critical transitional stage to a more complex arrangement of tokens, where the tokens, still largely indexical in relation to their objects, begin to arrange themselves in patterns of token-to-token combination. Importantly, these indexical combinations of token interactions still do not relinquish their referential relationship to objects or actions – monarchs who exercized great power throughout Europe.

As we move to the fully symbolic level, the relationships that matter shift to those between the tokens themselves, and only by abstraction to their physical or pragmatic referents. At this point one could thematize the nature of kingship, or, perhaps, the archetype of the king. Without elaborating on this in detail, I would suggest that Deacon's approach is a useful illustration of what Jung was beginning to observe in the word association test, and in his work with *dementia praecox* and the discourse of the psychotics. These observations formed the foundation for what would eventually be his theory of archetypes.

Another brief comment to set the stage for what is perhaps the most difficult part of my argument. We are all aware of the importance of Jung's close relationship with Wolfgang Pauli and their extended discussion of synchronicity in relation to quantum physics. Pauli was, in many ways, the only interlocutor to whom Jung paid actual attention and even deference. Pauli reciprocated with an extraordinary level of engagement in topics that he was aware could easily marginalize him in some scientific circles.

That said, when we read their correspondence, it is clear that one of Pauli's greatest challenges in the exchange was getting Jung to understand the statistical nature of quantum mechanics. I raise this historical bit of the discussion of synchronicity because, with the advent of quantum mechanics, much of theoretical physics became statistical in nature, and that process has continued to the present. The aspect of this development that concerns us here is the application of the statistical methods developed by physicists, as well as some other disciplines such as economics, to domains beyond their normal purview, including the study of language and the nature of symbols. One way in which my own argument over the last several years might be framed is that Jung, as well as Pauli, with some of this research at hand, might have arrived at a very different understanding of archetypes and synchronicity.[2]

3   Zipf's Law

Had he known of them, Jung might have appreciated a set of mathematical formulations that have the important quality of describing a wide variety of phenomena with no intrinsic connection to one another. There is, if you

---

[2]This was entirely the point of my paper at the 2004 Congress at Barcelona, which I will now try to outline as clearly as I can.

will, something transcendental about them, in the philosophical sense of being universal conditions of the world, without reference to specific states of affairs. The ones I am particularly interested in, in relation to the symbol, are power-law distributions, particularly Zipf's law, but also including the fractal geometry of Mandelbrot (1981, 1997) and scale-free structures of networks. Let me note that all of these patterns involve, amongst other characteristics, a relationship to scaling phenomena, or what is usually referred to as scale invariance. This means that these patterns apply regardless of the scale at which the phenomena are analyzed.

Zipf's law is named after the American linguist George Kingsley Zipf (1902–1950). He was something of a polymath or a dilatant, depending on your point of view, who began by examining variations in the size of cities. He discovered that, within a given geographic area, the size of population concentrations, from small villages to large cities is governed by a so-called power-law distribution. In the case of cities the abundance (or frequency) of agglomerations of size s followed a deceptively simple $1/s$ distribution.

The outcome of this calculation looks like a graph schematically depicted in Figure 1a. However, when the results are converted to a doubly logarithmic graph (log-log plot), the chart looks like Fig. 1b, which exhibits the characteristic linearity of a power-law distribution. Zipf's next step, and the result for which he is remembered, was to examine the frequency of words in a text. He found that the frequency of words in a text, ranked according to their abundance, fell like $1/r$ as a function of their rank r. In a log-log plot, the slope of the resulting linear function is then minus one.

In addition to his work on the statistics of word frequency, Zipf proposed a model for the generation of lexicons, or symbol systems, that he referred to as the principle of least effort. Briefly, the idea here was simply that both the listener and the speaker in an exchange of signs would seek to minimize their expenditure of energy – that is, put in the least effort (Zipf 1949). This

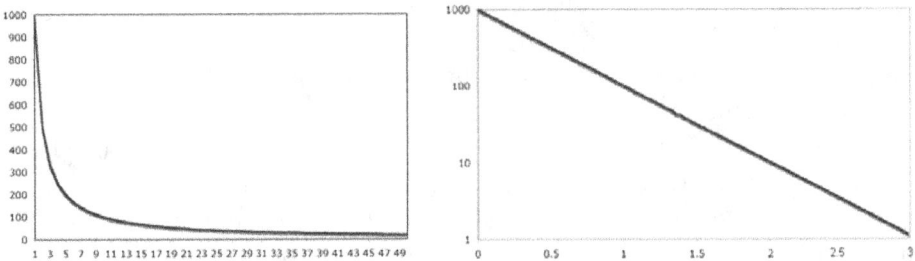

Figure 1: Schematic representations of a power-law distribution, plotted on a linear scale (1a) and on a doubly logarithmic scale (1b).

means that a kind of negotiation would take place between the parties of an exchange, in which each sought the greatest level of understanding for the least effort.

Needless to say, the simplest way to achieve this goal is to have a shared lexicon of exact one-to-one relations between the elements in the lexicon and the objects referenced by the lexicon. However, this approach entails massive memory requirements to insure the least ambiguity. It is the way in which most animals, other than humans, communicate. The monkey cry that designates the presence of a snake is distinct from the cry that designates an eagle. But while some animals can learn fairly large lexicons in captivity, and under well controlled conditions, we also know that in the wild the upper bound for say the bonobo chimpanzee, perhaps the most cognitively advanced primate short of humans, is on the order of about 40 "words", with little or no syntax. These lexicons are essentially indexical rather than symbolic, in the terms used by Peirce and Deacon.

Explicitly drawing on Deacon's and Peirce's distinction between symbols and indexes, Ferrer-i-Cancho and Solé (2003) simulated the development of a lexicon beyond the indexical level and concluded that Zipf's law was not simply a descriptive tool. Rather, it was actually a necessary emergent property of symbolic systems which, they also demonstrated, exist in what is known as a phase transition – a condition such as what happens as water turns to steam or freezes into ice.

However, the symbolic phase space in this instance has the added feature that the symbolic system proper remains in the phase space and does not resolve either into indexicals or into meaningless randomness. This feature, which entails a significant degree of referential ambiguity, was, they speculate, a likely contributing factor in the evolution of language, because it allows a limited lexicon to refer to a larger set of objects. In the presentation of their findings one can see how a phase transition emerges where the effort of the speaker becomes roughly equal to the effort of the listener (cf. Fig. 2).

The model of Ferrer-i-Cancho and Solé (2003) created a very abstract and idealized understanding of a language or symbol system. They have gone further in other papers, to examine the emergence of syntax and also to argue that semantic content may follow Zipf's law as well. This is a more controversial claim, but it has received some support from other researchers. For instance, Vogt (2004) adds a dimension to this discussion in that he enlarges the set of possible symbolic structures by examining the ways in which referential tokens can be aggregated into categories. Once again, the principle of least effort is at work, but the objective is to locate the category that best discriminates one reference from others.

Vogt refers to the conceptual structures of symbolic systems in terms of their density: symbolic density. He argues that in a search for appropriate categorical structures, the principle of least effort will motivate movement through a hierarchy of increasingly dense categories. Furthermore, this hierarchy of category density can be subsumed under a Zipf-Mandelbrot power law. This is exactly my argument regarding Jung's system; the complex, the archetype, synchronicity and even the notion of the Self are scale-invariant symbolic structures of increasing density that should, by virtue of their symbolic nature, as well as the curious fact of the scope of power-law like phenomena, fall under Zipf's law.

A question can be raised at this point, however. If synchronicities are simply at the high end of a power law, why do they carry the level of meaning and affective impact that they typically do? A large earthquake, after all, is, as Sornette (2003) has remarked, simply a small earthquake that keeps going. But, if the same principle applies to synchronicities, why do you have the experience of a "rupture of time" (Main 2004)?

Ironically, part of the answer is already available in the work of Ferrer-i-Cancho and Solé (2003), in their discussion of the emergence of language. As was mentioned, this process consists in the formation of a phase transition in which an entirely new and distinct regime emerges as symbols overwhelm the earlier simple indexical reference. This transition is, in no small measure, a catastrophe, in the technical sense of the word – and perhaps in practice as

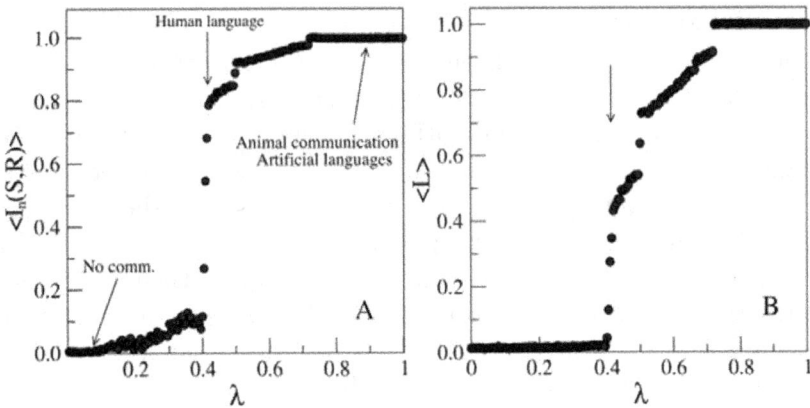

Figure 2: (A) $< I_n(S,R) >$, a measure for the accuracy of communication, as a function of λ, a parameter that weighs the effort for speaker and listener. (B) Average lexicon size $< L >$ as a function of λ. Both functions show a phase transition at $\lambda \approx 0.41$. For more details see Ferrer-i-Cancho and Solé (2003).

well, insofar as it likely catapulted the *genus homo* into an entirely different life-world to the general detriment of other organisms.

4 Dragon Kings

The term "dragon king" was originally coined by Didier Sornette at ETH Zurich, where he studies financial bubbles (Sornette 2003). Sornette began as an earthquake researcher, which involved him in the study of power laws, particularly Zipf's law and its variant, the Zipf-Mandelbrot law. While examining other systems that fail catastrophically, most notably the global financial system, he recognized the emergence of a separate class of events that should fall within the classic Zipf power-law distribution, but for some reason did not. One example, which uses Zipf's original research topic of city size, is the existence in some countries of one exceptional outlier city, such as Paris in France.

Sornette, in a series of papers, gives many examples of dragon-king phenomena, but they are, as I will discuss in a moment, united by one dynamic process. It is at this point that I believe we may begin to address the question of synchronicities as psycho-symbolic dragon kings. Sornette (2009) remarks that he chose the term dragon in part because dragons are rather mystical and mysterious creatures, much like a synchronicity. The notion of the king, on the other hand, refers to the characteristic vast wealth of kings, a feature that we might want to associate with the meaning-ladenness of a synchronistic experience.

So how do we address synchronicities in this context? In Sornette's discussion of the emergence of dragon-king phenomena, one critical element is the degree of coupled interactions within a synchronized system, and the impact degrees of coupling have on amplifying the characteristics of the system. In the case of a massive city such as Tokyo, for example, the great Shogun Tokugawa Ieyasu consolidated his power by forcing the other great lords to conduct elaborate progressions between Tokyo and their own fiefdoms, as well as maintain courts in both places. This arrangement also played to the competitiveness the great lords felt toward one another, and led to ever-greater concentration of resources in Tokyo.

As we will see, this became, as was the case in Paris, and to a degree London, a tightly coupled system of amplification with limited damping mechanisms that would regulate the system's development. Compare this to the United States, where a variety of early and explicit political decisions prevented any one city from attaining a similar level of cultural hegemony.

The looser coupling of the amplification and damping process resulted in a distribution of city populations that follows Zipf's law far more closely.

To work this part of the argument out in greater detail, we need to understand a bit more about the role played by amplification within complex systems, partly for its explanatory importance, but also because Jung made amplification, as opposed to reductive analysis, the cornerstone of his method. I want to be clear that I do not think of this similarity in terminology as a simple analogy; rather, it is a matter of substance. Jung's method, in other words, is directly associated with the perspective I am proposing. This takes us into questions related to emergent phenomena, an area to which I want to turn once again before concluding this paper.

My argument (Hogenson 2001) has been that archetypes, and related phenomena such as synchronicities, are emergent phenomena rather than preexisting structures. I even went so far as to argue that "the archetypes of the collective unconscious, as either modular entities in the brain or as neo-Platonic abstractions in some alternative ontological universe, *do not exist*, in the sense that there is no *place* where the archetypes can be said to be" (Hogenson 2001, p. 606). Jung's notion of the archetype-in-itself was, therefore, mistaken, and archetypal phenomena, as emergent, derived from the systemic interaction of brain, culture and narrative.

I have, since then, become even more convinced that when we talk about archetypal images we are dealing with complex dynamic systems within which the symbolic itself plays the most important role. Attempting to imbed the archetypal in evolutionary theory, neuroscience or developmental achievements is to put the cart before the horse. While all of these elements are part of the system taken as a whole, they have, if you will, become subsidiary to the workings of the symbolic environment that human beings inhabit – the environment to which we have become adapted (cf. Hendricks-Jansen 1997).

This means, and here I return to Deacon's use of Peirce's semiotic model, that amplification as a force within the emergent processes of the symbolic world can manifest in extraordinarily complex ways. This gives synchronicity, as a particular manifestation of archetypal emergence within the symbolic world, equally extraordinary scope. As Deacon himself comments in an important paper on emergence, to which I will refer in what follows, "a symbolic species such as Homo sapiens" occupies an ecological niche that is characterized by processes of "symbolic self-organization and by evolutionary processes that are quite different from those at lower levels" (Deacon 2005, p. 149).

5   Synchronicity

To flesh out the significance of this point of view, and more directly connect the symbolic and synchronicity to the ideas of amplification and dragon kings, I need to review, again drawing on Deacon, the fundamentals of emergent structures in complex systems. There are a number of ways in which emergent processes can be carved up and distinguished from one another. In many discussions of emergence, the notion of supervenience[3] plays an important role.

The emergent properties of water are an example. Simply put, the combination of two gases, hydrogen and oxygen, at normal temperatures and pressures form a liquid when a large group of molecules are aggregated. The "liquidness" of an aggregation of water molecules is an emergent, supervenient property of the aggregation. The fact that you need an aggregation of water molecules is important precisely because a single molecule does not possess the qualities of liquidness. Those properties emerge due to interactions among the molecules at the aggregate level. The thermodynamic property of liquidness supervenes on the behavior of the water molecules. But even at this level, some of the potential interactions are amplified by other interactions, while still others are damped. In Deacon's formulation of emergent processes, this combination of amplification and damping plays an important role.

Above the thermodynamic level of emergence, Deacon (2006) argues that we can see processes of what he calls morphodynamic emergence, such as crystal formation in supersaturated solutions. At this level of emergence the thermodynamic properties themselves are being amplified and damped such that new, higher-level structures, such as snowflakes, are formed. Moving still further up the scale of emergent processes, Deacon suggests that, as the morphodynamic processes stabilize into persistent molecular structures capable of self-replication, a new level of emergent amplification is established with such molecules as DNA. They shift the frame of time in the process of emergence by introducing a form of memory.

What has gone before is carried forward but, at the same time, this memory within the system creates a sense of movement toward some future state, and we begin to have what Deacon calls teleodynamic emergence. It is at this level that the symbolic itself emerges and, in turn, becomes part of the

---

[3]Generally speaking, supervenience is usually defined as a relation between two sets of properties A and B. Properties in A supervene on properties in B if and only if no two things can differ with respect to A-properties without also differing with respect to B-properties (cf. McLaughlin and Bennett 2011).

workings of complex systems with their own emergent properties. However, the symbolic in a sense detaches itself from the indexical and establishes the primary structures of symbolic systems within the symbolic domain itself. As a result, the potential for the formation of emergent structures is no longer bounded by reference to objects but rather by associative symbol-to-symbol reference.

Deacon captures this last point by reference to the Taoist metaphor of the empty hub of the spoked wheel, a metaphor that Jung also enlists. This passage from the *Tao-Te-Ching* reads, in Deacon's (2006, p. 119) rendering:

> Thirty spokes converge at the wheel's hub to an empty space that makes it useful. Clay is shaped into a vessel, to take advantage of the emptiness it surrounds. Doors and windows are cut into walls of a room so that it can serve some function. Though we must work with what is there, use comes from what is not there.

Richard Wilhelm, who rendered "empty" as "nothing," introduced Jung to this passage. Jung comments in the essay on synchronicity (Jung 1952, par. 919):

> "Nothing" is evidently "meaning" or "purpose," and it is only called Nothing because it does not appear in the world of the senses, but is only its organizer.

This Nothing, or emptiness, is what in complex systems theory would be called an affordance that provides the space within which emergence takes place. Deacon remarks, in relation to the passage from the *Tao-Te-Ching*, that (Deacon 2006, p. 120):

> The Western mind sees causality primarily in the presence of something, in the pushes and resistance that things offer. Here we are confronted with a different sense of causality, in the form of an "affordance": a specifically constrained range of possibilities, a potential that is created by virtue of something missing.

At this point it is appropriate to refer to Jung's conceptualization of the symbol as the best possible representation of something we do not understand (Jung 1971, par. 816): "The symbol is alive only so long as it is pregnant with meaning". We can play on this definition in the context of Jung's comment on the *Tao-Te-Ching* in his synchronicity essay to argue that the symbol is alive insofar as it references "nothing" or "emptiness," i.e. the affordance necessary for emergence to take place.

We are, however, not quite in a position to connect the scarab beetle incident to the emergence of dragon kings. To make this move I have to introduce a distinction that Sornette draws within the world of self-organizing systems. As emergent phenomena came to be more carefully studied, particularly after the development of chaos and complexity theory, it became evident that some self-organizing processes developed to a point referred to as self-organized criticality (Bak 1996). Bak compared the formation of self-organized criticality to the avalanches in a child's sand pile, in which small avalanches form as the pile grows, but at some point a critical level of sand will form a much larger avalanche – the pile has reached a point of self-organized criticality.

Importantly, the sand pile also illustrates the significance of loose coupling in the system. The large outlier avalanche in this example would fall near the top of a power-law distribution, within the frame defined by Zipf's law. It would be an unusual event, but not a dragon king. Researchers in complexity theory have come to refer to these events as black swans. To use the notion of dragon kings as a way of thinking about synchronicity, we need to understand their relationship to black swans. I do now believe that I was wrong in previous work (Hogenson 2004) to argue that synchronicities lay directly on the path of a power-law distribution.

Archetypes may be black swans in the symbolic domain, but, if Sornette's argument is correct, it may be better to see synchronicities as related, but different at critical points. It is here that some of what I said at the beginning

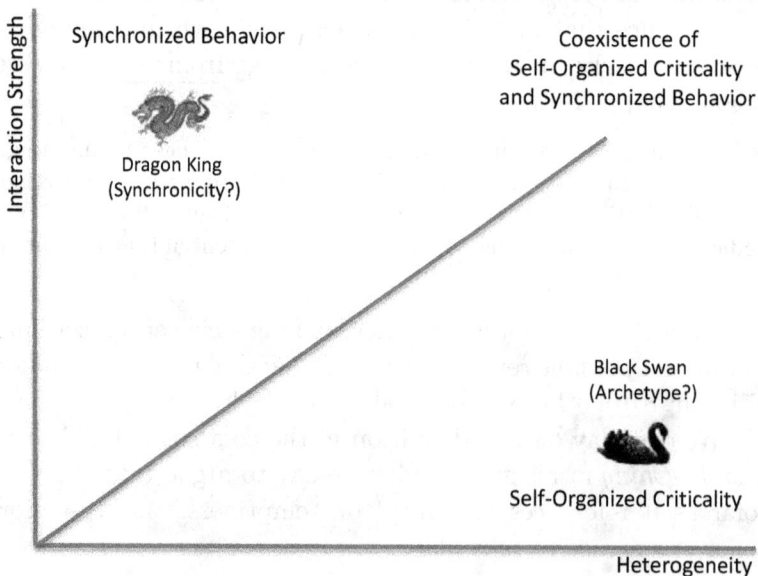

Figure 3: Schematic representation of dragon kings and black swans in complex systems of more or less interacting, more or less heterogeneous elements.

about the scarab beetle incident becomes important. To flesh this out let us look at the diagram in Fig. 3 after Sornette (2009).[4]

The horizontal axis of the graph in Fig. 3 is a measure of the heterogeneity of interacting elements in a self-organizing system, and the vertical axis measures the interaction strength of the elements of the system, their coupling. What appears to be the case in some systems – high-speed computer-driven stock market trading is one important example – is that the system develops into an almost entirely homogenous structure, where the parts are very tightly coupled. The collapse of the system takes the form of a dragon king. The system is basically driven to an extreme position. The black swan, on the other hand, forms in a more heterogeneous environment, with looser coupling, rather like the child's sand pile.

6 Conclusions

Let me finally return to the remarks I made about Jung's interest in scarabs, as well as the patient's dream and the advent of the chaffer. The scarab beetle event, in essence, resembles a tightly coupled, homogeneous system that is about to undergo a process of catastrophic self-organization: a synchronicity. As Sornette (2009, p. 11) remarks regarding the emergence of dragon kings: "The key idea is that catastrophic events involve interactions between structures at many different scales that lead to the emergence of transitions between collective regimes of organization." These interactive processes, as Sornette and Ouillon (2012) explain, create amplifying feedback loops in the aggregate system that push the system into a global phase transition.

In the case of the scarab, we have Jung and his patient evidently under considerable pressure – she is very rational, the work is not progressing, they are stuck. In this situation, the dream introduces a symbolism that captures not only the patient, but, importantly, Jung as well or even in particular. With the appearance of the chaffer, the aggregate system of Jung-patient-symbolism produces just such an undamped tightly coupled amplifying feedback loop that pushes the systems into a global phase transition. As Jung remarks, this rearranges the entire structure of the analysis. This is the process, I would suggest, that makes at least this synchronistic event, but perhaps others as well, into a dragon king.

---

[4]For clarity in the present context, I have rendered the diagram in a simplified form. The original may be found in the article by Sornette (2009).

References

Bak P. (1996): *How Nature Works: The Science of Self-Organized Criticality*, Copernicus, New York.

Bishop P. (2000): *Synchronicity and Intellectual Intuition in Kant, Swedenborg, and Jung*, Edwin Mellon Press, Lampeter.

Deacon T. W. (1997): *The Symbolic Species: The Co-Evolution of Language and the Brain*, W.W. Norton, New York.

Deacon T. (2006): Emergence: The hole at the wheel's hub. In *The Re-Emergence of Emergence*, ed. by P. Clayton and P. Davies, Oxford University Press, Oxford, pp. 111–150.

Ferrer-i-Cancho R., and Solé R.V. (2003): Least effort and the origins of scaling in human language. *Proceedings of the National Academy of Sciences of the USA* **100**(3), 788–791.

Hendricks-Jansen H. (1997): The epistemology of autism: Making a case for am embodied, dynamic, and historical explanation. *Cybernetics and Systems* **28**(5), 359–415.

Hogenson G.B. (2001): The Baldwin effect: A neglected influence on C.G. Jung's evolutionary thinking. *Journal of Analytical Psychology*, **46**, 591–611.

Hogenson G.B. (2004): Archetypes: Emergence and the psyche's deep structure. In *Analytical Psychology: Contemporary Perspectives in Jungian Analysis*, ed. by J. Cambray and L. Carter, Brunner-Routledge, Hove, pp. 32–55.

Jung C.G. (1971): *Psychological Types or the Psychology of Individuation*, ed. by H.G. Baynes, Harcourt Brace, New York.

Jung C.G. (1952): Synchronicity: An acausal connecting principle. In *Collected Works Vol. 8*, transl. by R.F.C. Hull, Princeton University Press, Princeton, pp. 419–519.

Main R. (2004): *The Rupture of Time: Synchronicity and Jung's Critique of Modern Western Culture*, Brunner-Routledge, Hove.

Main R. (2007): Ruptured time and the reenchantment of modernity. In *Who Owns Jung?*, ed. by A. Casement, Karnac, London, p. 19–38.

Mandelbrot B. (1981): Scalebound or scaling: A useful distinction in the visual arts and in the natural sciences. *Leonardo* **14**, 45–47.

Mandelbrot B. (1997): *Fractals and Scaling in Finance: Discontinuity, Concentration, Risk*, Springer, New York.

McLaughlin B. and Bennett K. (2011): Supervenience. In *Stanford Encyclopedia of Philosophy*, ed. by E.N. Zalta. This article is accessible at `plato.stanford.edu/entries/supervenience/`.

Rebeiro B.T. (1994): *Coherence in Psychotic Discourse*, Oxford University Press, Oxford.

Sornette D. and Ouillon G. (2012): Dragon kings: Mechanisms, statistical methods and empirical evidence. *European Physical Journal: Special Topics* **205**, 1–26.

Sornette D. (2009): Dragon kings, black swans and the prediction of crisis. *International journal of Terraspace Science and Engineering* **2**(1), 1–18.

Sornette D. (2003): *Why Stock Markets Crash: Critical Events in Complex Financial Systems*, Princeton University Press, Princeton.

Spitzer M. (1992): Word-associations in experimental psychiatry: A historical perspective. In *Phenomenology, Language and Schizophrenia*, ed. by M. Spitzer, F.A. Uehleinand, M.A. Schwartz and C. Mundt, Springer, New York, pp. 160–196.

Vogt P. (2004): Minimum cost and the emergence of the Zipf-Mandelbrot law. In *Artifical Life IX*, ed. by J. Pollack, M. Bedau, P. Husbands, T. Ikegami, and R.A. Watson, MIT Press, Cambridge, pp. 214–319.

Zipf G.K. (1949): *Human Behavior and the Principle of Least Effort*, Addison-Wesley, Cambridge.

Synchronicity and the Problem of Meaning in Science

Roderick Main

Abstract

This chapter examines C.G. Jung's attempt with his concept of synchronicity to address the seemingly negative impact of scientific rationalism on experiences of meaning in the modern world. Through close reading of Jung's texts and a critical re-evaluation of a recent debate on the nature of meaning in synchronicity, the first part of the chapter establishes that for Jung such meaning is multileveled – ranging from "quasi-linguistic" to "cosmic" – and is underpinned at all levels by the notion of the archetype as psychoid. Then, referring mainly to documents from Jung's collaboration with Wolfgang Pauli, the second part of the chapter draws out the radical implication that, as evinced by synchronistic events, such multileveled, archetypal meaning can imbue the physical as well as the psychological world. However, being able to access this dual psychic and physical meaning requires both a deep, transformative engagement with the unconscious and an ability to relate unconscious contents to contemporary knowledge.

1 The Impact of Scientific Rationalism on Experiences of Meaning

In his essay "Science as a Vocation", Max Weber (1918) famously characterized the modern world as disenchanted. By this he meant that as a result of the process of increasing intellectualization and rationalization, epitomized and driven by science, it was now, in contrast to pre-modern times, possible to order the natural and social worlds in "purely practical and technical" terms without recourse to "mysterious incalculable forces", "magical means", and the imploring of "spirits" (Weber 1918, p. 139).

One important concomitant of this process of rationalization and disenchantment, as the Canadian philosopher and social theorist Charles Taylor has elaborated, is a transformation of the way in which meaning is experienced (Taylor 2007, pp. 29–35). In the enchanted, pre-modern world, according to Taylor, meaning was experienced as residing not only in human minds but also in non-human subjects and in things (pp. 31-33). In this world, objects could be "charged", "magical", they could "impose meanings and bring about physical outcomes proportionate to their meanings", they had "influence and causal power" (p. 35). In the disenchanted, modern world, by contrast, meaning resides exclusively in the inward space of human minds (pp. 30f), objects are not "charged", and "the causal relations between things cannot be in any way dependent on their meanings, which must be projected on them from our minds" (p. 35).

For Weber, this "intellectualist rationalization" and concomitant transformation in the experience of meaning were "created by science and by scientifically oriented technology" (Weber 1918, p. 139). Against this background he reflects on the meaning and value of science itself, on "the vocation of science within the total life of humanity" (p. 140). He notes Tolstoy's judgment that, with its emphasis on endless progress, science has rendered death, and hence also civilized life, meaningless, because now "civilized man ... catches only the most minute part of what the life of the spirit brings forth ever anew, and what he seizes is always something provisional and not definitive" (p. 140). Traditionally, notes Weber, for the likes of Plato, Leonardo, Galileo, Bacon, and thinkers influenced by Puritanism, science was viewed as providing a way to "true being", "true art", "true nature", "true God", or "true happiness" (pp. 140–143). But these "illusions" have now been "dispelled" (p. 143). "Who", Weber (p. 142) asks,

> aside from certain big children who are indeed found in the natural sciences – still believes that the findings of astronomy, biology, physics, or chemistry could teach us anything about the *meaning* of the world?

The natural sciences, indeed, "are apt to make the belief that there is such a thing as the 'meaning' of the universe die out at its very roots" (Weber 1918, p. 142). This is not to say that science lacks meaning and value, but such meaning and value stem from extra-scientific commitments. Weber himself judges science to be "an objectively valuable 'vocation'" (p. 152), not only for its practical utility (pp. 144f) and generation of specialist knowledge that is "important in the sense that it is 'worth being known'" (p. 143) but also, and more fundamentally, for its "methods of thinking" (p. 150) and ability to bring about "self-clarification and a sense of responsibility" (p. 152). But he is forced to acknowledge that such meanings are based on presuppositions that themselves "cannot be proved by scientific means" but must be rejected or accepted "according to our ultimate position towards life" (pp. 143, 153).

As for the state of disenchantment which science has contributed so much to bringing about, this for Weber is "the fate of the times", which, for all its unpleasantness, must simply be borne "like a man" (p. 155) – for, as Weber bleakly writes elsewhere, the accompanying rationalization and bureaucratization of the social order have placed us in an "iron cage" (Weber 1904, p. 123) with little to look forward to but "a polar night of icy darkness and hardness" (Weber 1919, p. 128).

This grim assessment of the impact of scientific rationalism on experiences of meaning in the modern world, involving the separation from science of traditional sources of meaning, was broadly shared by the psychologist Carl Gustav Jung (1875–1961) and the physicist Wolfgang Pauli (1900–1958). For Jung, as he states in his late essay "The Undiscovered Self" (Jung 1957), science "is based in the main on statistical truths and abstract knowledge and therefore imparts an unrealistic, rational picture of the world" (Jung 1957, par. 498). This leads to a "leveling down" of "not only the psyche but the individual man and, indeed, all individual events whatsoever" (par. 499). The "statistical world picture" thus "thrusts aside the individual in favor of anonymous units that pile up into mass formations" – "organizations", "the abstract idea of the State" – in which the "goal and meaning of individual life (which is the only *real* life)" are submerged (par. 499).

The consequences of this "psychological mass-mindedness" (Jung 1957, par. 501) brought about by scientific rationalism show up, in Jung's view, both in individual pathology, where one-sidedly intellectual patients cut off from their instincts and emotions suffer a sense of "meaninglessness" (Jung 1934, par. 815; see also Jung 1951b, par. 982; Jung 1952b, par. 845), and in the social and political sphere where, he argues, mass-mindedness provides the conditions in which totalitarianism can flourish (Jung 1957, pars. 488–516; see also Main 2004, pp. 117–121, 135–138).

Pauli, too, as a natural rather than social scientist, was concerned about both the personal and collective consequences of the one-sided "rationalistic attitude of scientists since the eighteenth century" (Pauli 1952, p. 153) and the accompanying "de-animation of the physical world" (p. 156). While he greatly valued the knowledge that could be gained through science, the role of science in improving living conditions, and the personal satisfaction of achieving scientific insight (Pauli 1952, p. 152; see also Gieser 2005, p. 258), he felt that current science, which for him meant physics in particular, was incomplete because of its exclusion of feeling, value, psychological reality, and the realm of the non-rational and qualitative generally (Pauli 1952, pp. 206–208; Meier 2001, pp. 195f; Gieser 2005, p. 140).

At the individual level, this exclusion fostered a hypertrophy of reason such as, in Pauli's own case, had precipitated the personal crisis that led to his seeking treatment from Jung (Gieser 2005, pp. 142–154; Miller 2009, pp. 124–147). At a more social and political level, scientific rationalism resulted in a perilous dissociation of science from morality – a situation epitomized for Pauli by the direct and indirect involvement of physicists in the development of the atom bomb and their complicity thereby in mass murder (Gieser 2005, pp. 23, 323f; Miller 2009, p. 176).

While Weber believed that the science-driven rationalization and disenchantment of the modern Western world were an inevitability to which we would have to reconcile ourselves, Jung and Pauli believed that these processes might be tempered, if not reversed, by the development of a revised understanding of science – an understanding which somehow would reconnect science with excluded aspects of meaning in a fundamental way. The specific proposal for how this might be done was the concept of synchronicity, developed by Jung (1952b) with substantive contributions and much critical encouragement from Pauli (1952; Meier 2001).

With the concept of synchronicity Jung proposed that events not related to one another by efficient causation could be connected through shared meaning, through standing in a relationship of "meaningful coincidence" (Jung 1952b, par. 827) or, as Pauli preferred, "meaning-correspondence" (Meier 2001, p. 44). In the following I ask both what Jung means by "meaning" in relation to synchronicity and how, if at all, this understanding of meaning might help with the problem of the seemingly negative impact of scientific rationalism on experiences of meaning in the modern world.

2 The Meaning of "Meaning" in "Meaningful Coincidence"

Jung himself did not present a systematic account of meaning in synchronicity, nor has there appeared to be a clear consensus of understanding among subsequent Jungian writers on the topic. This lack of consensus is evinced by an exchange that took place in the pages of the *Journal of Analytical Psychology* between 2011 and 2012, in which two prominent Jungians, Warren Colman (2011, 2012) and Wolfgang Giegerich (2012), offered views that diverge radically not only from each other but also from Jung's own primary focus. I shall briefly summarize their debate as it highlights a number of important issues and provides a useful point of entry into the topic.

2.1 Constructed, Quasi-Linguistic, or Cosmic?

Colman (2011) argues that with his concept of synchronicity Jung tried to establish scientifically an objective principle of meaning in nature, a meaning which, as Jung puts it, "is *a priori* in relation to human consciousness and apparently exists outside man" (Jung 1952b, par. 942). Colman sees this as tantamount to Jung's trying to establish the reality of "the Self ... as ... a Greater Subject", the "Universal Mind", or "in short, God" (Colman 2011, pp. 472, 481–482). Colman himself proposes (p. 472)

> the contrary view that the meaning in synchronicity is a function of human meaning-making ... a phenomenon of human being in the world in which meaning is generated out of the interaction *between* mind and Nature.

He presents a sophisticated account of such meaning-making in terms of events being associated not causally and logically, as in science, but through congruent correspondence, as in primordial thinking and poetic metaphor, and their then being retroactively organized into narratives (Colman 2011, pp. 480–487).

In a challenging response, Giegerich (2012) takes issue with Colman's, and other authors' (e.g., Hogenson 2005, 2009), focus on the impact that synchronistic events may have subjectively on experiencers, or the role they can play in terms of human meaning-making. He sees this focus as stemming from misunderstanding of the German word translated into English as "meaningful" in the phrase "meaningful coincidence" (Giegerich 2012, p. 502):[1]

> The difference between *sinnvoll* and *sinngemäß* is crucial. When one reports what someone else said, adding that the report will be *sinngemäß*, one indicates that what follows is not a verbatim quotation, but merely "roughly the same", a repetition "faithful to" *(gemäß)* the basic *Sinn* (intended meaning) of what had been said, but now presented in the present speaker's own words or summary, as it were the speaker's version of the gist of it. Whether what you cite in a *sinngemäß* way also happens to be *sinnvoll* (meaningful) or not is another question entirely.

---

[1] "Meaningful coincidence" can be understood as *sinngemäße Koinzidenz* or as *sinnvolle Koinzidenz*. Jung himself used the term *sinngemäße Koinzidenz*, while Pauli occasionally (letter to Fierz of June 3, 1952; von Meyenn 1996, p. 634) proposed to replace coincidence by correspondence ("Entsprechung"), connection ("Zusammenhang"), or constellation ("Anordnung"). Another interesting characterization by Pauli refers to synchronicities as "corrections to chance fluctuations by meaningful and purposeful coincidences ("sinnhafte und zweckmäßige Koinzidenzen") of causally unconnected events" (von Meyenn 1999, p. 336). I am grateful to Harald Atmanspacher for providing these details about Pauli's proposals and usages.

In Giegerich's reading, what Jung means in calling an event a "meaningful coincidence" is merely that "the inner and outer event [of the coincidence] 'mean roughly the same thing'" (Giegerich 2012, p. 502). This use of meaning, he stresses, is "quite sober, down to earth, close to 'concept' or 'notion' ... roughly the same as when we speak of the meaning of a word" and "has absolutely nothing to do with Meaning with a capital M, with human *experiences* of meaning, with what is meaningful *for us* and makes existence meaningful, let alone with transcendent meaning" (p. 502). The impact of a synchronicity, such as the way the famous event with the scarab beetle allegedly had a positive effect on Jung's analysis of his patient, is to be seen merely as an "after-effect", "an additional piece of information about the serendipitous subsequent course of events and remains external to the synchronicity event itself" (p. 506). For Giegerich, Jung's aim with his concept of synchronicity was to address not the subjective interpretation of events but "an extremely puzzling, intellectually challenging *objective* problem: the problem given with the events themselves". It is "a strictly intellectual problem, a challenge for the scientific mind. And the solution offered by Jung is also a rational one" (p. 505).

In his reply to Giegerich, Colman (2012) notes that there are fluent German speakers, including Marie-Louise von Franz, the first main successor to Jung's work on synchronicity, who apparently do not share Giegerich's "low-key, quasi-linguistic" (Giegerich 2012, p. 503) understanding of the phrase *sinngemäße Koinzidenz*. So the issue may be one of different interpretations rather than of mistranslation (Colman 2012, p. 513; see von Franz 1992, p. 258). Colman also notes that the notion of rough correspondence, which Giegerich attributes to Jung, makes it virtually impossible to distinguish "where coincidence ends and synchronicity begins or *vice versa*" (p. 513). There is need for a factor beyond just rough correspondence that converts a mere coincidence into a meaningful coincidence.

For Jung, Colman observes, this is the connection of synchronicity to archetypes with their "highly numinous symbolic images" and "psychoid" nature, which point precisely to a kind of " 'cosmic' meaning" and "Grand Narrative" (2012, pp. 513f). Colman himself, however, remains "unpersuaded by any non-psychic explanation" of archetypes in terms of "self-subsistent meaning" and concludes by reaffirming his claim "that the meaning-making psyche is inextricably involved in the significant correlations of *sinngemäße Koinzidenz*" (2012, p. 516).

Despite the differences between Colman and Giegerich, they also have a notable area of agreement in their shared criticism of Jungian interest in the *unus mundus* (the idea of a unitary reality underlying the mental and the physical) and the possibility that synchronicity may provide "an opening to the sacred and to transcendence" (Giegerich 2012, p. 505; Colman 2012, pp. 512f). Both steer away from accepting any transcendental, cosmic, or religious interpretation of the meaning involved in synchronicities. Colman finds such an interpretation in Jung but rejects it. Giegerich denies that it exists in Jung's work on synchronicity, which he considers to be an exclusively scientific project.

In the remainder of this chapter I shall argue that the objective, transcendental interpretation of meaning is indeed present in Jung's work on synchronicity and that this can most accurately and also most helpfully be viewed as co-existing with, rather than being an alternative to, interpretations of synchronistic meaning in terms of either parallel content (à la Giegerich) or subjective impact and meaning-making (à la Colman).

In his work on synchronicity Jung's primary focus was indeed on science, but, far from being unconcerned with psychological and ultimately spiritual matters, his aim was to propose a revision of our understanding of science which would cease to exclude these from its world picture. The tensions between the understandings of Colman and Giegerich, and the divergence of both from Jung's position, can be lessened by recognizing that Jung implicitly referred to several levels of meaning in his work on synchronicity and that all of these levels are underpinned by the concept of the archetype understood as "psychoid".

2.2 Levels of Meaning

Robert Aziz, in his close examination of the concept of synchronicity in relation to Jung's psychology of religion (Aziz 1990, pp. 64–66, 75–84), identified "four interrelated layers of meaning" involved in synchronistic experiences. The first of these levels is simply the fact of two or more events paralleling one another. The paralleling is by virtue of a shared content or meaning, such as Jung's patient's dream and the appearance of the insect at Jung's consulting room window involving the same or very similar content of a scarab or scarabaeid beetle. This is the level of meaning that Giegerich stresses.

The second level of meaning identified by Aziz consists of the emotional charge or "numinosity" attending synchronistic events. It is a source of nonrational or pre-reflective meaning, suggested in Jung's example by the way in which, when the synchronicity occurred, his patient's "natural being could burst through the armor of her animus possession" (Jung 1952b, pars. 845f, 859f, 870). Colman seems to acknowledge this level when he writes of synchronistic experiences producing "an uncanny sense of what I can best describe as a feeling that the universe is alive" (Colman 2011, p. 475; Colman 2012, p. 514).

Aziz's third level of meaning is the significance of the synchronicity interpreted subjectively, from the point of view of the experiencer's personal developmental needs and goals, unconscious as well as conscious – in Jungian terms, their individuation. In Jung's example, this is expressed by the way in which, following the synchronicity, "the [patient's] process of transformation could at last begin to move" (Jung 1952b, par. 845), so that "the treatment could now be continued with satisfactory results" (Jung 1951b, par. 982). This is the kind of meaning with which Colman and many other analysts who write about synchronicity seem to be primarily concerned, but which Giegerich considers to be an "after-effect ... external to the synchronicity event itself".

The fourth and last of the levels of meaning identified by Aziz is the significance of the synchronicity objectively, that is, as the expression of archetypal meaning which is transcendental to human consciousness. From the symbolism of his example, Jung infers that the objective, transcendental meaning involved was that of the archetype of rebirth, since "the scarab is a classic example of a rebirth symbol" (Jung 1952b, par. 845). Giegerich and Colman both seem to be sceptical about this level of objective, transcendental meaning.

This kind of identification of levels of meaning is not idiosyncratic. As the psychologist Baumeister observes in his interdisciplinary study *Meanings of Life*, depth psychologists, literary critics, and even diplomats frequently recognize different levels of meaning (Baumeister 1991, p. 20). When they do so, what the levels usually refer to is, roughly, "the quantity and complexity of the relationships that are subsumed" (p. 20f):

> The simplest uses of meaning associate labels (such as names) to specific, immediate objects. These uses of meaning tend to be concrete and to be limited in time. In contrast, the highest levels of meaning may refer to complex, far-reaching relationships that transcend the immediate situation and may even approach timeless or eternal perspectives. ... In an important sense, higher level meanings refer to contexts for lower levels.

What interrelates the four levels of meaning identified by Aziz is the concept of the archetype. For while Aziz calls the fourth, objective level of meaning the "archetypal level" (Aziz 1990, p. 66), each of his other three levels of meaning also depends on the presence of the archetype. The shared meaning by virtue of which two or more events are taken to have parallel content and so to be in a synchronistic relationship derives from an archetype:

e.g., in Jung's paradigmatic example of synchronicity, underlying the scarab symbol in both its psychic and its physical appearances is the archetype of rebirth (Jung 1952b, par. 845). The numinous charge that Jung finds associated with synchronicities is something that, he argues, stems from the presence of an activated archetype (Jung 1952b, par. 841). And insofar as the subjective level of meaning is evaluated with reference to the developmental process of individuation, this will also be based on archetypes, since the activation of archetypes – shadow, animus/anima, self, and so on – is intrinsic to individuation for Jung (1928, pars. 266–406; 1951a, pars. 1-67).

This identification of archetypes as providing the basis of synchronicity is in fact explicitly made by Jung: "By far the greatest number of synchronistic phenomena that I have had occasion to observe and analyze", he writes (Jung 1952b, par. 912), "can easily be shown to have a direct connection with the archetype" (see also Jung 1952b, pars. 845f; Jung 1976, pp. 437, 447, 490). Consistently with Baumeister's observation, Aziz's fourth, archetypal level of meaning in synchronicity can be seen both as the highest level and as providing the context for the other levels.

2.3 The Archetype as Psychoid

In his synchronicity essay Jung (1952b, par. 840) writes that archetypes "constitute the structure of the collective unconscious", that they are "formal factors responsible for the organization of unconscious psychic processes: they are 'patterns of behavior'", and that "they have a 'specific charge' and develop numinous effects which express themselves as *affects*" (Jung 1952b, par. 841). This is all consonant with what Jung had been writing for years. But in the synchronicity essay he articulates some novel characteristics that reveal archetypes to be not just psychic but what he calls "psychoid" factors. In characterizing archetypes as "psychoid" he means that they cannot be fully represented psychically and therefore cannot simply be equated with "perceptible psychic phenomena" (par. 840). At an irrepresentable level, he suggests, archetypes can structure matter as well as psyche – and not just separately but at the same time in respect to the same pattern of meaning. Thus, of archetypes as "psychoid factors" he writes (Jung 1952b, par. 964):

> These are *indefinite*, that is to say they can be known and determined only approximately. Although associated with causal processes, or "carried" by them, they continually go beyond their frame of reference, an infringement to which I would give the name "transgressivity", because the archetypes are not found exclusively in the psychic sphere, but can occur just as much in circumstances that are not psychic (equivalence of an outward physical process with a psychic one).

For Jung, the psychoid nature of archetypes is most clearly evinced by number archetypes. In the synchronicity essay itself this is largely left implicit. Jung (1952b, par. 870) highlights the archetypal character of natural numbers but he primarily focuses on the psychic properties of number archetypes and their role in divination procedures (pars. 863–870). However, in a subsequent letter to Pauli (24 October 1953) he is more explicit about the broader significance of number archetypes as paradigmatic psychoid factors. As "the simplest and most elementary of all archetypes" (Meier 2001, p. 127), he writes, number archetypes can help "to locate and describe that region which is indisputably common to both [physics and psychology]". For numbers "possess that characteristic of the psychoid archetype in classical form – namely, that *they are as much inside as outside*" (p. 127). This, Jung suggests, is why "equations can be devised from purely mathematical prerequisites" and these equations later "will turn out to be formulations of physical processes" (p. 127). He thus concludes that "from the psychological point of view at least, the sought-after borderland between physics and psychology lies in the secret of the number" (p. 127).

In his essay Jung (1952b, par. 964) also characterizes archetypes as representing "psychic probability". This formulation highlights another analogy between physics and Jung's depth psychology, thereby again enhancing the plausibility of connecting the two. As Jung neatly summarized it in a letter to Pauli (13 January 1951): "In physical terms, probability corresponds to the so-called law of nature; psychically, it corresponds to the archetype" (Meier 2001, p. 70).

These developments in Jung's thinking about archetypes, each of which facilitates a rapprochement between psychology and physics, were all prompted by Pauli's criticisms of earlier drafts of Jung's essay. Pauli was concerned in particular about Jung's placing discontinuous phenomena in physics on the same level as synchronicity, since, as Pauli explained (Meier 2001, p. 56), "microphysics ... has no use for the concept of 'meaning'" – except to the minimal extent that "the term 'state' or 'physical situation' in quantum physics" might be "a preliminary stage for [the] more general term 'meaningful connection'" (Meier 2001, p. 56, n. 5). To allow for the "broader definition" of synchronicity that encompasses acausal phenomena in microphysics – the definition that Jung and eventually Pauli both favored (pp. 59–65) – it was necessary for the concept of the archetype, which was currently "inadequate" for this purpose, to undergo change (p. 65). This is the reason for its reformulation as "psychoid" and as representing "psychic probability" – and Pauli suspected that "more changes are in the offing" (p. 65).

2.4 Meaning in Synchronicity

Having clarified that for Jung the meaning in synchronicities is multileveled and is underpinned at all levels by the concept of the psychoid archetype, I would now like to return to the debate between Colman and Giegerich and to offer a few comments on their respective positions.

First, I think Giegerich is right to emphasize that the first of the four levels of meaning identified by Aziz, paralleling of content, is of much greater concern to Jung in his synchronicity essay (and indeed in most of the other places where he discusses synchronicity) than is the third, subjective level of meaning which Colman emphasizes. In the essay there are many references to meaning at the level of parallel content, several of them receiving italicized emphasis by Jung. For example, Jung refers simply to "meaningful coincidence", "a kind of *meaningful cross-connection*" (Jung 1952b, par. 827); to images standing "in an analogous or equivalent (i.e., meaningful) relationship to objective occurrences" (par. 856); to "the simultaneous occurrence of a psychic state with a physical process as an *equivalence of meaning*" (par. 865); to how "the connecting principle [in synchronicities] must lie in the *equal significance* of the parallel events; in other words their *tertium comparationis* [the third element in the comparison] is *meaning*" (par. 915); and to a "factor in nature which expresses itself in the arrangement of events and appears to us as meaning" (par. 962).

In contrast, there are no explicit references to the subjective level of meaning apart from Jung's briefly recalling in the "Foreword" of the essay "how much these inner experiences meant to my patients" (Jung 1952b, par. 816) and, *pace* Giegerich, Jung's recounting of the incident involving the scarab beetle, which, even with its scarcity of "relevant subjective information" (Aziz 1990, p. 65), does provide an outline of how the synchronicity fostered the patient's "process of transformation" (Jung 1952b, par. 845; Jung 1951b, par. 982).

Giegerich is also clearly right that in his writing on synchronicity Jung is more immediately concerned with science than therapy. The synchronicity essay begins and ends with lengthy discussions of subatomic physics and issues in the philosophy of science, and in between there are various forays into parapsychology and descriptive biology. By contrast, any references to the therapeutic implications of synchronicity are brief and primarily illustrative. Even outside of the essay it is surprisingly difficult to find places where Jung does discuss the subjective or therapeutic effects of synchronicity (see Main 2007, pp. 360–364).

However, Giegerich seems to overplay this hand. Even if Jung does not

evince much concern with Aziz's third, subjective level of meaning, there is
plenty of evidence of his concern with the fourth, objective level. In fact, the
greatest number of explicit references to meaning in Jung's essay concerns
the objective or archetypal level. Jung refers to meaning that can "exist
outside the psyche" (Jung 1952b, par. 915), "outside man" (par. 942), and
is "self-subsistent" (par. 944), "transcendental" (par. 915), "*a priori* in relation to human consciousness" (par. 942). He also devotes many pages to
setting out some of the considerations that, for him, point to the existence of
objective meaning. Chief among these considerations are synchronistic phenomena themselves, whether spontaneous or generated, with their apparent
ability to transcend the limitations of space and time to reveal "'absolute
knowledge' ... a knowledge not mediated by the sense organs ..., knowledge
of future or spatially distant events" (Jung 1952b, par. 948). In addition,
by way of cultural support, Jung adduces in Chapter 3 of the essay a range
of Chinese, Greek, medieval, and Renaissance forerunners of his idea of synchronicity – notions of Tao, the sympathy of all things, correspondences,
microcosm and macrocosm, and pre-established harmony – each of which
presupposes the existence of objective meaning (pars. 916–946). As further
indications, he refers to dreams whose content seems to suggest the idea
of self-subsistent meaning (Jung 1952b, pars. 945f); to the "'meaningful'
or 'intelligent' behavior of the lower organisms, which are without a brain"
(pars. 947–948); and to out-of-body-experiences or, as he refers to them,
"remarkable observations made during deep syncopes" (pars. 949–955).

In the light of all this counterevidence, it is clearly not the case that Jung
is exclusively preoccupied in the essay with the level of simple paralleling of
content. Furthermore, while Jung may have been primarily engaged with
a scientific problem, the nature of this problem was precisely how to revise
science in such a way as to open it up to include psychological factors – in
particular, factors associated with the concept of the archetype (Jung 1952b,
par. 962).

When Giegerich (2012, p. 502) writes that Jung's "low-key, quasi-linguistic" use of meaning "has absolutely nothing to do with Meaning with a
capital M, with human *experiences* of meaning, with what is meaningful *for*
us and makes existence meaningful, let alone with transcendent meaning",
he seems to suggest that the quasi-linguistic level of meaning is of such a
radically different kind as to preclude connection with the higher levels of
meaning he enumerates. Yet this is not necessarily so. As Baumeister (1991,
p. 16) notes:

> The meaning of a life is the same kind of meaning as the meaning of
> a sentence in several important respects: having the parts fit together

> into a coherent pattern, being capable of being understood by others,
> fitting into a broader context, and invoking implicit assumptions shared
> by other members of the culture. ... Meanings of life are a special usage
> of meaning, not a special kind of meaning.

For Jung, too, as suggested above, there is continuity among levels of
meaning. On the one hand, he can claim that myths, which for him consist
of culturally elaborated archetypal motifs, can help us to "frame a view
of the world which adequately explains the meaning of human existence in
the cosmos" (Jung 1963, p. 373). On the other hand, he can note that all
interpretations "make use of certain linguistic matrices that are themselves
derived from primordial images [i.e., archetypes]" (Jung 1934/1954, par. 67).
For Jung, the archetype provides the "source" of not just the cosmic but also
the linguistic meaning. He refers to the "equivalences" in synchronistic events
– the equivalences that Giegerich (2012, p. 502) finds so "sober" and "down
to earth" – as *"archetypal* equivalences" (Jung 1952b, par. 964, emphasis
added). In view of the more transcendental directions in which Jung goes
elsewhere in his essay, it would seem that his low-key, quasi-linguistic uses
of meaning are intended to facilitate its discussion at the most basic level
in relation to problems of science. Having done this, he can then show –
or imply and leave for others to show – how, through the concept of the
archetype, this entails the inclusion of richer and more complex levels of
meaning as well.

Colman, in his reply to Giegerich, seems to me right to argue that, if
the meaning in synchronicity referred to nothing more than "rough correspondence", we would have no way of distinguishing meaningful from mere
coincidences (Colman 2012, p. 513). For even series of events that Jung in
the end judges not to be synchronistic, such as his run of experiences involving fishes, do have roughly corresponding, parallel content (Jung 1952b,
pars. 826–827; Jung 1951b, pars. 969–971). Colman is also, I think, clearly
right that the factor which for Jung enables this distinction to be made is the
archetype understood as numinous and psychoid. Colman's overall assessment, in his original paper, of Jung's aim as being to establish scientifically
an objective principle of meaning in nature, with far-reaching religious implications, also seems to me correct and is well supported by the evidence
in the essay itself, as we have seen. That Jung's concern with objective
meaning might indeed lead him into religious territory, as Colman suggests,
is indicated by, for example, Jung's willingness, following Richard Wilhelm,
to equate the kind of meaning he has in mind with Tao, "one of the oldest
and most central ideas [in Chinese philosophy], which the Jesuits translated
as 'God' " (Jung 1952b, par. 917; Jung 1935, par. 143).

While recognizing this focus of Jung's, Colman himself develops implications of synchronicity at Aziz's third level of subjective meaning. Despite
Jung's own lack of attention to this level, developing it is, as we have seen,
not inconsistent with what Jung writes in his synchronicity essay. And outside the essay, there are other cases, albeit not many and not very detailed,
where Jung interpreted synchronistic events in relation to the experiencer's
subjective or therapeutic concerns (Main 2007, pp. 362–364). But perhaps
most telling is that, when this aspect of synchronicity was picked up by other
analysts, Jung was quick to approve. For example, Michael Fordham in his
book chapter "Reflections on the Archetypes and Synchronicity" discusses
in some detail the therapeutic aspects of both Jung's scarab incident and a
synchronicity from his own clinical experience (Fordham 1957, pp. 42–50).
Jung read this in manuscript and praised it in a letter to Fordham (3 January 1957) as "the most intelligent thing that has been said hitherto about
this remote subject" (Jung 1976, pp. 343f). From this it is clear that Jung
did recognize the subjective, psychological, meaning-making aspect of synchronicity and affirmed the value of exploring it. There thus seems to be a
misplaced emphasis when Colman writes that he is presenting "the *contrary
view* that the meaning in synchronicity is a function of human meaning-
making" (Colman 2011, p. 472, emphasis added). Jung is acutely aware
of the human contribution to the emergence of meaning and struggles with
this awareness when attempting to make his case for the existence also of
objective meaning. He acknowledges that, as usually understood, "meaning
is an anthropomorphic interpretation" (Jung 1952b, par. 916); that "what
that factor which appears to us as 'meaning' may be in itself we have no
possibility of knowing" (par. 916); and specifically that "we have absolutely
no scientific means of proving the existence of an objective meaning which is
not just a psychic product" (par. 915). In a later letter to Erich Neumann
(10 March 1959), he wrote that (Jung 1976, p. 495)

> meaningfulness always appears to be unconscious at first, and can there-
> fore only be discovered *post hoc*; hence there is always the danger that
> meaning will be read into things where actually there is nothing of the
> sort.

If, despite his recognition of these difficulties, Jung continues to argue
for the existence of objective meaning, this is not to the detriment of the
subjective level of meaning. For Jung, the transcendental view of meaning is not incompatible with a view that sees meaning as "a phenomenon
of human being in the world" involving "the interaction *between* mind and
Nature" (Colman 2011, p. 472). In synchronistic experiences transcendental,

archetypal meaning can enter the lives of individuals through the dynamic of unconscious compensation of their conscious attitude. The meaning expresses itself through archetypal imagery inflected by the experiencers' circumstances. If integrated, it promotes the experiencers' individuation (Aziz 1990, pp. 80–84; Mansfield 1995, pp. 16-19; Mansfield 2002, pp. 124–128).

I have argued in detail elsewhere (Main 2013a,b) that Jung's psychology is purposely framed so as to respect both immanent, material and transcendental, spiritual viewpoints equally, without allowing one to eclipse the other. I have suggested that this intrinsic doubleness can help to foster productive engagements of analytical psychology both with contemporary social and political problems (Main 2013a) and with other social theories (Main 2013b). The wide range of kinds of meaning that can be identified in synchronicity – with their extreme poles, quasi-linguistic and cosmic – may be another expression of this dual secular and religious perspective. Indeed it can be argued that the concept of synchronicity was introduced precisely to help to maintain this perspective (see Main 2004, pp. 100–114). Accordingly, to reject the notion of "self-subsistent meaning" and the underpinning "non-psychic", transcendental interpretation of archetypes, as both Giegerich and Colman do, seems to me to risk impoverishing analytical psychology by collapsing one of the deepest cultural tensions held by Jung's thought: the tension between the secular and the religious.

3   Shaking the Security
of Our Scientific Foundations

In a letter to Richard Hull (24 January 1955), one of the main translators of Jung's works, Jung (1976, p. 217, emphasis added) reported:

> The latest word about "Synchronicity" is that it cannot be accepted because it shakes the security of our scientific foundations, *as if this were not exactly the goal I am aiming at.*

Pauli, for his part, considered that the final chapter of Jung's essay might provide "a glimpse into the future of natural philosophy" (Meier 2001, p. 65). How, then, might synchronicity as a principle of acausal connection through meaning, understood to be multileveled and archetypal, shake up and revise the kind of scientific rationalism criticized by Jung and Pauli? And how might synchronicity thereby help with the crisis of meaning in modernity that Jung, Pauli, and others, such as Weber, observed?

3.1   Meaning in Matter, Psychology and Religion in Science

Jung characterized scientific rationalism – the science which had been as-
cendant since at least the 18th century – as "triadic", based on the three
principles of time, space, and causality. His radical proposal was to introduce synchronicity, and thereby also meaning, as a fourth principle (Jung
1952b, pars. 961–963; see also Pauli 1952, pp. 174f, 204f, 226–236). This,
he argued, would make possible "a whole judgment" (Jung 1952b, par. 961)
and "a view which includes the psychoid factor [i.e., the archetype] in our
description and knowledge of nature – that is, an *a priori* meaning or 'equivalence'" (par. 962).

There are at least three important implications of this proposal. First,
meaning is here recognized as a factor able to connect events that would
not otherwise be connected. This allows for the perception of other sets of
relationships than causal ones. Events which might be disregarded from a
causal point of view because they are unique, irrational, creative, or outright
anomalous can be grasped from a synchronistic point of view in terms of the
patterns of meaning in which they are woven.

The tendency of scientific rationalism to disregard the unique and rare in
favor of the statistical average taxed Jung and provided one of the launching
points for his essay (Jung 1952b, par. 821). In a letter to Fordham written
some years later (24 January 1955), he bemoaned (Jung 1976, p. 216) that
"wherever a philosophy based on the sciences prevails ..., the individual man
loses his foothold and becomes 'vermasst', turned into a mass particle, because as an 'exception' he is valueless". He then asserted that his wish to do
something to forestall the perilous social and political consequences of "this
blind and dangerous belief in the security of the scientific Trinity [of time,
space, and causality]" – above all, the dangers of de-individualization, mass-
mindedness, and totalitarianism – was actually "the reason and the motive
of my [synchronicity] essay" (Jung 1976, p. 216).

Second, Jung's proposal that the archetype is psychoid and can therefore
structure or arrange physical as well as psychic events implies that matter is
not, as in the view of scientific rationalism, fundamentally inert and meaningless. Meaning is not only something that the human psyche projects onto
matter but, rather, matter can be inherently imbued with meaning. Jung
states this explicitly to Pauli (7 March 1953): "In cases of synchronicity",
he writes, "they [i.e. archetypes] are arrangers of physical circumstances, so
that they can also be regarded as a characteristic of Matter (as *the feature
that imbues it with meaning*)" (Meier 2001, p. 101, emphasis added). With
this claim Jung suggests that synchronicity can – at least to some extent

(Main 2011) – reverse the process identified by Weber as "the disenchant-ment of the world", by Pauli as "the de-animation of the physical world", and by Jung himself as "the historical process of world despiritualization" (Jung 1938/1940, par. 141).

Third, because the meaning archetypes imbue is multileveled, ranging from basic levels of ordering, paralleling, and signifying to the kinds of higher levels that inform individual transformation and the framing of meanings of life, the meaning in material circumstances may be complex enough to connect with psychological and religious meanings. This seems to be implied when Jung, explaining in a letter to Pauli (3 March 1953) his reasons for publishing his religious essay "Answer to Job" (Jung 1952a) simultaneously with his synchronicity essay, writes that "by making the assumption [in the synchronicity essay] that 'being is endowed with meaning' (i.e., extension of the archetype in the object)" he was attempting "to open up *a new path to the 'state of spiritualization' [Beseeltheit] of Matter*" (Meier 2001, p. 98, emphasis added). This potential for an intimate connection of natural science with psychology and religion opens the prospect for a truly holistic form of understanding.

These were thoughts to which Pauli was broadly receptive. In the letter (27 February 1953) to which Jung was responding, Pauli had written that he now believed in "the possibility of a simultaneous religious and scientific function of the appearance of archetypal symbols" (Meier 2001, p. 87). In relation to this, he referred (p. 87, n. 5) to the conclusion of his essay on "The Influence of Archetypal Ideas on the Scientific Theories of Kepler", where he argued that "not only alchemy [as propounded by Robert Fludd] but the heliocentric idea [held by Johannes Kepler]" each proved "the existence of a symbol that had, simultaneously, a religious and a scientific function" (Pauli 1952, p. 212).

In a later letter to Jung (23 December 1953) Pauli wrote of his impression that (Meier 2001, p. 130, emphasis added)

> compensatorily from the unconscious, the tendency is being developed to bring physics much closer to the roots and sources of life, and that what is happening is ultimately *an assimilation of the psychoid archetype into an extended form of physics.*

Later again (23 October 1956), he wrote of certain dreams that seemed to him to be addressing "the problem ... of getting right to the archetypal source of the natural sciences and thus to a new form of religion" (Meier 2001, p. 150).

3.2 Transformation and "Double Vision"

It is more than incidental that the last two statements of Pauli's draw their content from expressions of the unconscious. Pauli and Jung were attempting to articulate the basis of a more holistic form of science, one which would include the unconscious as well as consciousness, psyche as well as matter, and the functions of feeling and intuition as well as those of sensation and thinking. But the model they were seeking could not be arrived at simply consciously, outwardly, and intellectually. It also required attention to the unconscious, to the inner world, and to emotions. As Jung asserted (7 March 1953), "only from his wholeness can man create a model of the whole" (Meier 2001, p. 99).

Pauli (27 May 1953) fully agreed with this statement, adding with emphasis (Meier 2001, p. 124):

> *it is impossible for me to find this correspondentia between physics and psychology just through intellectual speculation; it can only properly emerge in the course of the individuation process in the form of accompanying objective statements.*

At the conclusion of his study of the conflict between the esotericist Fludd, with his qualitative approach to knowledge, and the early modern scientist Kepler, with his new quantitative approach, Pauli (1952, p. 212) formulated this idea as follows:

> The process of knowing is connected with the religious experience of transmutation undergone by him who acquires knowledge. This connection can only be comprehended through symbols which both imaginatively express the emotional aspect of the experience and stand in vital relationship to the sum total of contemporary knowledge and the actual process of cognition.

Again, this statement has several important implications. One is that scientific insight can be fostered by engaging with the unconscious. Both Jung and Pauli put this implied route to knowledge into practice. In Jung's case, a major source of his psychological insights and theories was his period of imaginative and reflective engagement with his unconscious and its personifications, as recorded in his *Red Book* (Jung 2009). Notably, he later even considered it helpful to posit a specific "archetype of meaning" (Jung 1934/1954, par. 66), which in personified form could be encountered as the image of the Wise Old Man, "the superior master and teacher, the archetype of the spirit, who symbolizes the pre-existent meaning hidden in the chaos of life" (Jung 1934/1954, par. 74).

Pauli, for his part, attended closely to his dreams, particularly when they involved concepts from physics or included figures who seemed to symbolize either a critical attitude towards the prevailing science or the possibility of a new, more holistic science (see Meier 2001, passim; Gieser 2005, pp. 180, 319). He also regularly engaged in active imagination (see, e.g., Meier 2001, pp. 39f; Pauli 2002), and he even wrote an essay titled "Modern Examples of 'Background Physics' " (Meier 2001, pp. 179-196), in which he reflected at length on "the appearance of quantitative terms and concepts from physics in spontaneous fantasies in a qualitative and figurative – i.e., symbolic – sense" (Meier 2001, p. 179).

A second point to highlight is Pauli's statement that the symbols through which the connection between knowing and the religious experience of transmutation can be comprehended need to "stand in vital relationship to the sum total of contemporary knowledge" (Pauli 1952, p. 212). This implies that inner psychological transformation alone will not lead to holistic understanding. As Pauli remarks (Meier 2001, p. 180), *"the purely psychological interpretation* only apprehends *half of the matter. The other half is the revealing of the archetypal basis of the terms* actually applied in *modern physics"*. For an effective integration of depth psychology with physics, or indeed with any other discipline, the flow or understanding and influence must be two-way. Gieser (2005, p. 177) summarizes the way Pauli made this point in a letter to von Franz (16 October 1951):

> A truly symmetrical relationship requires that the concept of *introjection*, in other words information from the outside world to the psyche, be accorded as much importance as the concept of *projection*, contents flowing from the psyche to the world.

Finally, Jung's and Pauli's view of the relationship between personal transformation and knowledge implies that a fully holistic understanding of the world would only be possible for someone at an advanced stage of individuation. On 23 October 1956 Pauli sent Jung a series of dreams together with his own commentary. In his reply (December 1956) Jung added his own interpretations of some of the dreams, including two (dreamed on 1 October 1954 and 26 December 1955) which concerned "the difference between Danish and English" (Meier 2001, p. 143). For Pauli, Danish symbolized everyday language, while English symbolized dream language (p. 146).

In the first dream the difference between the two languages is said to correspond to *"the difference between v and w"* (Meier 2001, p. 143), and Pauli included in his letter a lengthy narration of his attempts to solve the linguistic conundrum with which his dream had presented him (pp. 143–146).

In the second dream a king tells Pauli "with great authority" that he (Pauli) has "an apparatus that enables you to see both Danish and English" (p. 152). Before commenting on the imagery in these dreams, Jung tells Pauli that he is "most impressed by your forays into linguistics" (p. 156). He then turns to the symbolism (p. 156f, emphasis added):

> The important thing about the dream of 26 December 1955 is the double vision. This is a distinctive characteristic of the human being who is at one with himself. He sees the inner and outer oppositeness, not just V = 5, which is a symbol of the natural person who, with his consciousness based on perception, becomes ensnared in the world of sense perception and its vividness. W (double V), by way of contrast, is the One, *the whole person who, although himself not split, nevertheless perceives both the external sensory aspect of the world and also its hidden depths of meaning.* Thus the split is based on the one-sided ensnarement in one or the other aspect. But if man has united the opposites within himself, there is nothing to stop him perceiving both aspects of the world in an objective manner. The inner psychic split is replaced by a split world-picture, and this is inevitable, for without this discrimination, conscious perception would be impossible. It is not in actual fact a split world, for facing the person who is united with himself is an *unus mundus*. He has to split this one world in order to be able to perceive it, always bearing in mind that what he is splitting is still the *one* world, and that the split has been predetermined by consciousness.

This is a far-reaching set of statements. Above all, the passage articulates the kind of perspective that Jung believed might result from the inclusion in our scientific picture of the world of a principle of acausal connection through meaning. In addition, with its claim that one can perceive "both the external sensory aspect of the world and also its hidden depths of meaning", the passage suggests the dual secular and religious perspective that is arguably intrinsic to analytical psychology (Main 2013b).

The references in the passage to an ontological unity ("the *one* world") encountered through an epistemological duality ("a split world-picture") suggest the philosophical position of dual-aspect monism that seems particularly fruitful for understanding Jung's and Pauli's work on synchronicity (Seager 2009, Atmanspacher 2012).

Finally, a major part of the argument of this chapter has been that the seemingly low-key, quasi-linguistic usage of "meaning" upon which Jung attempts to build his concept of synchronicity in fact connects, via the concept of the psychoid archetype, with higher levels of meaning, reaching all the way up to the most "transcendental" and "cosmic". It therefore seems apt that Jung's far-reaching statements about "the double vision" should emerge here specifically in relation to reflections on linguistics.

References

Atmanspacher H. (2012): Dual-aspect monism à la Pauli and Jung. *Journal of Consciousness Studies* **19**(9/10), 96–120.

Aziz R. (1990): *C.G. Jung's Psychology of Religion and Synchronicity*, University of New York Press, Albany.

Baumeister R. (1991): *Meanings of Life*, Guilford, New York.

Colman W. (2011): Synchronicity and the meaning-making psyche. *Journal of Analytical Psychology* **56**(4), 471–491.

Colman W. (2012): Reply to Wolfgang Giegerich's "A serious misunderstanding: Synchronicity and the generation of meaning". *Journal of Analytical Psychology* **57**(4), 512–516.

Fordham M. (1957): Reflections on the archetypes and synchronicity. In *New Developments in Analytical Psychology*, Routledge and Kegan Paul, London, pp. 35–50.

Franz M.-L. von (1992): *Psyche and Matter*, Shambhala, Boston.

Giegerich W. (2012): A serious misunderstanding: Synchronicity and the generation of meaning. *Journal of Analytical Psychology* **57**(4), 500–511.

Gieser S. (2005): *The Innermost Kernel: Depth Psychology and Quantum Physics – Wolfgang Pauli's Dialogue with C.G. Jung*, Springer, Berlin.

Hogenson G. (2005): The self, the symbolic, and synchronicity: Virtual realities and the emergence of the psyche. *Journal of Analytical Psychology* **50**(3), 271–284.

Hogenson G. (2009): Synchronicity and moments of meeting. *Journal of Analytical Psychology* **54**(2), 183–197.

Jung C.G. (1928): The relations between the ego and the unconscious. In *Two Essays on Analytical Psychology, Collected Works Vol. 7*, Routledge and Kegan Paul, London 1966.

Jung C.G. (1934): The soul and death. In *The Structure and Dynamics of the Psyche, Collected Works Vol. 8*, Routledge and Kegan Paul, London 1969.

Jung C.G. (1934/1954): Archetypes of the collective unconscious. In *The Archetypes and the Collective Unconscious, Collected Works Vol. 9/1*, Routledge and Kegan Paul, London 1968.

Jung C.G. (1935): The Tavistock lectures. In *The Symbolic Life, Collected Works Vol. 18*, Routledge and Kegan Paul, London 1977.

Jung C.G.(1938/1940): Psychology and religion. In *Psychology and Religion: West and East, Collected Works Vol. 11*, Routledge and Kegan Paul, London 1969.

Jung C.G. (1951a): *Aion: Researches into the Phenomenology of the Self, Collected Works Vol. 9/2*, Routledge and Kegan Paul, London 1968.

Jung C.G. (1951b): On synchronicity. In *The Structure and Dynamics of the Psyche, Collected Works Vol. 8*, Routledge and Kegan Paul, London 1969.

Jung C.G. (1952a): Answer to Job. In *Psychology and Religion: West and East, Collected Works Vol. 11*, Routledge and Kegan Paul, London 1969.

Jung C.G. (1952b): Synchronicity: An acausal connecting principle. In *The Structure and Dynamics of the Psyche, Collected Works Vol. 8*, Routledge and Kegan Paul, London 1969.

Jung, C.G. (1957): The undiscovered self. In *Civilization in Transition, Collected Works Vol. 10*, Routledge and Kegan Paul, London 1970.

Jung C.G. (1963): *Memories, Dreams, Reflections*, ed. by A. Jaffé, transl. by R. and C. Winston, Fontana, London 1995.

Jung C.G. (1976): *Letters Vol. 2: 1951-1961*, ed. by G. Adler and A. Jaffé, transl. by R.F.C. Hull, Routledge and Kegan Paul, London.

Main R. (2004): *The Rupture of Time: Synchronicity and Jung's Critique of Modern Western Culture*, Brunner-Routledge, New York.

Main R. (2007): Synchronicity and analysis: Jung and after. *European Journal of Psychotherapy and Counselling* **9**(4), 359–371.

Main R. (2011): Synchronicity and the limits of re-enchantment. *International Journal of Jungian Studies* **3**(2), 144–158.

Main R. (2013a): In a secular age: Weber, Taylor, Jung. *Psychoanalysis, Culture & Society* **18**(3), 277–294.

Main R. (2013b): Secular *and* religious: The intrinsic doubleness of analytical psychology and the hegemony of naturalism in the social sciences. *Journal of Analytical Psychology* **58**(3), 366–386.

Mansfield V. (1995): *Synchronicity, Science, and Soul-Making: Understanding Jungian Synchronicity through Physics, Buddhism, and Philosophy*, Open Court, Chicago.

Mansfield V. (2002): *Head and Heart: A Personal Exploration of Science and the Sacred*, Quest Books, Wheaton.

Meier C.A., ed. (2001): *Atom and Archetype: The Pauli/Jung Letters 1932–1958*, Routledge, London.

Miller A. (2009): *Deciphering the Cosmic Number: The Strange Friendship of Wolfgang Pauli and Carl Jung*, Norton, New York.

Pauli W. (1952): The influence of archetypal ideas on the scientific theories of Kepler. In *The Interpretation of Nature and the Psyche*, by C.G. Jung and W. Pauli, Pantheon, New York 1955, pp. 147–240.

Pauli W. (2002): The piano lesson: An active fantasy about the unconscious. Translated by F.W. Wiegel, H. van Erkelens, and J. van Meurs. *Journal for Jungian Studies* **48**(2), 122–134.

Seager W. (2009): A new idea of reality: Pauli on the unity of mind and matter. In *Recasting Reality: Wolfgang Pauli's Philosophical Ideas and Contemporary Science*, ed. by H. Atmanspacher and H. Primas, Springer, Berlin, pp. 83–97.

Taylor C. (2007): *A Secular Age*, Harvard University Press, Cambridge.

von Meyenn, K., ed. (1996): *Wolfgang Pauli. Wissenschaftlicher Briefwechsel Band IV, Teil 1: 1951–1952*, Springer, Berlin.

von Meyenn, K., ed. (1999): *Wolfgang Pauli. Wissenschaftlicher Briefwechsel Band IV, Teil 2: 1953–1954*, Springer, Berlin.

Weber M. (1904): *The Protestant Ethic and the Spirit of Capitalism*, transl. by T. Parsons, Routledge, London 2001.

Weber M. (1918): Science as a vocation. In *From Max Weber: Essays in Sociology*, ed. by H. Gerth and C. Wright Mills, Oxford University Press, New York 1946, pp. 129–156.

Weber M. (1919): Politics as a vocation. In *From Max Weber: Essays in Sociology*, ed. by H. Gerth and C. Wright Mills, Oxford University Press, New York 1946, pp. 78–128.

Investigating Synchronistic Events in Psychotherapy

Christian Roesler

Abstract

According to Jung's original definition of synchronicity, psychotherapy can be considered as an outstanding field where synchronistic events can be expected to appear. Even though synchronicity has been discussed intensively in recent years, up to now there has been no attempt to observe and document synchronistic events in psychotherapy in any systematic fashion. There are several collections of reports of synchronistic events, either documented in a more or less anecdotal way or resulting from broader investigations of so-called anomalous phenomena. The psychotherapeutic setting presents the opportunity to observe synchronistic events systematically and prospectively, and to link them with other data concerning the patient, his/her psychodynamics, and interpersonal and other conditions. Moreover, it offers the option to test Jung's assumption that synchronistic events are connected to the patient's process of individuation. A systematic research frame is proposed for Jungian psychotherapy in order to collect corresponding data. First results from a pilot study are reported.

1 Introduction

A central theoretical product of the Pauli-Jung-dialog is the concept of synchronicity. Jung published his article "Synchronizität als ein Prinzip akausaler Zusammenhänge" (Jung 1952) together with a treatise by Pauli on Kepler in their joint volume "Naturerklärung und Psyche" (Jung and Pauli 1952) after a previous publication on the topic "Über Synchronizität" (Jung 1951). According to Jung, synchronicity was defined as a coincidence of an inner psychological state or event with an external or objective event without any causal connection between the two. Instead, their connection seems, at least for the experiencing individual, as meaningful. Both Jung and Pauli struggled to find a theoretical model for synchronicity based on analogies with quantum theory (see Gieser 2005 for more details).

Jung's (1951) classic example for a case of synchronicity occurred in the context of psychotherapy. A female patient of Jung's presented a dream in a therapy session in which she had received a golden scarabeus as a present. Right at that moment they heard a noise tapping at the window. When Jung stood up and opened the window a beetle flew into the room which was the closest relative to a scarabeus occurring in central Europe, a so-called rose beetle. The patient was deeply moved by this experience.

Before this event the therapeutic process had become difficult and made no progress. Through the synchronistic experience it became possible for the patient to change her inflexible identification with a rational orientation of her consciousness and to begin a process of psychological transformation. Jung saw the archetypal symbolism of the scarabeus beetle in relation to the mystery of death and rebirth and in analogy to the psychological situation of his patient. She had to give up a too one-sided orientation concerning rationality and control of the ego and move towards a new balance between consciousness and the unconscious.

The example makes clear that from the beginning Jung connected the concept of synchronicity strongly with both the process of psychotherapy and the individuation process. Individuation is here seen as a spontaneous process developing out of the unconscious psyche, moving the individual towards his or her potential wholeness. In this process the unconscious confronts the conscious ego with symbols, as for example in dreams, to foster a constructive dialog between consciousness and the unconscious.

In Jungian theory the situation of analytical therapy with its special interpersonal relationship is seen as an arena where this internal dialog is promoted. During the spontaneous production of symbols from the unconscious, the likelihood for synchronistic events to appear is increased. The reason behind this is the constellation of collective unconscious and archetypal material.

Archetypes, which structure the unconscious, are organized in opposites which can be related to the concept of complementarity in quantum theory.[1] Synchronistic events are meaningful and can be interpreted, in the context of psychotherapy, like other symbolic material as, for example, dreams, images etc. (Fordham 1957, Main 2007, Hopcke 2009). Several authors have developed methodologies for utilizing synchronistic events in psychotherapy (Bolen 1979, Kreutzer 1984).

_____

[1]Compare Atmanspacher *et al.* (2002), Walach (2003), von Lucadou *et al.* (2007); see also Fach (2011) for a detailed description of a generalized quantum theory as an explanatory model for the appearance of synchronistic and other anomalous events.

2   State of Empirical Research

Even though there have been many publications since the time when Jung and Pauli formulated the concept, there have been only few studies on synchronicity from the background of analytical psychology using systematic empirical research methods (Coleman and Beitman 2009). Most studies have been single case studies with no coherent methodology based on free interpretations in the sense of a general psychoanalytic approach (Williams 1957, Bender 1966, Keutzer 1984, Wharton 1986, Hopcke 1990, Kelly 1993, Guindon and Hanna 2002).

But there are also studies applying systematic scientific methodology: Hanson and Klimo (1998) conducted a systematic analysis of reports on coincidences with negative consequences where 56% of the subjects interviewed reported synchronistic events. Hill (2011) developed a study of synchronistic events in mourning and showed that these synchronistic experiences have a healing function for grief. Meyer (1998) investigated the correlation between the proneness to experience synchronistic events and personality factors. He found that synchronistic experiences abound for introverted feeling types, and that they occur especially frequently in stressful life situations.

Several studies conducted in Germany investigated the occurrence of synchronistic experiences and exceptional experiences in a descriptive sense. In a representative survey investigating the frequency of exceptional experiences it was found that 36.7% had precognitive dreams and 18.7% experienced extrasensory perceptions in relation to death or crises (Schmied-Knittel and Schetsche 2003). In another nation-wide representative telephone survey in Germany with 1510 participants, 40.3% stated that they had at least once the experience of a meaningful coincidence which was incompatible with chance expectations (Deflorin 2003).

Temme (2003) found 36.7% of interviewees in a representative study saying they had at least one precognitive dream. The same study also showed that the content of the dreams circled around a limited number of topics: especially death and existential crises as well as outstanding changes in the life of the subjects (e.g., first meeting of their spouse).

The most common form of synchronistic experiences (Sannwald 1959) are dreams and visions (47.9%), premonitions (26.7%) and dreams and visions of a rather symbolic nature (15.1%). Precognitive dreams are often experienced as especially clear, emotionally intensive and easy to remember (Schredl 1999). In a data base collected by the counseling department of the Institute Frontier Areas of Psychology (IGPP) in Freiburg, Germany, containing 1465 cases of exceptional experiences, 6% were identified as "meaningful coincidences" (Atmanspacher and Fach 2013). The same percentage was found in a recent survey of the general population in Switzerland (Fach *et al.* 2013).

All these studies show that synchronicities are fairly well documented at a descriptive level. However, the small number of systematic empirical studies investigating the connection between inner and external events shows the need for more research connecting reports of synchronistic events with context data.

A key problem in studies of synchronicity is that chance expectations can never be exactly computed (or even excluded). The reason is that the base rate for the occurrence of single events is fundamentally inaccessible (Diaconis and Mosteller 1989). Similarly, causal connections between synchronistic events can hardly be excluded with certainty – they could just be too complex to be identified or hidden as common causes for the two events observed (Primas 1996). As a consequence, the difficulties of investigating synchronicities in an experimentally well-confined study design are extreme indeed.

3   Synchronicity in Psychotherapeutic Settings

This situation led the author to design a study on synchronicity using empirical research methods and placing it in the field of psychotherapy. Psychotherapy as a research field has several advantages:

- Psychotherapy is a highly standardized and reduced situation (referring to the setting, persons involved, time frame, space, topic, etc.).

- Context information (patient's pathology and psychodynamics, biography, transference, course of therapy etc.) is available after and prior to the synchronistic event.

- Dreams are documented right after their occurrence (this is the case at least in Jungian psychotherapy where clients are usually asked to document their dreams regularly).

- The therapeutic relationship is usually stable over several years.

- Follow-up investigations of the development after the synchronistic event are possible.

This means that analytical psychotherapy is the designated field for empirical research on synchronicity. It allows the systematic documentation of synchronistic events and offers the possibility of interpreting the events in context, i.e. in relation to the biography and the psychodynamics of the patient as well as with his life situation and his situation in therapy. In this way, the problem of a retrospective reinterpretation or manipulation of the original data can be reduced. Follow-up studies of the impact of the event on further developments are possible.

4   Design of the Study

The general idea of the study is to create conditions under which it is possible to collect data about synchronistic events in psychotherapy and corresponding context data in a systematic way. The goal is to interpret connections between individual psychological conditions and the occurrence of synchronistic events. A necessary step in this project is to establish a documentation scheme. We start with a selected collection of case reports from publicly accessible literature. The following sources have been exploited:

- Ryback and Sweitzer (1990): 23 cases of precognitive dreams,

- Tart (1990): nine cases of transcendent experiences,

- Demoll *et al.* (1960): two cases of spontaneous experiences,

- Bauer and Schetsche (2003): six cases of supernatural events.

This collection of 40 case reports concerning synchronistic experience was analyzed via *qualitative content analysis* (Mayring 2010), an interpretive method that seeks categories inherent in the empirical material. This analysis produced a system of categories presented in the following table.

context	**change**	1. general 2. crisis 3. growth
	stability	1. general 2. conflict
	specific	1. couple relationship 2. family 3. other social relationship 4. work 5. other
	psychological	1. hope 2. fear 3. personal affective relation

	topic, content, symbolism	negative affect positive affect
inner state	type of experience	1. dream 2. hallucination 3. vision 4. premonition (emotion, spontaneous behavior, physiological reaction, physical effect, information) 5. inner voice 6. illusion 7. statement
	focus person(s)	1. self 2. other: familiar, unfamiliar, anonymous
coinciding event	topic, content, symbolism	negative affect positive affect
	manifestation	1. psychological state 2. external event
	focus person(s)	1. self 2. other: familiar, unfamiliar, anonymous
coincidence	subjective explanation	1. god/higher being 2. magic causality 3. transcendental reality 4. unexplainable/anomaly
	consequences	1. topic/focus 2. subjective changes (world concept, self concept, social relations) 3. persistence (temporary/ongoing) 4. dynamic (beginning of, end of, part of development)
relations	time	1. synchronic 2. asynchronic
	space	1. coinciding 2. distant
	focus person(s)	1. participant 2. observer (w/o focus person) 3. representative (active/passsive)

		1. realistic
coincidence type		2. symbolic
	subtype	1. precognition
		2. telepathy
		3. clairvoyance

The results of this qualitative analysis show some first systematic structures inherent in the empirical material. As a first observation, synchronistic experiences occur under special conditions, especially in life situations that are characterized by rapid change, crises or even illness and death. Secondly, synchronistic experiences are typically organized around a so-called focus person which is connected with the change situation. This can be either the reporting person or a proxy person. In many cases the experience leads to changes in the world concept or the self concept of the person, or is part of a dynamic which changes psychological or interpersonal conditions.

5 Case Example

To illustrate the application of the category system, the following case example from Tart (1990) will be analyzed and put into the scheme.

My best friend, Mike, was in a car accident and for approximately a month was in a coma. One night I dreamed that he came to my parent's house. The dream was extremely vivid. We sat and talked for what seemed about an hour, about all kinds of subjects. Mike told me about the wreck, that his girlfriend had not died instantly (like the papers had reported) but that she was okay now, and that he was fine and would see me again one day. The odd thing about the dream was that it was completely real, but not surreal like most of my "vivid" dreams. It really felt like reality. When Mike got up to leave, he mentioned that he wouldn't see me again for a long time, but that I wasn't to be upset, because he was fine. As he walked out the door, he looked back and said that his mom was about to call, and to let her know everything would be okay. I awoke with a start from the dream, and sat up in my bed. About one minute later, at around five in the morning, the phone rang. I had a room downstairs that had been a family room, and it had a phone. I got to the phone before the third ring and answered it. It was Mike's mother. She simply said Mike had died earlier that morning. I was still quite groggy from my sudden awakening, and all I could think of to say was, "I know. He told me". She started crying and hung up the phone. The thing that struck me about this incident was that at the time, it did not seem odd at all. It was simply a fact that Mike and I had talked prior to his leaving. It did not surprise me that Mike had died, because we had talked about that in our conversation, and Mike had told me that his mom would call, so the call did not even seem notable. I did notice a sudden change in my attitude after this event. Prior to Mike's death, I had been consumed by fear of death, often crying myself to sleep worrying about dying, even though I was brought up in a church environment that taught that death was not to be feared. After this incident, I lost my fear of death, but more than that, I gained a love of life, the absence of which had stifled my childhood. I never considered this a case of transcendental experience, in part because it was so normal and natural. However, had I not had this experience, I don't believe I would have had the courage to follow my creative scientific thoughts that lead to my leading an R&D team.

If we put the information from the case report into the category system the following description results:

context	stability	crisis: car accident
	specific	other social relationship: close friend
	psychological	fear of death
inner state	topic content symbolism	positive affect assurance about well-being of friend, goodbye, departing in hope phone call by mother
	type of experience	dream
	focus person(s)	self other: familiar
coinciding event	topic content symbolism	phone call by mother information about Mike's death
	manifestation	external event
	focus person(s)	self other: familiar

coincidence	subjective explanation	none (transcendental?)
	consequences	topic/focus: lost fear of death, gained love of life, courage to follow his creative scientific thoughts; subjective changes of self concept and emotions; persistence ongoing; dynamic: beginning of development
relations	time	asynchronic
	space	distant
	focus person(s)	participant observer (with focus person)
coincidence type		realistic
	subtype	precognition

This first step of the study was not primarily designed to gain insight into the conditions and consequences of synchronicities. A documentation scheme for further data collection, especially adapted to psychotherapeutic settings, has to be added. It is presented in the following table.

patient	psychopathology	e.g., depression, trauma
	biographical background	
	psychodynamics	1. complexes 2. conflict(s) 3. defense mechanisms 4. interpersonal relationships
	personality	1. typology: introvert/extravert 2. psychodynamics: anancastic/hysteric
	earlier synchronistic/ anomalous experience	
	external life situation	e.g., divorce, crisis

therapist	personality	1. typology: introvert/extravert 2. psychodynamics: anacastic/hysteric
	earlier synchronistic/ anomalous experience	
	external life situation	e.g., divorce, crises
psychotherapy	transference – countertransference	
	development	1. therapeutic goals 2. course of therapy 3. current issues, situation
	consequences	1. for life of patient 2. for psychotherapy 3. for therapeutic relation

The second part of this documentation scheme is designed for the psychotherapist to provide the necessary context data from therapy and diagnosis. Hypotheses about the meaning of the synchronistic event and interconnections with the psychodynamics and the course of therapy and the development of patient (individuation) are to be noted in addition.

In psychodynamic psychotherapy in general and especially in Jungian psychotherapy many context data should be available already before the occurrence of a synchronistic event. It is necessary to collect detailed data about the therapist and the patient because in Jungian understanding the synchronistic event occurs in an interpersonal unconscious sphere which is influenced by unconscious conditions from both partners of the relationship.

6   Future Course of the Study

The documentation scheme presented above will be circulated in the German Jung Society (DGAP) inviting participation. In Germany, psychotherapy is integrated into the legal healthcare system, and psychotherapists have to provide extensive information about diagnosis, biography and psychodynamics, and the personality of the patient. They have to develop a schedule for the therapy including prognosis in order to apply for funding of the therapy. This entails that most of the information required for the second part of the documentation scheme is available.

In a later stage we pan to circulate translated versions of the documentation scheme in the international society of Jungian therapists (IAAP) and invite participation there as well. This way it will be possible to create a corpus of cases over time which can then be analyzed both interpretively (qualitatively) and statistically (quantitatively). Based on such analyses, it is our hope to gain more insight into the structure and conditions of the occurrence of synchronistic events in psychotherapeutic settings. This should greatly help to move forward toward an empirically grounded theory of synchronicity.

References

Atmanspacher H., Römer H., Walach H. (2002): Weak quantum theory: Complementarity and entanglement in physics and beyond. *Foundations of Physics* **32**, 379–406.

Atmanspacher H. and Fach W. (2013): A structural-phenomenological typology of mind-matter correlations. *Journal of Analytical Psychology* **58**, 219–244.

Bauer, E. and Schetsche M., eds. (2003): *Alltägliche Wunder: Erfahrungen mit dem Übersinnlichen*, Ergon, Würzburg.

Bender H. (1966): The Gotenhafen case of correspondence between dreams and future events: A study of motivation. *International Journal of Neuropsychiatry* **2**, 398–407.

Bolen J.S. (1979): *The Tao of Psychology: Synchronicity and the Self*, Harper and Row, New York.

Coleman S.L., Beitman, B.D., Celebi E. (2009): Weird coincidences commonly occur. *Psychiatric Annals* **39**(5), 265–270.

Deflorin R. (2003): Wenn Dinge sich verblüffend fügen. Außeralltägliche Wirklichkeitserfahrungen im Spannungsfeld zwischen Zufall, Unwahrscheinlichkeit und Notwendigkeit. In *Alltägliche Wunder: Erfahrungen mit dem Übersinnlichen*, ed. by E. Bauer, M. Schetsche, Ergon, Würzburg, pp. 121–147.

Demoll R., Oliass G., Schumacher J. (1960): Berichte über spontane Erlebnisse. *Zeitschrift für Parapsychologie und Grenzgebiete der Psychologie* **3**(2/3), 184–191.

Diaconis P., Mosteller F. (1989): Methods for studying coincidences. *Journal of the American Statistical Association* **84**, 853–861.

Fach W. (2011): Phenomenological aspects of complementarity and entanglement in exceptional human experiences. *Axiomathes* **21**, 233–247.

Fach W., Atmanspacher H., Landolt K., Wyss T., Rössler W. (2013): A comparative study of exceptional experiences of clients seeking advice and of subjects in an ordinary population. *Frontiers in Psychology* 4:65, 1–10.

Fordham M. (1957): Reflections on the archetypes and synchronicity. In *New Developments in Analytical Psychology*, ed. by M. Fordham, Routledge and Kegan Paul, London, pp. 35–40.

Gieser S. (2005): *The Innermost Kernel. Depth Psychology and Quantum Physics – Wolfgang Pauli's Dialogue with C.G. Jung*, Springer, Berlin.

Guindon M.H., Hanna F.J. (2002): Coincidence, happenstance, serendipity, fate or the hand of God: Case studies in synchronicity. *The Career Development Quarterly* **50**(3), 195–208.

Hanson D., Klimo J. (1998): Toward a phenomenology of synchronicity. In *Phenomenological Inquiry in Psychology: Existential and Transpersonal Dimensions*, ed. by R. Valle, Plenum, New York, pp. 281–308.

Hill J. (2011): *Synchronicity and Grief: The Phenomenology of Meaningful Coincidence as it Arises During Bereavement*, Dissertation, Institute of Transpersonal Psychology, Palo Alto.

Hopcke R. (1990): The Barker: A synchronistic event in analysis. *Journal of Analytical Psychology* **3**, 459–473.

Hopcke R. (2009): Synchronicity and psychotherapy: Jung's concept and its use in clinical work. *Psychiatric Annals* **39**, 287–296.

Jung C.G. (1951): Über Synchronizität. In *Die Dynamik des Unbewussten, Gesammelte Werke, Bd. 8*, Rascher, Zürich 1967, pp. 579–591.

Jung C.G. (1952): Synchronizität als ein Prinzip akausaler Zusammenhänge. In *Die Dynamik des Unbewussten, Gesammelte Werke, Bd. 8*, Rascher, Zürich 1967, pp. 475–577.

Jung C.G. and Pauli W. (1952): *Naturerklärung und Psyche*, Rascher, Zürich. English translation: *The Interpretation of Nature and the Psyche*, Routledge and Kegan Paul, London 1955.

Kelly S. (1993): A trip through lower town: Reflections on a case of double synchronicity. *Journal of Analytical Psychology* **38**, 191–198.

Keutzer C. (1984): Synchronicity in psychotherapy. *Journal of Analytical Psychology* **29**, 373–381.

Main R. (2004): *The Rupture of Time: Synchronicity and Jung's Critique of Modern Western Culture*, Brunner-Routledge, Hove.

Mayring P. (2010): *Qualitative Inhaltsanalyse: Grundlagen und Techniken*, Beltz, Weinheim.

Meyer M.B. (1998): Role of cognitive variables in the reporting of experienced meaningful coincidences or "synchronicity". Dissertation, Saybrook Institute, San Francisco.

Primas H. (1996): Synchronizität und Zufall. *Zeitschrift für Parapsychologie und Grenzgebiete der Psychologie* **38**, 61–91.

Roesler C. (2010): *Analytische Psychologie heute. Der aktuelle Forschungsstand zur Psychologie C.G. Jungs*, Karger, Basel.

Roesler C. (2012): Archetypen - Ein zentrales Konzept der Analytischen Psychologie. *Analytische Psychologie* **170**, 487–509.

Ryback D., Sweitzer L. (1990): *Wahrträume: Ihre transformierende und übersinnliche Kraft*, Droemer Knaur, München.

Sannwald G. (1959): Statistische Untersuchungen an Spontanphänomenen. *Zeitschrift für Parapsychologie und Grenzgebiete der Psychologie* **3**, 59–71.

Schmied-Knittel I. (2003): Todeswissen und Todesbegegnungen: Ahnungen, Erscheinungen und Spukerlebnisse. In *Alltägliche Wunder: Erfahrungen mit dem Übersinnlichen*, ed. by E. Bauer, M. Schetsche, Ergon, Würzburg, pp. 93–120.

Schmied-Knittel I. and Schetsche M. (2003): Psi-Report Deutschland. Eine repräsentative Bevölkerungsumfrage zu außergewöhnlichen Erfahrungen. In *Alltägliche Wunder: Erfahrungen mit dem Übersinnlichen*, ed. by E. Bauer, M. Schetsche, Ergon, Würzburg, pp. 13–38.

Schredl M. (1999): Präkognitive Träume. Überblick über die Forschung und Zusammenhang zum Traumerleben. *Zeitschrift für Parapsychologie und Grenzgebiete der Psychologie* **40/41**, 134–158.

Tart C.T. (1981): Causality and synchronicity: Steps toward clarification. *Journal of the American Society of Psychical Research* **75**, 121–141.

Tart C.T. (1990): TASTE: The Archives of Scientists' Transcendent Experiences. Accessible at www.issc-taste.org.

Temme T. (2003): Ich sehe was, was Du nicht siehst. Wahrträume und ihre subjektive Evidenz. In *Alltägliche Wunder: Erfahrungen mit dem Übersinnlichen*, ed. by E. Bauer and M. Schetsche, Ergon, Würzburg, pp. 65–92.

von Lucadou W., Römer H., Walach H. (2007): Synchronistic phenomena as entanglement correlations in generalized quantum theory. *Journal of Consciousness Studies* **14**(4), 50–74.

Walach H. (2003): Generalisierte Quantentheorie (Weak Quantum Theory): Eine theoretische Basis zum Verständnis transpersonaler Phänomene. In *Auf dem Weg zu einer Psychologie des Bewusstseins*, ed. by W. Belschner, L. Hofmann, H. Walach, BIS, Oldenburg, pp. 13–46.

Wharton B. (1986): Deintegration and two synchronistic events. *Journal of Analytical Psychology* **31**, 281–285.

Williams M. (1957): An example of synchronicity. *Journal of Analytical Psychology* **2**, 93–95.

Complementary Aspects
of Mind-Matter Correlations
in Exceptional Human Experiences

Wolfgang Fach

Abstract

Exceptional experiences of humans are discussed from the perspective of mind-matter correlations within a dual-aspect framework of thinking as proposed by Pauli and Jung. An essential implication of their conjecture is a classification of exceptional experiences which matches a large body of documented client cases and empirical data from the general population. Another implication is that mind-matter correlations typically have a robust (structural) and an evasive (induced) component. Ways in which these components can be interrelated are analyzed with respect to the psychodynamics of exceptional experiences depending strongly on biographic and systemic conditions.

1 Introduction

Individuals reporting exceptional experiences (EE) like extrasensory perceptions, magical influences, apparitions, or poltergeist phenomena are under the impression that scientific principles of causation and laws of nature are suspended. A number of studies (Gallup and Newport 1991, Greeley 1975, Haraldsson 1985, McClenon 1994, Newport and Strausberg 2001, Belz 2009, Schmied-Knittel and Schetsche 2012) indicate that the frequency of EE occurrences lies between 30% and 50% in the population of Western countries and more in other cultural contexts. A representative survey in Germany showed that 50% up to 70% of the general population report at least one EE in their own lifetimes (Schmied-Knittel and Schetsche 2012).

In general, EE differ from diagnozed mental disorders even though the two occasionally overlap (Cardeña *et al.* 2013, Belz 2009, Belz and Fach 2012, Fach *et al.* 2013). Moreover, their status as "anomalous" or "paranormal" seems inappropriate or at least arguable. For reasons to be discussed later in this article we prefer the notion of a deviation from less exceptional (ordinary) experiences.

Moreover, we restrict our investigations to the way in which EE are phenomenally represented in the mental system. For such studies it is not relevant if EE are solely subjective events in the sense of attribution fallacies, delusions, or mental disorders, or if they are objective events violating established scientific knowledge. We use current ideas in the philosophy of mind to propose a systematic phenomenological classification of EE without addressing their diagnostic or ontological status.

The dual-aspect framework of thinking outlined by Wolfgang Pauli and Carl Gustav Jung entails empirical implications for EE which could be successfully validated by a large body of empirical material from clients seeking advice and from the general population. In addition, the Pauli-Jung conjecture allows us to interpret EE as acausal mind-matter correlations induced by psychosocial contexts. Phenomenological characteristics of different patterns of EE and their psychodynamic background will be discussed.

2 Phenomenological Classification of EE

2.1 Mental Representations

Even if patterns of EE appear tremendously varied (see Cardeña *et al.* 2013), a systematic and concise classification has been developed on the basis of a few key postulates of Metzinger's (2003) theory of mental representations (Fach 2011, Belz and Fach 2012). Following Metzinger, human beings create a mental *reality model* as an internal description of parts of reality. This model consists of two fundamental components:

- The *world model* contains representations of the material world including the subject's own physical body. As a matter of principle, the referents of these representations are observationally accessible to other individuals as well, so that intersubjective, sometimes called "objective" "third-person" knowledge about them is possible.

- In the *self model* internal states of the human organism such as sensations, cognitions, volitions, affects, motivations, and inner images are represented. Knowledge about these states is private; it can be experienced only by the subject itself based on "first-person" accounts.

Subjects are able to differentiate internal states from external events in their environment because mental states induced by external sensory stimuli differ from states generated by internal processing. Therefore, touching a hot stove as an event in the physical world can be distinguished from experiencing the resulting pain in the self model. Although the two submodels are separated within the overall reality model, they are ordinarily correlated in a strong mutual psychophysical relationship. The visual and palpable physical body represented in the world model is usually experienced as connected with internal bodily sensations represented in the self model.

The dichotomy of the submodels of the reality model resembles the Cartesian distinction between *res cogitans* and *res extensa*, but in contrast to Descartes's ontologically conceived dualism, Metzinger's distinction is explicitly epistemic. In his basically naturalistic point of view mental processes are understood as a result of physical activity of the brain. We use his approach as a tool only for a systematic classification of EE, without sharing his philosophical underpinnings. From this perspective, EE can be defined as deviations in the reality model of individuals and/or their social surrounding without any commitment concerning their ontological status.

2.2   EE as Deviations in the Reality Model

In the conceptual framework of the reality model, four different phenomenological classes of deviations can be deduced. As a logical consequence of its two components, four classes of deviations are possible: deviations in the self model, deviations in the world model, and two types of deviations in their relation, constituted by connected (coincidence phenomena) or disconnected (dissociation phenomena) elements of both models:

- *External phenomena* are perceived in the world model. This class comprises sensual perceptions of visual, auditory, tactile, olfactory, and kinetic phenomena or inexplicable changes of physical objects. Such phenomena are subjectively seen as a violation of laws of nature or of conventional cause-and-effect relations.

- *Internal phenomena* are perceived in the self model. They include somatic sensations, unusual cognitions, moods, feelings, and inner images. As with external phenomena, the subject is convinced that familiar explanations are suspended. The experiences often appear as ego-dystonic and as an influence of foreign forces.

- *Coincidence phenomena* refer to meaningful connections between ordinarily disconnected elements of the self model and the world model that are not based on the regular senses or bodily functions. Spatiotemporal restrictions may appear as inefficacious, as in several kinds of "extrasensory perception".

- *Dissociation phenomena* exhibit disconnections of ordinarily connected elements of the self model and the world model. For instance, subjects are not in full control of their bodies, or experience autonomous behavior not deliberately set into action. In out-of-body experiences, the mental self is experienced as located outside the body.

3   Empirical Data

3.1   Common Patterns of EE

An extensive body of data about the prevalence and the characteristics of EE has been collected by the counseling department of the Institute for Frontier Areas of Psychology and Mental Health (IGPP) at Freiburg, Germany (Bauer *et al.* 2012). IGPP offers professional help and advice for subjects reporting EE, and all counseling cases have been documented using a classification scheme developed by IGPP personnel.

Based on the theoretical framework described above, our classification is explicitly phenomenological. It offers a systematic perspective on EE which is not entirely in line with the historically evolved catalogues of disorder symptoms, especially in the DSM (American Psychiatric Association 2013) and the ICD (World Health Organization 2010) as standard psychiatric diagnosis manuals. We will show that our empirical material is in striking agreement with our theoretically deduced classification.

A factor analyses of 1465 cases arising from our documentation between 1996 and 2006 (Fach 2011, Belz and Fach 2012) rendered six typical patterns of EE. External and internal phenomena appear uniquely mapped to the scheme presented in Sec. 2.2, while dissociation and coincidence phenomena split into two subclasses. These subclasses can be delineated by a slight dominance of external or internal features, respectively. The six factors (cf. Fig. 1), ordered by decreasing relative frequency of occurrence, can be described in the following way.

1. *Poltergeist and apparitions* (32%) fit into the class of external phenomena. They comprise unexplained movements or changes of objects, sensory perceptions without identifiable sources like acoustic phenomena and mimicry sounds (e.g., raps, steps, voices), visual appearances (lights or shapes etc.), tactile and olfactory phenomena. The phenomena are often ascribed to influences of ghosts or deceased people.

2. Extrasensory perceptions (25%) refer to experiences of coincidences of events without causal connection, but related by some common meaning. They are reported between the inner, mental state – which can be

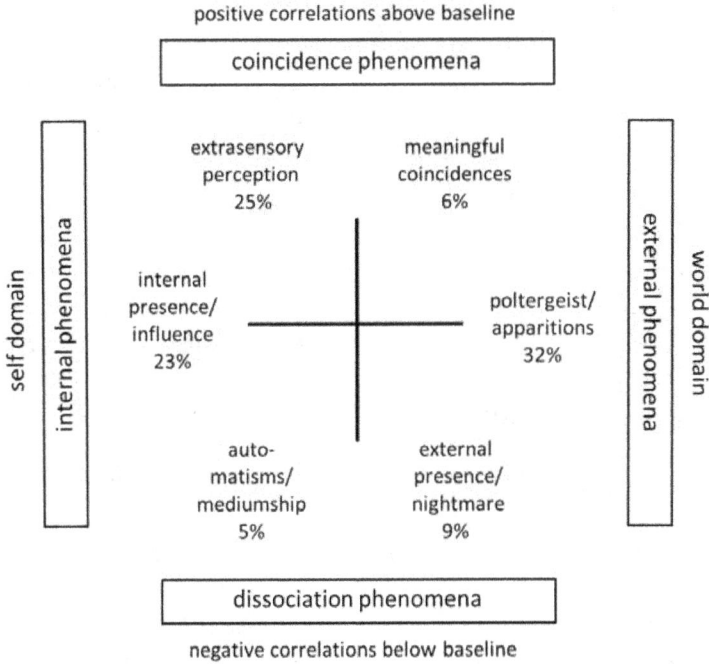

positive correlations above baseline

coincidence phenomena

self domain / internal phenomena

external phenomena / world domain

extrasensory
perception
25%

meaningful
coincidences
6%

internal
presence/
influence
23%

poltergeist/
apparitions
32%

auto-
matisms/
mediumship
5%

external
presence/
nightmare
9%

dissociation phenomena

negative correlations below baseline

Figure 1: Six patterns of EE within four classes. Percentages are relative frequencies of occurrence, for further details see text.

an ordinary mental state or an internal phenomenon – of the affected subject and inner states of others ("telepathy") or external physical events past or present ("clairvoyance") or in the future ("precognition").

3. *Internal presence and influences* (23%) belong to purely internal phenomena like somatic experiences (energy flux, pain) without medically established explanation, thought insertion, inner voices, strange ideas, and inner visual impressions. Subjects affected by such phenomena often assume magical influences or believe that they are possessed by external powers, ghosts, or demons.

4. *External presence and nightmare* (9%) are dissociative phenomena with a tendency toward external phenomena. An invisible entity-like presence is felt by "atmospheric" or even tactile sensations (nightmare) often accompanied by the inability to perform bodily movement (sleep paralysis). Experiences of bodily attacks and sexual assaults are commonly attributed to black magic and demons historically known as *succubi* and *incubi*.

5. *Meaningful coincidences* (6%) refer to coincidences between "objective" external events (e.g., accumulation of accidents, appearance of specific patterns or numbers) among which no causal relation is available or seems plausible. Subjects relate them to one another by attributing salient meaning to them, often in terms of fateful influences, higher powers or conspiracies.

6. *Automatism and mediumship* (5%) are based on psychophysical dissociation with a tendency toward internal phenomena. Subjects experience coordinated and spontaneous bodily movements (e.g., automatic writing, channeling) not voluntarily set in action or controlled by their will. Such phenomena are often interpreted as internal contacts with external forces or entities like ghosts or angels.

3.2   EE in the General Population

In addition to the documentation system used for the counseling cases, a questionnaire (PAGE-R) has been developed at IGPP to survey the clients. The questionnaire is based on the same conceptual framework as the documentation system. For each of the four basic classes of EE, the frequency of their occurrences is assessed for eight items with five possible ratings between "never" and "very often" (see Fach *et al.* 2013 for more details).

The questionnaire was also used in a survey across the general population in Switzerland with a total number of 1580 participating individuals (Fach *et al.* 2013). All factors extracted by suitable statistical analysis are consistent with both the theoretical framework and the results described in the preceding section. However, both frequency and intensity of EE over the entire lifespan of individuals were found to differ considerably.

Frequencies were rated significantly (by about 50%) higher for IGPP clients than for the Swiss sample. The difference in intensity, measured by the question of how deeply previously experienced EE keep preoccupying individuals until today, is larger in the IGPP sample by a factor of about two. Furthermore, IGPP clients show a distinct ambivalence in their valuation of EE. They consider them as "positive and enriching" and as "negative and burdened" at the same time, both on a much higher level than the Swiss sample from the general population.

For both samples, EE occur predominantly in the waking state and mostly spontaneously. This is to say that mental techniques, drugs, contact with occultism and healers, or contexts of extreme situations do not play major roles.

Altogether, there is significant evidence that the phenomenology of EE is organized according to one basic structure consisting of four classes of deviations. The consistency of the results for different samples confirms this assumption. The fact that EE occur with different intensity and frequency in different samples supports that they are distributed along a continuum (as has been proposed for mental disorders by van Os *et al.* 2000).

4 EE and Mind-Matter Correlations

4.1 Mind and Matter as Dual Aspects

The good agreement of the extensive empirical material with the theoretically derived four basic classes of EE raises the suspicion that the phenomenology of EE is grounded on fundamental ordering principles. This resonates strongly with the framework of thinking developed by the psychiatrist Carl Gustav Jung and the physicist Wolfgang Pauli in the mid 20th century (cf. Atmanspacher 2012).

With respect to synchronistic phenomena, i.e. acausally connected meaningful coincidences, Pauli and Jung proposed a fundamental holistic background reality without a separation of mind and matter. Although this overall picture as a whole is entirely at variance with Metzinger's account, his notions of self model and world model may be mapped onto the mental and the material aspects of the Pauli-Jung conjecture. According to their proposal, the mental and the material are conceived as dual aspects emerging from a psychophysically neutral reality which Jung later called the *unus mundus*, the one world (Jung 2006, p. 148):

> Since psyche and matter are contained in one and the same world, and moreover are in continuous contact with one another and ultimately rest on irrepresentable, transcendental factors, it is not only possible but fairly probable, even, that psyche and matter are two different aspects of one and the same thing. The synchronicity phenomena point, it seems to me, in this direction, for they show that the nonpsychic can behave like the psychic, and vice versa, without there being any causal connection between them.

Pauli and Jung saw the role of measurement as a link between local and nonlocal (holistic) domains of reality in physics as mirrored by the act in which subjects become consciously aware of "mental objects", as it were, arising from holistic unconscious contents in psychology. The link between holistic and local realms in both mental and material domains is conceived as *bidirectional*. Unconscious contents can become conscious, and simultaneously this very transition changes the unconscious left behind. Analogously, physical measurement necessitates a decomposition of the holistic realm, and simultaneously this very measurement changes the state of the system left behind. For more details see Atmanspacher and Fach (2013).

4.2 Structural and Induced Mind-Matter Correlations

Conceiving the mind-matter distinction in terms of an epistemic split of the *unus mundus* implies correlations between mind and matter as a direct and generic consequence. These correlations are remnants of the wholeness that is lost due to the distinction made and they are not due to causal interactions in the conventional sense of efficient causation between the mental and the material. Additional correlations may be contextually induced by interventions in the mental or material domain. The Pauli-Jung conjecture suggests a distinction between two basically different kinds of mind-matter correlations (Atmanspacher and Fach 2013):

- *Structural correlations* are the consequence of archetypal ordering factors giving rise to mental and material events at the same time. They arise due to the epistemic split of the *unus mundus* and define a baseline of ordinary, persistent, and empirically reproducible mind-matter correlations (e.g. mind-brain correlations or psychosomatic correlations).

- *Induced correlations* are the consequence of back-reactions that changes of consciousness induce in the unconscious and, via unconscious archetypal activity, in the physical world as well. They depend on all kinds of contexts, occur only occasionally, are evasive and not (easily) reproducible. Induced correlations are represented by deviations above or below the baseline of structural mind-matter correlations.

4.3 Complementarity and Meaning

Jungian synchronicities may be regarded as special cases of induced mind-matter correlations. When ordinarily disconnected elements of self and world appear connected in coincidence phenomena, this represents a deviation from ordinary correlations, i.e. above the baseline of structural mind-matter correlations. The connection manifests itself by the experience of meaning.

Pauli and Jung saw the concept of meaning as a constructive way to characterize acausal mind-matter correlations, and they considered meaningful correspondence and efficient causation as complementary ways to interpret correlations. The experience of meaning, although being subjectively ascribed by the subject concerned, is not completely arbitrary. It depends on the situation as a whole, including conditions that are not consciously available to the subject. According to Jung, synchronistic events arise due to constellated archetypal activity, which limits the range of possibly attributable meanings.

While coincidence phenomena or synchronicities per definition are meaningful correlations between the mental and the material, such correlations are not obvious in purely internal or external phenomena. For a long time, Jung insisted that the concept of synchronicity should be reserved for cases with an experience of meaning that takes on existential dimensions. In later years, Jung opened up toward the possibility that the notion of synchronicity could be conceived more broadly (Jung 2006, pp. 167-168):

> As soon as a psychic content crosses the threshold of consciousness, the synchronistic marginal phenomena disappear, time and space resume their accustomed sway, and consciousness is once more isolated in subjectivity. We have here one of those instances which can be best understood in terms of the physicist's idea of "complementarity": When an unconscious content passes over into consciousness its synchronistic manifestation ceases; conversely, synchronistic phenomena can be evoked by putting the subject into an unconscious state (trance). The same relationship of complementarity can be observed just as easily in all those extremely common medical cases in which certain clinical symptoms disappear when the corresponding unconscious contents are made conscious. We also know that a number of psychosomatic phenomena, which are otherwise outside the control of the will, can be induced by hypnosis, that is, by this same restriction of consciousness.

This quote stresses the important role of a complementary relation between consciousness and the unconscious. It makes clear that the insight into unconscious complexes can lead to a disappearance of induced mind-matter correlations as a consequence of that same insight which, needless to say, manifests itself as an experience of meaning.

4.4 Bonding and Autonomy

Investigations of EE patterns suggest that the associated mind-matter correlations are induced by conflicting complementary human needs, especially bonding and autonomy. Following the attachment theory of Bowlby (1969), two behavioral control systems are most basic and important for human survival and procreation: the attachment system and the exploration system. Their functioning and interplay in child development has been well confirmed by empirical research, and nowadays both systems play an important role in cognitive psychology, developmental psychology, and psychoanalysis.

The attachment system propels the infant into close proximity with its caregiver to get protection, whereas the exploration system drives it to investigate, manipulate, and master the environment. If the exploration system motivates a child to behavior that is sensed as too risky, the attachment system becomes activated and the child returns to the caregiver.

An unimpeded development and differentiation of both control systems usually enables the adult to counterbalance bonding behavior by preserving autonomy for self-reliance and individuation. Individuals who have experienced impaired bonding and attachment in their childhood may have strong feelings of insecurity and difficulties in forming emotional relationships. Unsuccessful experiences of exploration and deficient learning and integration of autonomy-related skills will manifest in avoidance-oriented behavior.

In psychodiagnostics, the tremendous conflicts that can derive from problems with bonding and autonomy find their expression in the "Operationalized Psychodynamic Diagnosis" (OPD) which has been developed by German psychoanalysts, psychiatrist, and specialists in psychosomatic medicine (OPD Task Force 2008). The OPD extends the established diagnosis systems (DSM and ICD) by describing seven basic conflicts and possibly resulting symptoms and mental disorders. Three most important conflict patterns are "dependence versus autonomy", "submission versus control", "desire for care versus autarchy". There is an obvious relation to our notions of bonding and autonomy.

Already Koestler (1972), speculating about the source of synchronistic events à la Pauli and Jung, postulated the tendencies of differentiation and integration as ubiquitous in all domains of life. In human emotive behavior, he saw them reflected in self-assertion, competitiveness, and aggression on the one hand and in adaption, cooperation, and altruism on the other.

As implied above, empirical findings indicate that a number of socially and clinically relevant variables are significantly correlated with EE-patterns (Belz and Fach 2012). Each pattern occurs under specific social and psychodynamic conditions and corresponds with different amounts of bonding and autonomy in satisfaction of needs. If one of both tendencies becomes dominated by the other, it can be repressed by psychodynamic defense mechanisms.

The phenomenology of EE seems to manifest the repressed aspect, but because it is unconscious it cannot be properly interpreted by the subject. In accordance with Jung, counseling practice shows that the conscious realization of unconscious aspects involved in EE implies that the EE will disappear.

5 EE and Psychodynamics

5.1 External and Internal Phenomena

The relation between complementary characteristics of EE and their psychodynamic background can be demonstrated best by comparing individuals experiencing external versus internal phenomena. For instance, the occurrence of external poltergeist phenomena is unpredictable and elusive, whereas the pattern of internal presence becomes manifest as reliable, permanent, and highly personal. Aside from such phenomenological differences, there are a number of correlations with psychosocial and psychodynamic aspects (Belz and Fach 2012). Table 1 summarizes and compares some complementary aspects of internal and external phenomena.

	external type	**internal type**
social situation	family dominant bonding hidden conflicts	single dominant autonomy overt conflicts
characteristics of EE	external physical objective novelty elusivity diffuse threat	internal psychosomatic subjective confirmation persistence concrete threat
social behavior	adapted approving	unadapted challenging
defense mechanism	repression	projection

Table 1: Complementary aspects of external and internal EE phenomena.

External phenomena, especially poltergeist phenomena, often occur in families whose members are characterized by an exceptionally intense need for bonding and reliability. In addition, a thorough exploration of such families usually identifies one family member (the focus person) – often an adolescent in puberty – with a strong but not outspoken need for autonomy. Due to subtle structures of dependence and relationship among the family members, due to psychological immaturity and due to a lack of strategies for coping with conflicts, the focus person represses the desire for autonomy which is incompatible with the desire for attachment.

From a systemic point of view, the repressed internal autonomy manifests itself externally through physical objects that start to "act of their own accord". In this sense, the phenomenology of poltergeist phenomena is complementary to the powerful bonding interaction among the family members. The unconsciously repressed autonomy reappears dislocated in the external world. Neither the focus person nor the family members are aware of the externalized conflicts and the underlying meaning of the EE. Because of the enormous threat imposed by EE, the family bonding gets stronger and the phenomena increase in frequency and intensity.

In contrast, the internal type is socially less integrated, often living on their own, and frequently unemployed. Such individuals spend a lot of time and energy fighting against others which they "identify" as the external source of influence and internal phenomena. In the majority of cases, close or intimate relationships with the suspected source of influence preceded the occurrence of EE. Internal phenomena seem to express an ambivalent desire for attachment which is incompatible with the prevalence of autonomy in the conscious attitude and thus projected onto others.

Apart from the fact that both types of clients externalize the cause for the EE and misplace the responsibility for their problems exclusively into the world domain, the psychodynamics point in opposite directions. Figure 2 shows this circular process: The external type represses his conflicts within the self domain into the unconscious from where they reappear in the world domain, while the internal type projects from the self domain directly into the world domain. Via the unconscious, the EE manifest themselves complementary to the direction of defense because the repressed or projected contents are incompatible with the respective target domain. Repression of autonomy into the unconscious coincides with manifestations of autonomy within the world domain whereas projection of a desire for attachment coincides with manifestations of bonding within the self domain.

The special role of the unconscious in the manifestation of EE is corroborated by observations during counseling interventions exactly as Jung describes in his quote above: If the individual becomes aware of the repressed autonomy or the projected bonding, the corresponding phenomena tend to disappear.

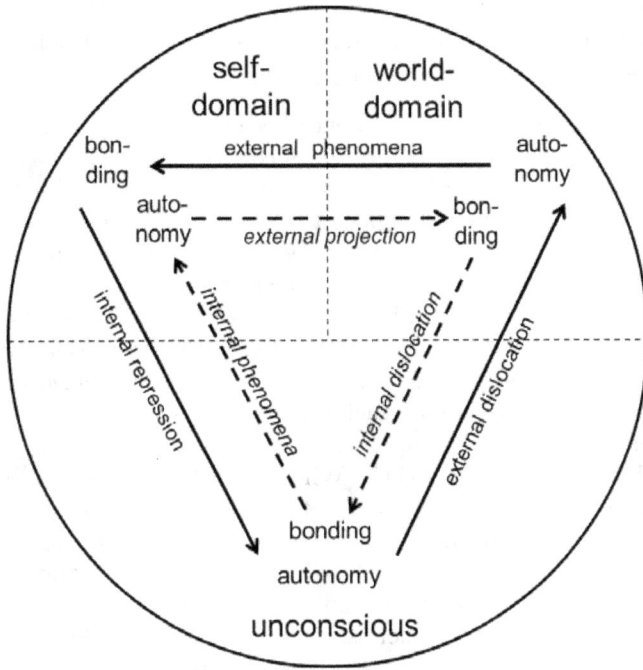

Figure 2: Clockwise and counterclockwise psychodynamics of internal (dashed lines) and external (solid lines) phenomena.

5.2 Coincidence Induced by Dissociation

Even if the former explanations may suggest so, not all patterns of induced mind-matter correlations are based on psychodynamic defense mechanisms. For instance, there are widely known patterns of coincidence phenomena that take place without obvious internal conflicts but under critical external life circumstances. In such cases, typically one of two persons who share an intense and emotional relationship experiences exceptional phenomena while the other undergoes a critical life event. Such cases of "crisis telepathy" seem to reflect the closeness of subjects who react as a system as a whole rather than as two individuals – even if they are spatially separate. As soon as an externally induced issue leads to an existential threat to the dyadic system, its parts are meaningfully associated in their reactions. EE of this kind are singular events and do not repeat, so the afflicted subject typically sees no reason for seeking help or advice.

By contrast, typical coincidence phenomena in the counseling setting are more frequent and persistent, not controllable and therefore burdened. Lasting psychological and social dysfunctioning leads to "chronic" EE as unsolvable dilemmas. Here is a report that illustrates this situation:

> A 50 year-old woman asked for advice because she needed urgent help "to cut through the energy band" which connected her with an internet acquaintance she had met in a grief forum after the death of her father. Via chat, phone and email she spent almost every night for several hours with him and a "crazy closeness" developed: "We could mutually call each other telepathically. I could feel him and knew what he was doing. He also was able to feel me physically". The client reported that they had made experiments confirming their telepathic link. While he had tried to initiate a personal meeting with her she had avoided this and therefore they had never seen each other face to face.
>
> Long before she had met him on the internet, a therapist gave her the idea of being a "survivor twin". Whether there really was a brother who had died during birth could not be clarified. Nevertheless, she believed to have found in this relationship her "missing part" which she had always longed for. In the difficult period of separation from her ex-husband, her internet acquaintance had accompanied her with support and caring. Nevertheless, when she told him about her new and current partner, he responded with a sudden and totally incomprehensible breakup of contact. However, the "energy band" remained and she could still feel him. He, also experiencing similar phenomena, had written her angry emails, accused her of maintaining the connection, and demanded that she cease it. He refused her desire to clarify or discuss the situation.

The counseling process revealed that the client's childhood was marked by an intense feeling of abandonment and subjection. She reported that her father had often been absent for business reasons, and therefore often unavailable when she had needed support. Her mother was unpredictable and violent in her reactions. Often she was locked up by her mother in a dark basement and experienced extreme anxiety. Because of these experiences of chronically threatened bonding and control needs she could not develop genuine trust and experiences emotional dependence on others as a major threat and loss of control. Her strong desire to addict herself to a partner in whom she can fully trust is in conflict with a need for preserving personal autonomy and controlling relationships to feel not helpless. Over her lifetime, the client has developed a pronounced empathy to control others. Numerous "telepathic" and "precognitive" experiences have occurred to her.

Against this background it is understandable that she organizes her partnerships either by emotional restraint or by physical distance, both proportional to the intensity of her feelings. She lives together with her ex-husband in "platonic relationship", has a sexual relationship with a man living separate from her, and experiences her most intense emotional closeness through extrasensory perception of an internet acquaintance she never met personally.

From both a biographic and a systemic point of view her EE express a combination of conflicting needs of autonomy versus bonding and trust versus control in relationships. Searching for intimate closeness while lacking trust and frightened to lose autonomy, the extrasensory perception enables "nearness at a distance" and also gives control over the environment.

In a circular model, the dynamics can be seen as a dissociation of her bonding needs into two aspects. One of them is that she transfers her imaginary twin, her "missing part" with whom she wants to be unified again, onto the person who wants to be in love with her. The other aspect is that, at the same time, she represses her own feelings of infatuation and desire for devotion into the unconscious (Fig. 3). This creates an antidromic dynamic of both bonding components: The brother transference is incompatible in the world domain, because her internet acquaintance wants to be her lover

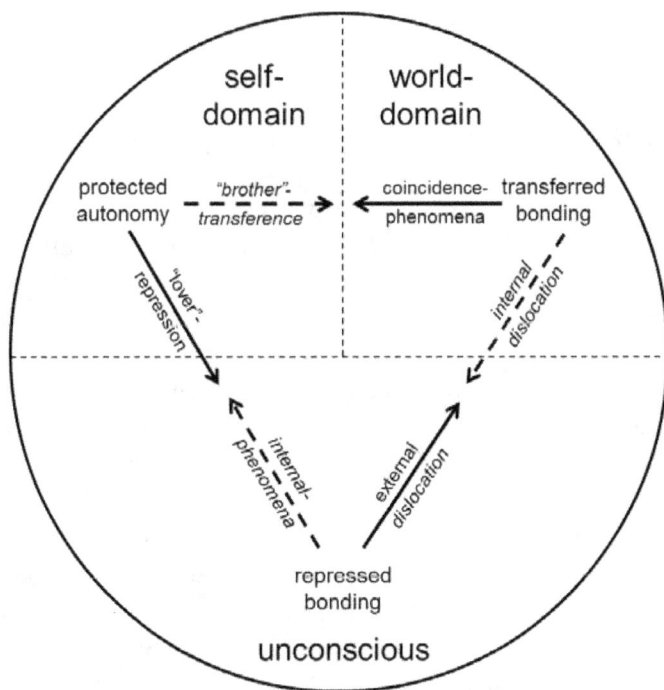

Figure 3: Antidromic psychodynamics of bonding aspects in coincidence phenomena.

rather than her brother. In the other direction, her repressed unconscious bonding component is dislocated to the world domain. Even if her feelings are sensed by her internet acquaintance and compatible with his desire, they are not confirmed by herself. In turn, he represses his frustration and reactions to the unwanted brother-transference and avoids confronting her. Thus, the interaction is very complex and consists of unconscious reciprocal aspects dissociated by defense mechanisms of both individuals, influencing both their self domain and their world domain.

As a consequence of these dynamic transmissions between self and world, meaningful coincidences are experienced. The client interprets these coincidences to confirm her belief in a "crazy closeness" with her supposed twin-brother. She cannot realize her own psychodynamic involvement in the configuration as a whole. The transference might have been resolvable by a personal meeting and confrontation. After a few counseling sessions, she could realize that their telepathic bond is dissolvable only if she takes on responsibility for herself and dares to get involved with a real rather than virtual relationship. In this way, an increase of trust in relationships might weaken the dominance of control and autonomy and allow her more emotional bonding and, ultimately, a decline of her extrasensory perceptions.

6   Conclusions

The approach described in this article proposes a link between exceptional experiences and the dual-aspect monism outlined by Pauli and Jung. Four classes of possible deviations from ordinary experiences have been systematically derived from the Pauli-Jung conjecture, which are consistent with key features of Metzinger's theory of mental representations. Assuming the mind-matter distinction as fundamental for the human reality model, internal phenomena in the self domain and external phenomena in the world domain have been predicted. Concerning the relationships between these two domains, dissociation phenomena and coincidence phenomena have to be expected as well.

Empirical studies with clients seeking advice and with the general population yielded six typical patterns of EE pertaining to the four classes: pure internal and pure external phenomena, and two patterns each for dissociation and coincidence phenomena. The fact that all EE patterns occur not only in clients but also in the general population gives reason to suppose that exceptional experiences are a widespread and inherent part of human life conditions. The distribution of EE over the different patterns is almost identical for clients and for the general population, and the distribution over frequency and intensity forms a continuum.

For these reasons it is implausible to generally assign EE as mental disorders – although there are clearly overlaps. While Metzinger's reality model pictures mind-matter dualism as an epistemic distinction embedded within a naturalistic view, our approach is different. In the Pauli-Jung version of dual-aspect monism EE appear as acausal mind-matter correlations induced by psychosocial contexts and psychodynamic processes via a domain that is neutral to the mind-matter distinction. Hence, induced mind-matter correlations are to be expected as accompanied by unconscious conflicts and symptoms of disorder which should not be too readily pathologized.

The systematic classification and the analysis of complementary relations in the phenomenology and psychodynamics of EE permits specific intervention strategies and has a strong potential for the future development of counseling and therapy of clients reporting EE. Studies of the occurrence of EE in everyday life are significant for both a deeper understanding of human nature and the problem of mind-matter correlations.

7 Acknowledgments

I would like to thank Harald Atmanspacher for his invaluable editorial help with the preparation of the manuscript. His support is greatly appreciated.

References

American Psychiatric Association (2013): *Diagnostic and Statistical Manual of Mental Disorders: DSM-5*, American Psychiatric Association, Arlington.

Atmanspacher H. (2012): Dual-aspect monism à la Pauli and Jung. *Journal of Consciousness Studies* **19**(9/10), 96–120.

Atmanspacher H. and Fach W. (2013): A structural-phenomenological typology of mind-matter correlations. *Journal of Analytical Psychology* **58**, 219–244. A commentary by David Tresan and a reply by the authors are published subsequent to the article.

Bauer E., Belz M., Fach W., Fangmeier R., Schupp-Ihle C., Wiedemer A. (2012): Counseling at the IGPP – An overview. In *Perspectives of Clinical Parapsychology*, ed. by W.H. Kramer, E. Bauer, and G.H. Hövelmann, Stichting Het Johan Borgman Fonds, Bunnik, pp. 149–167.

Belz M. (2009): *Außergewöhnliche Erfahrungen*, Hogrefe, Göttingen.

Belz M., Fach W. (2012): Theoretical reflections on counseling and therapy for individuals reporting ExE. In _Perspectives of Clinical Parapsychology_, ed. by W.H. Kramer, E. Bauer, and G.H. Hövelmann, Stichting Het Johan Borgman Fonds, Bunnik, pp. 168–189.

Bowlby J. (1969): _Attachment and Loss_, Penguin Books, Harmondsworth.

Cardeña E., Lynn S.J., Krippner S., eds. (2013): _Varieties of Anomalous Experience: Examining the Scientific Evidence_, second, largely revised edition, American Psychological Association, Washington DC.

Fach W. (2011): Phenomenological aspects of complementarity and entanglement in exceptional human experiences. _Axiomathes_ **21**, 233–247.

Fach W., Atmanspacher H., Landolt K., Wyss T., Rössler W. (2013): A comparative study of exceptional experiences of clients seeking advice and of subjects in an ordinary population. _Frontiers in Psychology_ **4**, 1–10.

Gallup G.H., Newport F. (1991): Belief in paranormal phenomena among adult Americans. _Skeptical Inquirer_ **15**, 137–146.

Greeley A.M. (1975): _The Sociology of the Paranormal. A Reconnaisance_, Sage, London.

Haraldsson E. (1985): Representative national surveys of psychic phenomena: Iceland, Great Britain, Sweden, USA and Gallup's multinational survey. _Journal for Psychical Research_ **53**, 145–158.

Jung C.G. (2006). _On the Nature of the Psyche_, Routledge, London. Also contained in _The Structure and Dynamics of the Psyche. Collected Works, Vol. 8_, Princeton University Press, Princeton 1969, pp. 159–234.

Koestler A. (1972): _The Roots of Coincindence_, Random House, New York.

McClenon J. (1994): Surveys of anomalous experience. A cross-cultural analysis. _Journal of the American Society for Psychical Research_ **88**, 117–135.

Metzinger T. (2003): _Being No One: The Self Model Theory of Subjectivity_, MIT Press, Cambridge.

Newport F., Strausberg M. (2001): Americans' belief in psychic and paranormal phenomena is up over last decade. _Gallup Poll News Service_, 8 June 2001.

OPD Task Force, eds. (2008): _Operationalized Psychodynamic Diagnosis OPD-2. Manual of Diagnosis and Treatment Planning_, Hogrefe & Huber, Cambridge.

Schmied-Knittel I., Schetsche M. (2012). Everyday miracles: Results of a representative survey in Germany. _Mind and Matter_ **10**, 169–184.

Van Os, J., Hanssen, M., Bijl, R.V., and Ravelli, A. (2000): Strauss (1969) revisited: A psychosis continuum in the general population? *Schizophrenia Research* **45**, 11–20.

World Health Organization (2010): *International Statistical Classification of Diseases and Related Health Problems, 10th revision*, World Health Organization, Geneva.

Business Leadership, Synchronicity, and Psychophysical Reality

David Laveman

Abstract

The political, social, and economic context of today's world indicates a drastic imbalance of material success and qualitative inner values which has led close to truly disastrous situations more than once. A key role in this development is played by leaders in general and business leaders in particular. This contribution argues that more awareness toward synchronistic coincidences in their professional routines and organizational transformation initiatives may be capable of inducing a complementary worldview conducive for a more beneficial future. This will be illustrated by three examples: the significance of serendipities in drug development, the case of Apple founder Steve Jobs, and a historically relevant excursion to Hermann Melville's novel Moby Dick.

1   Introduction

In the second half of the mid 20th century, boundaries between traditionally separate domains of physical, psychological and cultural inquiry were breaking down at an accelerated pace. Established disciplines such as the natural sciences, anthropology, sociology, and religion were beginning to notice connecting patterns of behavior and thought across cultural domains. Their findings raised significant epistemological questions that challenged the materialist worldview dominant in science since the 17th century.

Investigations into the nature of the most fundamental particles of the material world revealed a different order of reality where the previously "universal" Newtonian laws of physics did not apply. The laws of quantum theory, most illustratively Heisenberg's uncertainty principle, gave convincing voice to the limitations of what we can know. Simultaneously, C.G. Jung and his colleagues were discovering previously unsuspected parallels between dream imagery and mythological symbols. The nature of the psyche, the inner life of the individual appeared to have the ability – under certain circumstances – to transcend local realities of time, place and personal history.

Phenomena of this kind dissolve the seemingly unbridgable gap between the mental and the material. Connections between those two realms now became seriously addressed in the psychiatrist Jung's work together with the physicist Wolfgang Pauli. Such connections do not conform to the ordinary understanding of cause and effect, but nevertheless spark an extremely compelling interest to those who experienced them. Such phenomena have subsequently become widely known under the term "synchronicity": a "meaningful coincidence" of events not explainable by a conventional causation of one factor causing the other.

The original formulation of synchronicity provided by Jung and Pauli, as revealed in their correspondence, can be dated from fall 1948 to early 1951. (Their entire written correspondence covers the time from 1932 to 1957.) It culminated in their joint publication *The Interpretation of Nature and the Psyche* (Jung and Pauli 1952) in which Jung, clearly influenced by Pauli's suggestions and criticisms, published his synchronicity essay. Subsequent letters, building upon their insights, speculated about the relationship between psychology and physics, science and religion, and the nature of a reality that produced such phenomena (Gieser 2005, p. 6). The seminal Jung-Pauli correspondence is a key example of a modern, well-documented, cross-disciplinary attempt to bridge the abyss between mind and matter.

Pauli and Jung, each in their own way, embodied many of the crosscurrents and conflicting tensions of their age within their own personalities. Their respective training and creative minds made them excellent candidates to launch this seminal investigation. At age 21 Pauli published an extensive review of the theory of relativity (which remains relevant until today). Albert Einstein gave this laudatory summary (Gieser 2005, p. 17):

> One does not know what to admire most, the psychological grasp of the development of the ideas, the assurance of the mathematical deduction, the deep physical insight, the capacity for lucid and systematic presentation, the knowledge of literature, the technical integrity, the confidence of criticism.

Einstein's extraordinary praise recognizes Pauli's blending of keen intellect and clear exposition of ideas, combined with an unusual (for a physicist) "psychological grasp" and "knowledge of literature". As Einstein clearly perceived, Pauli's mind was not limited to physics alone.

Jung showed this same characteristic multivalent quality of mind. Rowland (2012), in *C.G. Jung in the Humanities*, provides an extended meditation on Jung's importance for the 21st century as a creative writer about "imagination, myth, symbolism, poetics, and literature" (p. 1). Although he was often approached as a psychotherapist, Jung could not divorce the problems of his patients from the larger cultural issues within which they lived. Because Jung's work was inclusive of nature, body, myth, culture, and relationship, "they became spaces where the human sciences, the humanities" could find a "new breath of being" (p. 1f).

The stage was set where Jung and Pauli, by virtue of their transdisciplinary interests, their desire for precision and clarity in communication, and their shared intuitions approached differently, could come into sustained contact that was transformative to both their investigations and themselves.

2 The Shaping Factor of the Social Context

Both Pauli and Jung, highly original and creative in their own right, were deeply influenced by the era in which they lived and worked. The social context within which their ideas developed and found adherents, had been gaining momentum since the beginning of the 20th century. It was dramatically accelerated by the revelations of the period from the beginning of World War I to the end of World War II. Gieser's chapter on *Niels Bohr and the Copenhagen School* (Gieser 2005, pp. 55–89) gives a compelling account of how the "Zeitgeist" of the times beginning at the end of the 19th century shaped fierce debates.

Forman (1971) argued that during this period it became clear that the larger cultural background of a specific era influenced what was considered "objective" science. Physicists and mathematicians were under "the anti-intellectual spirit of the Weimar Republic, which in turn accommodated anti-determinism" (Gieser 2005, p. 57). On one hand this opened the way to an uncritical examination of the emotional release offered by Nazi propaganda; yet on the other, it loosened up a strict, constraining reliance on a purely deterministic and reductionist science.

Unexamined and often conflicting philosophical premises provide a major input into the development of a background matrix from which a foreground social context emerges. Max Planck, for instance, who introduced the quantum of action into physics, remained a believer in causality all his life and rejected the positivism his observations implied. On the other side of the spectrum, Pascual Jordan, a pioneering quantum physicist himself, "embraced positivism wholeheartedly and even saw it as a new, open acausal worldview" (Gieser 2005, p. 59).

From such contradictory viewpoints, Gieser asserts, emerged the famous uncertainty principle: one can give an exact description of something only by limiting it to a partial aspect. A complete description demands not only multiple perspectives, but often even mutually exclusive ones, apparently confounding to the rational mind (Gieser 2005, p. 67). This entailed a shift toward the human condition, including that evanescent notion of "consciousness", and a radical question of what the new evidence of quantum physics meant for the previously straightforward notion of "objective" science.

On a larger scale, the role of language, psychology, philosophy, epistemology, and history became relevant tools of a science that sought to discover the fundaments that constitute reality. Notwithstanding the controversies about how this was to be applied, in what manner, with what definitions, and what agendas – consciousness and the unconscious became subjects of vigorous discussion and debate in the mid 20th century.

Regarding the formation of the idea of synchronicity, the development of quantum physics and especially the uncertainty principle set in motion a receptive intellectual climate for the radical notion of highly unique but valid acausal mind-matter correlations. It showed that one could not split off the subjectivity of the observer from the objectivity of the observed without initiating bidirectional interdependence that altered observer and observed alike. "Meaning", a troublesome issue for a rigid objective science, became important for the very reasons it was once shunned. It spoke directly to the human dimension of psychophysical reality and thus must be taken into account for any theory that aimed at claiming holism.

The collaboration of Pauli and Jung proved fruitful as each person's strength compensated, at least partially, for weaknesses in the other. Both shared enough commonality about the importance of empirical observation and coherent theory, the importance of the unconscious, and a postulated realm outside time and space that had a formative, ordering impact on that which occurred within time and space, to make productive communication possible. In particular, Pauli's brilliance in conceptual thinking and his sensitivity for logical inconsistencies helped Jung translate his superior experience with the unconscious, and his extensive knowledge of alternative approaches to apprehending a reality that existed in other cultures, into a communicable language that sought ascertainable evidence. Jung, in turn, helped Pauli understand the extraordinary fecundity of his unconscious mind, expressed through his highly symbolic dreams, and achieve a better integration of his inferior feeling function.

While Jung and Pauli searched for an empirically based, scientifically coherent "both-and" (Gieser 2005, p. 142) to express their views on indiosyncratic synchronicity phenomena, the larger social context of the mid 20th century illustrated a collective uncertainty about traditional values. This, in turn, created a background context where organized camps emerged to promote "either-or" positions. A few mid-century examples, occurring simultaneously in the larger environment make this point clear.

- The Manhattan project of the 1940s demonstrated that the new insights of modern physics were double-edged. The remarkable accomplishment of splitting the atom made highly efficient and relatively inexpensive energy potentially available. However, the significant difficulties of safe disposal of radioactive waste, and the massive dangers associated with the ever-present threat of a nuclear annihilation created a sense of perpetual uncertainty about the future.

- The mass genocide revealed after World War II raised to global consciousness major questions about the nature of 20th century civilization. The power that mythologies have to shape perception became widely apparent. The human misery associated with destructive mythologies, e.g. Aryan racial superiority and Jewish inferiority, became a poster child for the coercive power of unconscious chthonic impulses.

- "Alienation" became a buzzword of the 1950s to describe the sense of "meaningless" ennui that was understood as a byproduct of a bureaucratic, secular industrialized mass society which lost its connection to authentic sources of wisdom and meaning.

- Political rivalry assumed global proportions and divided themselves into competing, and often mutually exclusive camps fighting for supremacy: Cold war stalemates erupted into periodic hot wars. "Communist" versus "free world" ideologies resulted in hot wars in Korea and Vietnam. Socialist and capitalist camps promised adherents better lives and more equitable prosperity. Sometimes rivalries turned racial, thus South Africa adopted apartheid to secure advantages for the ruling white class against a far larger but disadvantaged black population.

While it is impossible to say how these events directly effected both Pauli's and Jung's thinking, there is abundant evidence that Pauli, by virtue of his Jewish heritage, and Jung, by virtue of his temporary proximity to Nazi ideology, personally felt the power of the larger world in their lives and work. Jung and Pauli (1952) published their joint book precisely because this represented a new view of science that allowed room for a human, non-deterministic creative element in science to emerge.

Quantum physics was seen as incomplete by some, notably Einstein, because it could not preserve a determinism based on the idea of local realism. Pauli, like Bohr, felt that insisting on an overly strict determinism was an indication of its incompleteness. Acausality and synchronicity created a more robust and complete worldview, augmenting the cause-and-effect relations in time with the inconstant connection of the unique quality of each moment and the linking factor of "meaning" for acausally connected coincidences. Gieser sums up the importance of the joint book (Gieser 2005, p. 297):

> Their being published together symbolized the first attempt at a uni-
> fied world view, where the physicist goes into psychology in order to
> understand the development of his own science and the psychologist is
> forced into the world of physics to find parallels with his own discipline's
> discovery of psychophysical connections.

With their publication a sustained interest in synchronicity has been part of the intellectual climate ever since. Its definition, meaning, and implications have been the subject of numerous studies by physicists, depth psychologists, contemporary writers, scholars of religion and spirituality. They all document its value and relevance to diverse issues significant for humankind. Among the disciplines, synchronicity provides insights into topics such as emergence, complexity, chaos theory, quantum mechanics, consciousness studies, neuroscience, psychodynamic therapy, mythology, mystical experience, cultural insights of ancient Eastern traditions, and trickster mythologies of indigenous American and African cultures. Aside from these more scholarly fields, synchronicities have been widely experienced by large numbers of individuals in ordinary population.

However, despite its relative ubiquity and subject of continuous interest for a long time, there is one overarching area that has seen no active research, sustained inquiry, and only passing reference to synchronicity. This area is the field of leadership, and business leadership in particular. Leadership studies, monographs and publications run in the thousands over the past decades. Their sponsors are leading business schools and consulting firms, yet virtually none of them have looked at how synchronicity can be thoughtfully applied. By looking at the dramatic change in the social context since the time Pauli and Jung put forth their original work, the purpose of this contribution is to argue for the importance and relevance of synchronicity research for business leadership.

3 A Turn in the Road

Significant changes in the social context has taken place since Jung and Pauli first formulated their ideas on synchronicity and psychophysical reality. A brief summary of those changes indicates their far-reaching implications:

- *Scientific Breakthroughs*: Chaos, complexity, ecology and neuroscience did not exist as developed sciences in the era of Pauli and Jung. They provide strong evidence for numerous linkages, interconnections and cyclic patterns not well understood sixty years ago.

- *Cultural Confluence*: The mass introduction of yoga, meditation, mindfulness, and Eastern religion and philosophy to the West along with a parallel rise in the availability of organic foods and alternative health regimes suggest a widespread desire for new ways of living and working as a counter-measure to high stress levels of modern life.

- *Economic Turmoil*: The 2008 global financial crisis demonstrated how fragile and interdependent the world economy has become.

- *Democratization of Information*: The invention of the internet, the worldwide web, social media, and widely available personal computing power, makes available once hard-to-access information increasing the potential for unanticipated organization and disruptive innovations that can upend the status quo.

- *Environmental Degradation*: Multiple environmental threats have been identified and tracked that were not widely understood in mid 20th century. Among them are deforestation, melting of the polar icecaps, potential significant natural resource shortages, energy, clean water and food production. Greater accuracy is available of reaching tipping points of no return in the near future if current trends continue.

- *Unintended Risk*: The unintended nuclear disaster at Chernobyl and more recently at Fukushima created a global realization of the potential for wide-scale disaster of what was formerly thought to be safe. The emergence of terrorism as a tactic that could penetrate the most sophisticated defenses put normally non-combatant civilian populations at risk.

- *Chronic Stress and Psychological Distress*: The emergence of fundamentalist ideologies among religions seeks refuge from modernity by abandoning scientific evidence and resorting to literal interpretations. Medication based healthcare, despite some notable successes, also raises concerns that an over-reliance on such mask significant social issues and an epidemic of depression that have a major influence on creating a "meaningful life".

- *Unprecedented Organized Research and Technical Innovation*: The development of cyberspace, artificial realities, social media, nano-technologies, robotics and genetic engineering create new opportunities and dangers faster than they can be assimilated and understood.

- *Rapid and Accelerating Change*: The most fundamental shift since the times of Pauli and Jung, demonstrated in each of the examples above, is the accelerating rate of change. Some large-scale stark examples include the unanticipated rapid collapse of the Soviet Union, the reunification of Germany, regime changes in Eastern Europe, the collapse of apartheid without the anticipated violence, the emergence of the "Arab Spring" with its still to be determined consequences reshaping the Islamic world. All of these changes demonstrate that long-standing social orders that were thought to be relatively stable change quickly and dramatically.

Michael Spence, winner of the 2001 Nobel Prize in economics, provides thought provoking observations when comparing the world of the 1950s to the current era. In his recent book *The Next Convergence* he remembers conversations he had in the 1950s about the gap between rich and poor countries. The "natural inclination was to ask why? How could differentials of that magnitude exist?" (Spence 2011, p. 17). He identifies the human propensity to view the world as a "snapshot" reality rather than a frame in a motion picture.

In today's world this is no longer the case (Spence 2011, p. 16f): "Dynamics and thinking about rapid, accelerating and permanent change is conceptually harder and more than slightly unsettling for most of us." This is a critical fact that encapsulates in a simple statement the change from the time of Jung and Pauli to now.

Its implications are enormous. The escalating rate of change is the theme that runs through all the specific issues noted in the list above. The larger social context of the 1950s saw large-scale polar oppositions come into open conflict with one another. Now, the polar oppositions are still there, but in a more fragmented, chaotic way. Small cells of fundamentalist terrorists can wreak havoc on large-scale modernity, any number of "perfect storms" can erupt, such as the 2008 global financial crises. The unprecedented economic crisis demonstrated how a number of formerly unrelated trends can suddenly coalesce into unanticipated crises. In regards to the 2008 boondoggle the following trends and entities all negatively reinforced one another. The result was near catastrophe.

- Global financial interdependency created opportunities by moving large sums of capital across former geographic and national boundaries with unprecedented speed. The 2008 economic crisis revealed that it could spread panic with equal speed.
- Quickly evolving new financial instruments (e.g. CDSs, credit default

swaps), thought to be understood and theoretically "safe" assuming a stable economic environment, were now revealed to be highly risky in a jittery one.

- Regulators ceased to function as regulators as it became a competitive necessity for them to earn income, by offering additional services to the very companies they had a responsibility to regulate.

- Wall Street made "cheap money" available to suspect companies which felt compelled to keep investing it in additional questionable products, otherwise risk a significant negative impact on share price and reputation.

- Unscrupulous lenders did not fully disclose the risks of back-end mortgages they were selling; while consumers, who were poor credit risks, were only too willing to buy a home they could hardly afford, assuming when it came time to refinance the housing market would continue to rise, thereby increasing their equity.

The outcome was near disaster for the world economy. It was driven by an escalating pace of change that was distinctly different from anything seen in the 1950s. Given the significant change in the "Zeitgeist" and social context of today's world versus that of the mid 20th century, and given the radical implications of synchronicity and its underlying picture of reality, there are two central questions:

> *Is there a way of conceiving synchronicity that maintains its fundamental insight into meaningful acausal connectivity while simultaneously making it applicable to significant 21st century problems associated with the escalating rate of change and increasing fragmentation?*
> *What arena of endeavor is most promising for corresponding research and development?*

The questions themselves presuppose that synchronicity can and should have an applicable value beyond its theoretical and empirical relevance. For some the notion of applicability is fraught with problematic issues, which will be discussed subsequently. The assertion that will now be addressed next is that research is needed into a more precise understanding of the "worldview" implications of synchronicity. It will be argued that leadership in general, and business leadership in particular, is the appropriate and needed arena for the next wave of sustained inquiry and active research into synchronicity.

4   The Rising Power of Global Business and the Role of Leadership

The contextual shifts outlined above have been accelerated by an exponential increase in the emergence, power and influence of the institution of "business". A recent analysis of global business done by a wealth management firm listed core S&P industries[1] they follow and the anticipated percentage of revenue from US and international sources over the next year. First, such revenue breakdown becomes a key indicator of where one puts investment dollars. Secondly and perhaps most importantly the industries that comprise the S&P 500 touch every aspect of ones life (Glenmede Trust 2013): utilities, telecom, financials, health care, staples, industrials, materials, energy, and consumer discretionary (e.g. restaurants, auto-parts, cable TV etc.).

Statistical substance to these facts derives from an analysis of the top Global 2000 companies in Forbes Magazine (DeCarlo 2013). Forbes, a highly prestigious business oriented periodical has been in existence well over seventy years. It uses multiple metrics to determine rankings of the world's biggest companies, equally weighting sales, profits, assets and market value. The findings indicate that "big business" is rapidly expanding into more geographic regions around the world. There is the obvious emergence of China and India as major economic powers.

Less obvious is the fact that 63 countries entered the Global 2000 in 2013 (versus 51 in 2004). In sheer economic clout, the Global 2000 account for $38 trillion in revenue, $2.43 trillion in profit, $159 trillion in assets and $39 trillion market capitalization, and they employ 87 million people worldwide (DeCarlo 2013). This reality alone, independent of millions of smaller businesses around the world, is startling for its concentration of real world economic power.

Cultural observer and depth psychologist James Hillman accurately portrayed the role of business entering into the 21st century. "Business is where the daring and most challenged mind is at work and where power is most central" (Hillman 1995, p. 1). It cannot be ignored for its sheer ubiquity in shaping cultural, economic, organizational, and ecological change. Hillman then states that (p. 1) "the drama of business, its struggles, challenges, victories and defeats forms the fundamental myths of our civilization". It is a short step for him to conclude that "business supplies the ideas that shape our lives, their values, and their ambitions" (p. 2).

---

[1]Standard & Poor (S&P) industry services track numerous industries worldwide and indicate their performance.

With this real-world power there is also the possibility of a darker vision of how current trends, driven by the trance-like organizational and motivational power of business, could develop. Glen Slater (2006, p. 172f) sketches such a possibility:

> If certain trends in technology, psychology and society at large maintain their present direction, thirty to forty years from now a new species of human being will stand before us. Stepping straight from the pages of science fiction, the cyborg (shorthand for cybernetic organism), the human-machine hybrid, will be realized. The nature of being human, human nature will be forever altered – the psyche as we know it will cease to exist. The instinctual and archetypal roots of existence will be detached. The ability to perceive and experience life in a manner that reflects the evolution of the earth and our adaptation to its environment will fade away. Leaving behind these ties to nature we will enter a cycle of development governed by values of efficiency, longevity, rational intelligence and adaptation to technological surrounds. The dawn of a second creation, a remaking of ourselves in our own industrialized and egocentric image, will be upon us.

Judged by its size, reach, organizational power, and influence, the business enterprise of the 21st century is a central determinant of the future into which we will live. A central feature of business, and especially of the largest and most powerful ones, are that they are run by oligarchies, e.g. boards of directors, C-suite executives, strategic business unit heads, and key functional senior management such as IT. In the hands of this relatively small group of institutional leaders there is a vastly disproportionate amount of power to make decisions, direct resources, communicate values, inspire or deflate large employee groups for which they are responsible. They can make superior or inferior products, start or cease to fund research programs, take new directions or follow the herd, use their resources wisely or foolishly, decide to be an agent of transformation or perpetuate the status quo.

To become a leader of such an enterprise there is a high value placed on being action-oriented, having superior communication skills, and being an extraordinary results producer in increasing revenue, profit, market share, product development, and managing change. Often they are scrutinized for long periods of time in a number of different business situations before being given the reigns of power.

Senior executive leaders are facing unprecedented complexity and accelerating, continuous change. Roland Smith, senior faculty member at the Center for Creative Leadership, describes the new environment in which business leaders must thrive as one of perpetual whitewater. His notion of increased turbulence is supported by an IBM study of over 1,500 CEOs who identified their number one concern as the growing complexity of their environments, with the majority of them saying that their organizations are not equipped to cope with this complexity. This theme was consistent among many of the interviewees in the study by Petrie (2011), some of whom used the army phrase VUCA to describe the new environment in which leaders must work:

- V – Velocity: change happens rapidly and on a large scale.
- U – Uncertainty: the future cannot be predicted with any accuracy.
- C – Complexity: challenges are complicated by many factors and there are *few single causes or solutions*.[2]
- A – Ambiguity: there is little clarity on what events mean and what effect they may have.[3]

5   A Role for Synchronicity in Business Leadership

Gieser's summary of the Pauli-Jung dialog resulting in a theoretical study of synchronicity is a superb synthesis of how Pauli and Jung, masters of seemingly contradictory insights (psychological and physical), supported and challenged each other to create a new idea of reality. Their emerging view connected the external, verifiable objective world in a "meaningful" way with the personal subjective factor of the experiencing individual. Elevating the validity and importance of subjective factors, dreams, intense feelings, areas of compelling interest and involvement, and even psychosomatic symptoms (Gieser 2005, pp. 273–298) became legitimate subjects of concern.

Pauli and Jung proposed a worldview in which conscious observers were not disregarded but assumed an important role. Germane for our focus is the continuum between lawful causal connections between events on one side and contingent connections between events by their joint meaning on the other (Gieser 2005, p. 294). As we move toward reproducible causal relations (including statistical causation), the role of the observer becomes less significant. However – and this is the critical point for locating synchronicity within a more comprehensive worldview – as we move in the opposite

---

[2]Italics indicate that there is receptivity to an approach that is possibly meaningful but non-causal. Most CEOs resort to multi-determinism which often sets in motion a plethora of change initiatives which overwhelm the organization, each trying to work on an identified cause with little attention paid to interconnectivity.

[3]This ambiguity about meaning provides an additional opening to the relevance, understanding, and validity of the "meaningful" coincidences that underlie aspects of the concept of synchronicity.

direction toward the unique, limit theorems become inapplicable, statistical analyses loose stability and reproducibility decreases together with a growing importance of "meaningful correspondences" between the mental and the physical.

Synchronicity then was a special case, possibly "induced" through non-causal though correlative mantic methods, e.g. divination. It would be wrong to denote synchronistic events as mere chance events: their uniqueness is due to their joint meaning which is not part of the laws of science. They impress, and sometimes overwhelm the individual (sometimes even groups) who experiences such psychophysical correspondence. Depending on the disposition of those who experience synchronicities, the events could be dismissed as simply good or bad luck on the one hand, or valued as yielding a compelling sense of the uncanny, inexplicable and even numinous on the other. To those inclined toward pondering their subjectively compelling quality, they reveal an *a priori* "order", in which outer and inner events are connected in a meaningful way – for reasons not understood so far (Gieser 2005, p. 296).

Here was a truly radical, revolutionary worldview that did not deny the power of science, but rather insisted, based on quantum physics and depth psychology, to develop it further to a holistic view of reality. Pauli and Jung speculated about the principle that orchestrates psychophysical correspondences. Their conjecture was to step – at least partially – outside the world of matter and psyche and postulate "not only a self-regulating principle encompassing the totality of all psychic phenomena but also a *superordinate organizing principle overarching psyche and world, beyond psyche and matter*" (Gieser 2005, p. 282, my italics). This superordinate organizing principle was conceived as fundamentally abstract (German: "unanschaulich") and required a neutral language for its depiction, devoid of associations with existing orthodoxies, to bring to conscious awareness and to communicate its nature.

From Jung's perspective, the "Self" as the organizing center of the collective unconscious seemed to fit this condition. From Pauli's perspective it was transcendent and symmetrical with respect to the mind-matter distinction. Both agreed about the assumption of "archetypes" as a psychophysically neutral ordering principle, nonexistent in matter, psyche, time and space. Amplifying the archetype with historical examples from the East (*Tao-Te-Ching, I Ching*) and West (Pythagoras, alchemy, Kepler-Fludd), they demonstrated that the idea of archetypes is ancient and universal.

Placing synchronicity in a larger psychophysical worldview points to its immediate relevance for leaders vested with significant power and responsibility. They are ideally positioned to address the escalating rate of change and increasing complex interdependency of today's world. The social context of mid 20th century, by today's standards, represented relatively stable, large-scale conflicting polar opposites. The social context of the current era is significantly different: there is a thawing, morphing and fragmentation of conflicting opposites well beyond the stable boundaries of 50 years ago. Significant tension between entrenched oppositions has leaped the centrally controlled, organized boundaries of mid 20th century, now insinuating itself into multiple areas of our environment, culture and everyday lives.

Today many large multi-national corporations transcend national boundaries. There is little central control over proliferation of weapons of mass destruction often aided by industry to help enable communications, logistics, resources etc., either knowingly or unknowingly. Global financial interdependency demands a coordinated effort from national and regional central bankers who are not infrequently in conflict with one another. Runaway industrialization in Asia has put a severe strain on natural resources and is remaking the balance of power. Ubiquitous democratization of information has undermined those power centers with a vested interest in its control and dissemination. Accelerating technological innovations are disrupting once stable companies and industries: the word "coopetition" has been coined to describe a set of relationships between normally competitive companies that have come to understand that cooperation serves their interests.

As paradoxes multiply, and formerly intact boundaries break down, the resulting fragmentation leads to an increased and deeply felt sense of chaos. It is not surprising that there is greater emphasis on how chaos and order are intimately connected, and how one may emerge from the other. This rapid and unpredictable change is self-evident, thus concern with "tipping points" and "black swan" events take hold, even in the popular imagination. The inescapable role of human consciousness in co-creating outcomes for better or worse is now accepted but not well understood. Neuroscience, quantum physics, depth psychology, and techniques of mindfulness and meditation are looked at with hope and anticipation.

But a word of caution is in order as well. Synchronicity can be misunderstood to directly confront or seek to invalidate established and successful conceptions of inner and outer world. If this remains unconscious, the knee-jerk response to synchronistic events is interpreted as a knack for being at the right or wrong place at the right or wrong time, a regression from the rigors of science to the infantile wish of "magical thinking". Or, conversely, it can lead to even more intense old-style determinism-oriented research to isolate the critical causes responsible for these uncanny coincidences.

I think that all such interpretations are insufficient to today's real-world challenges. It is my hope that taking synchronicity seriously, business leaders may catalyze a radically different worldview that includes highly relational self-world (mind-matter) representations. Synchronistic phenomena are then understood as an outcropping of an underlying holistic reality manifesting itself in a particular way, at a particular time, and with particular individuals.

Based on my 25 years of experience in corporate change and transformation initiatives with global, regional, and domestic businesses, ranging from among the largest in the world to among the smallest, I am convinced that business environments represent a unique challenge to the merits of a robust theory of synchronicity, not found in laboratory science, theoretical research, literary applications, spiritual and personal growth agendas or psychodynamic therapy. These areas represent current fields in which synchronicity has been studied. Some of their contributions have been life-changing on an individual level, but insufficient for the larger issues of today's world.

By contrast, business environments, their leaders and the organizations through which they operate, must be keen accumulators of any possible advantage that increases their odds for highly practical commercial success. Their environment is continuously demanding. It is extremely fast paced. It contains constant scrutiny, especially for publicly listed companies. It must address ever-impending competitive threats and endless possibilities of upending disruptive innovation. The black-and-white nature of "profit and loss" creates an overarching environment to which they must conform or risk severe loss of prestige, independence, and even the ability to survive. Thus there is a decidedly one-sided emphasis on what is pragmatic, efficient, and growth oriented. New projects and initiatives must demonstrate a clear-cut cost-benefit value in advance. For synchronicity to be an effective, alternative perspective for leadership, it must show how it can make a difference to all these pressures.

6 Issues for Synchronicity

Synchronicity suggests a compelling perspective on psychophysical relations transcending disciplinary boundaries and reductive physicalist hegemonies in the sciences until now. This includes, despite of their usefulness for particular specialized questions, modern directions of research in neuroscience and complexity studies as Harald Atmanspacher (2014) aptly notes:

> The core of psychophysical phenomena: a holism in which wholes do not
> consist of parts to begin with ... Pauli and Jung's daring ideas in their
> full scope may persuade us to believe that the repertoire of complex

dynamical systems is not deep enough. Similarly, brain science alone
will be unable to unveil the mysteries of psychophysical phenomena,
neither in the "decade of the brain" nor in decades to come. What is
needed is a new idea of reality, implying novel and refined metaphysical
structures. If we can make progress on this route, it will provide us,
and our culture, with a satisfactory and beneficial worldview.

This view contrasts sharply even with the most advanced business-centric
thinking. For example, recently the *Harvard Business Review* devoted an entire issue to "Strategy for Turbulent Times". In a spotlight article entitled
"What Is the Theory of Your Firm?", Todd Zenger, a senior business professor, concludes his advocacy for theories that develop value over competitive
advantage with a telling statement about the underlying paradigm that constitutes valid theory in the business world (Zenger 2013, p. 78):

> Theories define expectations about causal relationships. They enable
> counterfactual reasoning: if my theory accurately describes my world,
> then when I choose this, the following will occur. They are dynamic and
> can be based on the contrary evidence and feedback. Just as academic
> theories enable scientists to generate breakthrough knowledge, corpo-
> rate theories are the genesis of value-creating strategic action. They
> provide the vision necessary to step into uncharted terrain, guiding the
> selection of what are necessarily uncertain strategic experiments. A
> better theory yields better choices. Only when your company is armed
> with a well-crafted theory will its search for value be more than a ran-
> dom walk.

This statement epitomizes the issue and indicates the opportunities of
the concept of synchronicity if it is to be applicable to business leadership.
Clearly, conventional causal thinking is in effect here and provides a con-
straint on what may be considered "openness to feedback and evidence".
However, the role of theory to provide the vision and confidence needed to
step into uncertain and uncharted terrain demonstrates that theory is indispensible as a beginning step toward greater value and valid comprehension.
The old divide between something that has either causal plausibility or is
merely a random walk is still operative. Synchronicity proposes new options.

According to Main (2007, p. 14), a synchronistic experience is characterized by the following four points: (1) Two or more events parallel one
another through having identical, similar or comparable content. (2) There
is no discernible or plausible way in which this paralleling could be the result of conventional cause-and-effect relations. (3) The paralleling must be
sufficiently unlikely and detailed as to be notable. (4) The experience must
be meaningful beyond being notable.

The major issues of synchronicity that are central for its development in a business leadership context keep its connection to the original definition provided by Jung, and are developed further based on the additional experience with the concept since Jung first articulated it. The points outlined in the following are based on the work of Main (2007) and Atmanspacher and Fach (2013).

- *Context*: Astute scholars and researchers acknowledge that synchronicity and the associated concept of meaning are elusive and defy precise definition. Thus a strategy of broader general categorization may be useful within a more precise working definition (Main 2007, p. 11).

- *Coincidence*: Main (2007, p. 12) cites that the Chambers Twentieth Century Dictionary defines coincidence as "the occurrence of events simultaneously or consecutively in a striking manner but without any causal connection between them".

- *Meaningfulness*: The definition of "coincidence" is close to the core connecting principle of "meaning" that characterizes synchronicity. This is an example of the difficulties that Atmanspacher and Fach (2013) point out when they state that the notion of "meaning" is "notoriously difficult and is often used differently in different contexts".

Main (2007, p. 17) suggests to relax the need for simultaneity of two or more events, while acknowledging that their near simultaneity increases the likelihood that they will be notable. However, in some cases considerable time-lags reduce the possibility of a causal explanation and increases an understanding of the meaningfulness of an event. This may be particularly true in those synchronicities with long-range cultural ramifications.[4]

It is possible for two highly peculiar paralleling events to be solely subjective psychic events in two or more separate individuals with no obvious causal connection. Two individuals having the same dream, mass hallucinations, etc. could be examples of a synchronicity principle at work. If there is an inescapable psychic state in the person who experiences a synchronistic relationship with a connected external event, reflection on its meaning can become a stimulus for significant changes in representations of world and self, resulting in different priorities in decision making.

Acausality should not be regarded absolute. Later knowledge or even different worldviews may recognize causal relations not currently identified.

---

[4] An example of this category of synchronicity will be provided in the example of Moby Dick (Sec. 8.3 below). It demonstrates cultural, business and leadership "meaning" brought together by a confluence of remarkable coincidences.

Von Franz (1980) provides an excellent discussion of the distinction between modern worldviews based on linear time, central for causal thinking, and worldviews based on a "field" orientation to time. In this latter case, interpreting the meaning of the inner-outer circumstances present in a unique moment in time determines the meaning and its implication for action.

7   Inducing Synchronicity in Business

The notion that synchronicity phenomena in which meaning is the linking factor between coincidental events introduces numerous complications. Since the development of quantum theory, observations of external reality have become intimately linked with the nature of the observer. As Gieser (2005, p. 342) says: "Every observation is now seen as a unique *creative* act where it is necessary to choose a perspective on reality. It is in the meeting of subject and object that reality is created." However, choosing one's perspective on reality is a rational decision in the controlled setting of a scientific experiment. The wide range of subjective (conscious or unconscious) influences in the outside world is quite another thing.

In business leadership, the possibility of inducing synchronistic phenomena is of primary importance. It relates synchronicity to practical application, the *conditio sine qua non* for valid interest in business. However, the differentiations associated with practical application in a business setting need careful attention.

It is useful to begin with a business professor's view of the importance of mental models in the context of effective leadership and organizational transformation. Peter Senge, former director of *Systems Thinking and Organizational Learning* at MIT Sloan's School identifies mental models as (Senge 1990, p. 8)

> deeply ingrained assumptions, generalizations, or even pictures or images of how we understand the world and how we take action. Very often we are not consciously aware of the effects of our mental models or the effects they have on our behavior.

His assertion is supported by his report of the rise of Royal Dutch Shell from the weakest of the world's big seven oil companies in the 1970s to the strongest in the 1980s "in large measure from learning to challenge managers' mental models" (Senge, p. 8).

Senge also wrote the introduction to the only business oriented book of which I am aware that directly addresses synchronicity and leadership by Jaworski (2011): *Synchronicity – The Inner Path of Leadership*. This highly readable first-person account is filled with interesting anecdotes about Jaworski's personal transformation from being an attorney to the founding of the "American Leadership Forum". It clearly relates his interest in Jung's theory and a series of meaningful coincidences that supported him on his transformational journey.

It appears that the induction of Jaworski's synchronicity experiences depended strongly on his compelling and all-consuming belief in his idea about the importance of the "American Leadership Forum", his boldness to connect with people with whom he could learn, his capacity to concentrate deeply, and his burning desire to see his ideas become real despite the obstacles he would need to confront.[5] He also was consumed with a mission, distinct from selfish or narcissistic desire, to see his idea incarnate in time and space.

Circling back to Senge's business oriented notion of mental models, Atmanspacher and Fach (2013) remind us that contemporary philosophy of mind, in particular the ideas of Metzinger, assert that "a subject's model of reality as a whole is composed of two basic elements: a self model and a world model". The relationship between the self and world models are given by intentionality and context. When there is a stable connection between self model and world model, they occur as structurally related, such as brain-mind correlations. However, when there is an induced deviation from an ordinarily expected connection, this typically entails an increased intensity of phenomenally experienced meaning. Large deviations are usually infrequent in actual experiences and thus stand out as unique, uncanny and unpredictable.

Gieser (2012) expressed her general skepticism about induced synchronicity phenomena when it comes to motives of personal gain. However, she then succinctly outlined noteworthy observations of the conditions that might be favorable to their appearance:

> In my mind, and according to my ethics, you cannot manipulate synchronicity. But I think that you might prepare conditions in a company or a group that could be fertile for inviting synchronicities. In my mind, a prerequisite for that is genuine and deep commitment. This can come from a creative leader, and it can be in the form of an unconscious "possession" by an idea (in which case the leader might well become a victim of his creative force, finally being consumed by it).

---

[5] As we will see in Sec. 8.2, this conforms well with the case of Steve Jobs who had many such meaningful coincidences throughout his career.

8 Three Compelling Case Studies

8.1 Serendipity and Synchronicity in Drug Development

In his PhD thesis *On Fictions And Realities In Drug Development*, Boyer (2012) gives an informed account of how "rational drug design" developed through early successes, of the limits of such an approach, and of the continuing role "serendipity" plays in discovery and development.[6] Boyer (2012, p. 26) traces the early influence of rational drug design to the success Paul Ehrlich had in linking the

> localization of the foci disease from organs to the side-chains and finally to the receptors. Accordingly, the pharmaceutical focus moved from a systemic to a molecular level. Causes of disease were associated with the occurrence of particulate entities like microorganisms. This idea was subsequently expanded to the presence, absence or the functional state of certain biochemical structures such as specific receptors.

Since these early days there has been a virtual explosion of knowledge in systemic biological and molecular properties aided by significant advanced technologies such as screening methods and computer-based molecular design.

These advances led to a targeted and rational theory in which molecular structures not behaving according to the norm were correlated with a disease. A causal understanding of disease was thought to hold much promise for the development of key methods of contemporary drug development. However, this has not been proven out by the data. Drug releases to the market have declined in recent years while cost has skyrocketed exponentially. Thus the US Government Accountability Office has tracked the increase in research cost by a factor of 2.5 between 1993 and 2004. This has raised important questions for individuals, healthcare providers and political agendas.

Boyer's thesis follows this trend. Through qualitative interviews with expert drug developers he attempts to better understand how "the human is represented in pharmaceutical sciences concerning drug development"(Boyer 2012, p. 3). Boyer's thesis does not explore the implications for pharmaceutical leadership in developing new and efficacious medications, given the changed business environment and declining productivity. However, a hint toward leadership implications can be found in Boyer's next set of findings.

Starting with the attitudes and development of rational drug design, Boyer's observations reveal the greater than expected role serendipity plays

---

[6]See Merton and Barber (2003) for an introduction of the notion of serendipity from a sociology-of-science point of view.

in drug development. Unfortunately an awareness of serendipitous events and discoveries had been steadily pushed to the background as the knowledge of disease states increased and rational methods of discovery were refined. Terms like "innovation", Boyer, (2012, p. 45) notes, were used only retrospectively after marketplace success was demonstrated.

In other words, there was no real curiosity about the nature of true innovation. The dominant rational-causal paradigm invented metaphors like "magic bullet" and "lock and key" to rule public perception. Models were developed to illustrate causal links between macroscopic disease symptoms, specified "master switches" involved in the establishment of a disease and specific molecular modulators. However, there was always a something that did not make sense for all the rationality and selected precedence to support the propagation of the theory.

The real world of actual results, despite the dramatic increase in knowledge and the additional opportunities due to sophisticated technology, indicated a decline in the output of beneficial medications accompanied by a sharp increase in investment needed to produce the same or lesser results. The conclusion to be drawn was that "knowing more does not inevitably lead to increased output". This led to a host of question about application and about "how rational drug development really is" (Boyer 2012, p. 110).

Boyer notes that against the prevailing view toward the efficacy of rational drug design stood findings that clearly suggested that "chance – others may call it serendipity – has a share hardly underestimated in contributing to the ultimate effectiveness" (Boyer 2012, p. 46). The questions therefore still remain as to how one should select substances to be tested from a very large pool of candidates. Once selections are made, experience indicates they are only starting points for the many further transformations the chemical structure of the compounds will undergo. There was constant readjustment to predefined parameters.

Boyer's findings need to become of primary interest to pharmaceutical leadership. For example, he notes that an explanation for serendipity is that (Boyer 2012, p. 62)

> knowledge about a sub-system and the action of a compound upon it is necessarily restricted to the field of investigation. The compound's efficacy on a systemic level, however, cannot be deduced. Hence, finding the perfect match of compound, target and indication can be considered the result of pure luck rather then insightful, goal-oriented development.

As plausible as this statement seems, once one resorts to "pure luck" there is an admission that there is no guiding principle at work. "Pure luck" suggests that only endless trial and error may produce a result.

Cambray (2009) provides additional evidence that while serendipity and lucky accidents are important, they are not synonymous with synchronicity. He notes that it "usually takes time and research to discern whether or not a serendipitous occurrence contains a synchronicity" (Cambray 2009, p. 103). Interestingly, he then cites a pharmaceutical example of the discovery of penicillin to make his point. Without recounting the entire confluence of coincidentally meaningful events that eventually resulted in the development of penicillin during World War II after the discovery of its active agent years before. Once the drug "became commercially available, deaths by bacterial infections dropped dramatically"(p. 105).

On Cambray's account, serendipity may have been present. But since synchronicity emphasizes meaning as the linking factor between psychic state and objective events, the clustering of coincidences that finally resulted in penicillin becoming commercially available deserves closer attention. It was not only Fleming's original observation that a mold he had randomly chosen contained a "halo of inhibition" making it curiously free of bacteria.

In a more detailed account of the discovery, Cambray identifies in more detail the series of coincidences that makes this noteworthy: the mold chosen was extremely rare; the spores entered Fleming's petri dish at a critical moment (any time later, bacteria would have been able to overwhelm the spores); a heat wave in London had just broken when the mold was discovered (otherwise the bacteria would have suppressed the mold's effectiveness).

To turn the mold into "a medical miracle ... again some amazing coincidences were at play" (Cambray 2009, p. 104). The culture of the penicillin mold just happened to be in a building at Oxford where a group of researchers used the mold to isolate an enzyme on the surface of bacteria without a biomedical focus. Two of them, Florey and Heatly, could not create adequate supplies containing penicillin mold extract and went to the US where they were referred to a research center at Peoria, Illinois, because of its large fermentation laboratories. It happened to be the only lab where corn steep liquor would have been discovered. Corn steep liquor is the only nutrient upon which the mold thrived and thereby it could be made available in vast quantities beyond anything possible in Oxford.

And still, the meaningful coincidences do not end here. As the US entered the war, the need to fight infection was paramount and the army began an organized search for the mold, sending penicillin samples to military units around the world. However, it was lab aide Mary Hunt who (Cambray 2009, p. 105) "brought in a yellow mold she discovered growing on a rotten cantaloupe at a fruit market right in Peoria. This proved to be Penicillium Chrysogenum, a strain that produced 3000 times more pencillin than

Fleming's original mold!" With adequate supplies available, the commercial production of penicillin now became feasible.

This example distinguishes the clustering of numerous coincidences that are meaningful both individually and in total because they produced a much needed practical result responsible for saving countless lives. What is not at all clear is which state of mind operating in Fleming, Mary Hunt and others may have contributed to their lucky breaks. This is where leadership, taking the theory of synchronicity seriously, can orchestrate funding necessary for critical research into this mysterious realm.

8.2   The Instructive Case of Steve Jobs

Steve Jobs, founder and long-time CEO at Apple Inc., was well known for his ability to inspire teams of people (as well as enrage them), drive schedules no one thought doable, see possibilities where others could not, and create inventions previously not envisioned. There are numerous references throughout Isaacson's (2011) carefully researched biography to the ability of Steve Jobs to create what became known as a "reality distortion field" (p. 117). One of his co-workers described it in this way: "In his presence reality is malleable, he can convince anyone of practically anything. It wears off when he is not around" (p. 118). Apple co-founder Steve Wozniak, who knew Jobs very well, gives more color to the phenomenon (p. 118): "His reality distortion field is when he has an illogical vision of the future such as telling me that I can design the Breakout game in just a few days. You realize it cannot be true but he somehow makes it true."

Synchronistic experiences happen more frequently when strong emotions are involved. Gieser (2012) noted one has to "burn" for goals, be oriented toward a vision larger than self-interest, and be able to generate a level of concentration that brings mind and body together. Isaacson (2011, p. 561) observed that "the unified field theory that ties together Jobs' personality and products begins with his most salient trait: his intensity ..."

On the external level, Jobs was involved with the creation of a staggering amount of significant, culture changing inventions. He is best known for his role in the first personal computer for the non-geek, the iphone, ipod, ipad. But he was also responsible for revisioning retail experience and creating "the App store which spawned a new content creating industry" (Isaacson 2011, p. 566). Throughout all these inventions there was an overarching drive to produce highly stylized, captivating and technically perfect products. Such zeal, dedication, focus created great things – and came with a well documented shadow side.

Jobs' biography makes clear his unyielding belief that he could create reality the way he likes it. This had both extraordinarily positive and negative consequences in his personal and business life. Isaacson (2011, p. 119) notes that "if reality did not comport with his will he would ignore it". In the instance of the cancer that eventually killed him, Jobs' insistence on creating his own health regime, while delaying proven conventional treatment, cost him the opportunity of catching the disease early enough while there was still time for treatment. By contrast, in his business life, there are numerous well-documented examples in which his traits toward willfully bending reality his way worked in his favor (Isaacson 2011, p. 564): "Jobs' intensity was also evident in his ability to focus. He would set priorities and set his laser attention on them and filter out distraction. If something enraged him ... he was relentless."

Jobs changed the way people interact with one another. However, the interest in him with respect to synchronicity is a better understanding of the phenomenological and qualitative factors operating in his psyche that drew to him the timing, the people and the confluence of objective factors that created the results for which he is known. Jobs himself believed that his Zen training and love of simplicity were such factors. There are enough clues in Isaacson's biography to develop hypotheses about how inner psychic states of tensions may have correlated with corresponding circumstances that catalyzed scores of workers and transformed multiple industries.

Some individuals, for reasons not at all clear, produce synchronicity phenomena by their mere presence. The so-called Pauli effect is one example and Jobs' reality distortion field is another. The absence of a framework, by which to interpret and put in proper context synchronistic experiences that spontaneously happen to business leaders, leaves them ill-prepared to make use of their inner resources to constellate deeper levels of their own psychic endowment. This in turn may be correlated to material changes in external, objective coincidences forwarding or inhibiting significant corporate agendas.

In the case of Jobs, his enormous creativity and ability to come up with new ideas usually was interpreted as a one-off phenomenon having to do with a superficial view of his likeable and unlikable personality traits. For instance, Isaacson (2011, p. 118f) notes that

> a lot of people distort reality, when Jobs did so it was often a tactic
> for accomplishing something. ... At root of the reality distortion was
> Jobs' belief that the rules did not apply to him. ... Rebelliousness and
> willfulness were ingrained in his character. He had the sense he was
> special, a chosen one, an enlightened one.

Without an interest in developing a theory of synchronicity, especially for leaders with highly unusual aptitudes like Jobs, an opportunity is missed to better understand how our psyche and consciousness supports and/or inhibits creative emergence to occur. And additionally, how can negative and painful circumstances be adjusted by modifying something within our psyche with the same level of commitment that is given to designing effective external actions? A theory of synchronicity can give leaders a much needed alternative paradigm for managing today's rapid change in a complex ridden world.

8.3 Herman Melville and Moby Dick

Upon first impression, Herman Melville's novel Moby Dick appears as a great work of literature, a morality tale about man's disassociated relationship to nature and a study in madness and obsession in the person of Captain Ahab. However, it is the suggestive powers of the unusual synchronicities associated with the writing and publication of the novel that make this 19th century epic relevant to this essay. The nature of these synchronicities is documented by Tarnas (2006, p. 239ff).

The objective side of the synchronistic events begins with Melville's birth near Nantucket (Massachusetts) in 1819. Eleven days later, the whaling ship Essex leaves the nearby port of New Bedford. Fifteen months later, the Essex is sank by a white sperm whale in the South Pacific. Survivor Owen Chase writes an account of the incident in which he says the whale "rammed the ship furiously and repeatedly". So far this not dramatically unusual.

Much fiction owes their inspiration to real life events and it would appear Moby Dick is no different. However, we know something of Melville's state of mind after he coincidentally meets the son of Owen Smith, who relays to Melville the events around the sinking of the Essex. There is something about the story that is deeply compelling to him. In Jungian terms, the story awakened in him dormant archetypal energies constellating psychic forces larger than his ego identity. Melville is clearly possessed by extraordinary tensions begging for expression. He describes his subjective state while writing the epic as such (Tarnas 2006, p. 240):

> In the mere act of penning my thoughts of this Leviathan, they weary me, and make me faint with their outreaching comprehensiveness of sweep, as if to include the whole circle of sciences, and all the generations of whales, and men, and mastodons, past, present and to come, with all the revolving panoramas of empire on earth, and throughout the whole universe, not excluding its suburbs. Such, and so magnifying is the virtue of a large and liberating theme! We expand to its bulk. To produce a mighty book you must choose a mighty theme.

Given the intensity of Melville's inner state, it is plausible to assume that he was constellating archetypal energies when writing the epic. His description gives us a clue as to the more precise nature of the archetype. He uses images that suggest wholeness, and a breaking down of the normal subject-object or self-world split: "comprehensiveness of sweep", the "whole circle of science", all the generations of men, "past, present and to come", "the empire on earth and throughout the whole universe". It can be assumed that he was constellating the psychoid archetype (Jung's term), in which his personal internal state and that of the objective world correlate to one another in a meaningful way.

Further investigation tells us something about the quality of experience that may have induced the synchronicity as well as their larger social meaning. Melville's commitment to "produce a mighty book" was expressed in his portrait of the whaling industry, its revenge crazed captain, the numinous quality of the White Whale, and the ultimate tragedy that awaited the crew of the "Pequod". Melville indeed had chosen "a mighty theme".

If the story ended here, the most we could say is that Melville is unusually moved by the story. It unleashes a highly creative, manic, even obsessive need for expression but not necessarily a synchronicity. The coincidence of the departure of the Essex near his birthday and birthplace may be notable but not extraordinarily meaningful. What happens next makes this an example of the paradigm busting, potentially creative power of a true synchronicity. Tarnas (2006, p. 240) writes:

> Amazingly as Melville was completing his book in August 1851 ... the whaleship Ann Alexander was rammed and sunk by an enraged sperm whale it had been pursuing in the same waters in which the same fate had befallen the Essex over thirty years earlier – to this day the only two well documented cases of such an event. Melville was stunned when he learned of this great coincidence.

As compelling as this series of highly unusual coincidental events is, where is its relevance for business leadership in the 21st century? Investigation of the social context of the times in which Melville lived and wrote Moby Dick as well as his personal history give important clues to this relavance. Wagner's (2010) study *Moby Dick and the Mythology of Oil* provides the background. It is only a minor coincidence that Melville was writing Moby Dick almost exactly a century before the seminal book by Jung and Pauli was published with Jung's essay on synchronicity.

By the mid 19th century, America "experienced phenomenal growth and change in almost every aspect" (Wagner 2010, p. 37). Consider that in the 60 years since the founding of the Republic, the number of states increased from 13 to 31, the land mass went from primarily the northeast to spanning the entire range to the far west. The population increased from 3.9 million to 23.2 million and the gross domestic product from $190 million to $2.3 billion. And most importantly the way of life was undergoing a radical change. Between 1790 and 1850 there were thousands of inventions introduced – major and minor – which transformed manufacturing, transportation, mining, communications and agriculture. New Bedford whalers became the dominant players in the worldwide whaling industry. The emergence of economic cycles of boom and bust were happening in short succession. Between 1819 and 1847 there were four of them, causing disruption and dramatic changes in personal circumstances (Wagner 2010, p. 41).

Melville himself experienced the significance of these larger social changes in his own life. Born into a comfortable and socially prominent family, his father's business misadventures led to a dramatic change in family circumstances. Escaping creditors, they were forced to leave New York City at night and move to the Albany area. Melville's biographer Arvin suggests this experience had a traumatic effect on him after his idolized father died when he was still adolescent. Arvin states that he underwent "an emotional crisis from which he would never be free" (Wagner 2010, p. 54).

With these facts a parallel can be formed between the social context of Melville's time and ours: It was a period of extraordinary economic growth – then it was confined to the US, today it is a global phenomenon. As today, it was a time of tremendous disruptive technological innovation as the industrial revolution changed the way people lived. But it was a time of economic growth that revealed a shadow side of boom and busts. Today the accelerated global growth and its concomitant "financial engineering" led to the near global depression of 2008.

The underlying current of mid 19th century compared to the time of the founding of the American Republic was that of rapid, accelerating change, and a growing commercial rapaciousness that exploited human labor and dignity for material gain. Then it was expressed by a growing agitation about slavery, exploitation of immigrant populations for industrial expansion and the displacement of native American populations for access to the natural resources a rapidly growing economy needed. Now, over a century and a half later the issues are still exploitation, cheap labor, abuse of human rights and an explosive commercialism that has become a global phenomenon. Moby Dick, arguably ahead of its time as indicated by its anemic sales and inferior initial reviews, only came to be understood as highly relevant and meaningful in 1924, after D.H. Lawrence declared it as a book of "esoteric symbolism of profound significance" (Wagner 2010, p. 65).

All these suggested meanings are meant to trigger deep reflection; they are not to be understood as "answers" to anything. The example of synchronicities in Moby Dick is highly relevant to the theme of the global dominance of business and its technological handmaidens in the 21st century and the importance of its leadership to steer the ship of business toward stewardship or turn it toward the mad pursuit of dominance. I am suggesting that a robust and informed attention to synchronicity phenomena can serve as a counter-measure to the global obsession with unrestrained commercial growth at any cost.

9 Toward an Organizational View of Psychophysical Reality

Behind every synchronistic occurrence is an implied nature of reality. This nature demonstrates that leaving "meaning" out of the equation, ignoring meaningful correspondences as insignificant, and being just amused by quirky coincidences may miss very important clues as to how to interpret dilemmas business leaders face and have to solve.

Most business organizations overvalue the outer, objective reality and undervalue inner subjective reality. This happens on an individual and collective level simultaneously. Insights made possible by careful attention to inner-outer correlations are missed. Knowledge encouraging favorable synchronicities is not sought. An understanding of the fundamental nature of the organization remains primarily outer-oriented. Thus, most interventions rely on material rewards and consequences to achieve greater outer-world mastery. The overwhelming precedence of this orientation entails that a disciplined inquiry into outer-inner correlations typically is pejoratively labelled as "luck", " chance", or even "nonsense".

The implication of synchronicity research with corporate leaders is that conducting it in an organized and rigorous way can bring to consciousness questions not normally asked. If the basic nature of organizations is more of a psychophysical field rather than a collection of discrete, self-existent entities, then heroic efforts to "control" events may not be needed. Small but authentic shifts in consciousness and a knowledge of unconscious factors may yield substantial results. None of this is guaranteed. But the abundance of anecdotal evidence suggests it is possible.

In conclusion, the argument put forward here is that business leadership, by virtue of its dramatic rise to developing, organizing and distributing vast global resources, manpower, and innovation has to radically remake itself "inside out" and "outside in" to adequately address issues of the 21st century. The modern sixty plus years of synchronistic experiences, both in its popular and more rigorous scientific form, have provided a platform upon which this revisioning of leadership can take place.

Because of its philosophically holistic framework, its access to empirical rigor, its valuing of both conscious and unconscious subjective experience, and its bold assertion of "meaning" as a connecting principle, synchronicity complements a 21st century social context that has at its core paradoxical confluence and accelerated fragmentation of opposites. If we conceive of 21st century civilization as a living organism, then its health, more than ever before, is based on the homeostatic functioning of its global economically interdependent culture.

Although information and communication infrastructure makes economic parity possible, there is a dissociated opposite at work. It emerges symptomatically in the form of relatively small organized groups who have access to this same infrastructure with disproportional power to disrupt its functioning, just as rogue cancer cells must be effectively addressed or they will kill their host environment.

It is worth keeping in mind Jung's remarks to Pauli in a letter of 15 December 1956, in which he emphasizes the importance of discrimination without dissociation (von Meyenn 2001, p. 800):

> As soon as an individual has managed to unify the opposites within himself, nothing stands in the way of realizing both aspects of the world objectively. The inner psychic dissection becomes replaced by a dissected worldview, which is unavoidable because without such discrimination no conscious knowledge would be possible. In reality, however, there is no dissected world: for a unified individual there is one "unus mundus". He must discriminate this one world in order to be capable of conceiving it, but he must not forget that what he discriminates is always the one world, and discrimination is a presupposition of consciousness.

The clear implication for leadership is that their primary task in the current era needs to be revisioned toward unifying the opposites within themselves. What makes this inside-out challenge more difficult is that a leader's rise to the top of a business organization is an expression of their demonstrated mastery of the outer environment. Their bias is toward action and not reflection, toward problem solving and not introspection, toward technical innovation and not alchemical transformation.

However, it would be a mistake to take Jung's remark out of context. Its recipient was Wolfgang Pauli, the physicist, its subject was the inextricable link between inner and outer world, and its intention was discrimination and the indispensible role of consciousness. Thus, in this short concluding quote, we have recapped in miniature that the synchronicity experience – though not explicitly stated – points toward a potentially transformational approach uniquely adapted to the complex concerns of the current era.

Acknowledgments

This work could not have been accomplished without the support of Harald Atmanspacher, chief editor of the present volume. His enthusiasm for the project was invaluable in helping this previously unexplored area come into being. Nathan Schwartz-Salant has continuously modelled the fierce compassion needed to burn through obstacles. My friend and coach Tony Freedley, a former Navy Seal, demonstrates the warrior spirit needed to play to win in a big game. My daughter Cara makes sure to remind me that my failures are no big deal and my successes never go to my head.

References

Atmanspacher H. and Fach W. (2013): A structural-phenomenological typology of mind-matter correlations. *Journal of Analytical Psychology* **58**, 218–243.

Atmanspacher H. (2014): Notes on psychophysical phenomena. In *The Pauli-Jung Conjecture and Its Impact Today*, ed. by H. Atmanspacher and C. Fuchs, Imprint Academic, Exeter.

Boyer M.R. (2012): *On Fictions and Realities in Drug Development – An Account on Rationality in Drug Development*, PhD Thesis, ETH Zurich.

Cambray J. (2009): *Synchronicity: Nature and Psyche in an Interconnected Universe*, Texas A&M Press, College Station.

Corlett J.G. and Pearson C.S. (2003): *Mapping The Organizational Psyche*, Center For Application of Psychological Type, Gainesville.

DeCarlo S. (2013) The world's biggest companies. *Forbes Magazine*, April 17, 2013.

Forman P. (1971): Weimar culture, causality, and quantum theory: Adaptation by German physicists and mathematicians to a hostile environment. In *Historical Studies in the Physical Sciences, Vol. 3*, pp. 1–115.

Gieser S. (2005): *The Innermost Kernel*, Springer, Berlin.

Gieser S. (2012): private communication October 17.

Glenmede Trust (2013): Market Roundup, Glenmede Factset S&P 500.

Hillman J. (1995): *Kinds of Power*, Doubleday, New York.

Isaacson W. (2011): *Steve Jobs*, Simon and Schuster, New York.

Jaworski J. (2011): *Synchronicity: The Inner Path of Leadership*, Berrett-Koehler, San Francisco.

Jung C.G. and Pauli W. (1952): *Naturerklärung und Psyche*, Rascher, Zürich. English translation: *The Interpretation of Nature and the Psyche*, Routledge and Kegan Paul, London 1955.

Main R. (2007): *Revelations of Chance*, SUNY Press, Albany, pp. 11–37.

Merton R.K. and Barber E. (2003): *The Travels and Adventures of Serendipity*, Princeton University Press, Princeton.

Petrie N. (2011): Future trends in leadership development: A white paper. *Integral Leadership Review*, December 2011, 1–10.

Rowland S. (2012): *C.G. Jung in the Humanities*, Spring Journal Books, New Orleans.

Senge P. (1990): *The Fifth Discipline*, Doubleday, New York.

Slater G. (2006): Cyborgian drift: Resistance is not futile. *Spring* **75**(1), 171–195.

Spence M. (2011): *The Next Convergence*, Farrar, Gross, Giroix, New York, pp. 11–17, 140–147; 270–273.

Tarnas R. (2006): *Cosmos and Psyche. Intimations of a New World View*, Penguin, New York.

von Franz, M.-L. (1980): *On Divination and Synchronicity*, Inner City Books, Toronto.

von Meyenn K., ed. (2001): *Wolfgang Pauli. Wissenschaftlicher Briefwechsel Band IV/3*, Springer, Berlin.

Wagner, R.D. jr. (2010): *Moby Dick and the Mythology of Oil: An Admonition for the Petroleum Age*, PhD Thesis, Pacifica Graduate Institute, Santa Barbara, pp. 18–74, 187–203.

Zinger T. (2013): What is the theory of your firm? *Harvard Business Review*, June 2013.

Time and Tao in Synchronicity

Beverley Zabriskie

Abstract

From their different perspectives, the depth psychologist C.G. Jung and the physicist Wolfgang Pauli sought to observe and speculate on the underlying patterns of an interactive world. Their meeting might have remained a "frozen accident", a random coming together of two powerful intellects. Instead, their collaboration became meaningful. Through their explorations of acausal connections through contingency, equivalence, and meaning, they extended mind's participation in the universe with the surprising theory of synchronicity. In their discussions, they engaged the profound questions of the role of the mind in the perception of and amidst the matters of space and time.

The garden of forking paths is an incomplete but not false, image of the universe as Ts'ui Pen conceived it. In contrast to Newton and Schopenhauer, your ancestor did not believe in a uniform, absolute time. He believed in an infinite series of times, in a growing, dizzying net of divergent, convergent and parallel times. This network of times which approached one another, forked, broke off, or were unaware of one another for centuries, embraces all possibilities of time. We do not exist in the majority of these times; in some you exist, and not I; in others I, not you; in others, both of us. In the present one, in which a favorable fate has granted me, you have arrived at my house; in another, while crossing the garden, you found me dead; in still another, I utter these same words, but I am a mistake, a ghost.

Jose Luis Borges (1962, p. 28)

1   A Timely Meeting

When the eminent psychiatrist C.G. Jung met the brilliant young scientist Wolfgang Pauli, the potential synchronicities between them were not apparent. The two men were of different generations, disciplines, contexts. At age 57, Jung was grounded in a cohesive life of family, research, and practice.

Then 32 years old, Pauli was rootless and fragmented. While his theories had solved key issues in physics, his personal life was dissolved in emotional chaos.

By 1932, Jung had integrated many facets of his theory, yet was yeasty in his intellectual ferment as he re-layered and re-framed the psyche. He had posited the four psychological types, the constant conjunctions and disjunctions of consciousness and the unconscious in dreams and fantasies, the imagistic configurations of psychic attitudes in shadow, anima, animus, and the presence of archetypal patterns constellated when an individual meets existential challenges. Jung was impacted by William James' perception of complementary kinds of dynamics in the psyche (James 1890, p. 206) between the "upper self" and the "under self". James had also posited a complementarity between psychology and physics, going beyond earlier ideas of psychophysical parallelism.

Jung had re-emerged from the emotional turmoil and imaginal tempest he recorded in his journal, *The Red Book*. In its words and images, he had tested the Jamesian margins between conscious and unconscious fields of knowing, between magic and mystery, art and science, the centro-verting signals of dreams and the extroverting mandates of a relational life. He was also mourning the death of a valued colleague, the sinologist Richard Wilhelm.

Jung had already tried to add the paradigms of physics to his psychological, philosophical and religious approaches to psychology. Despite his awkward ignorance of mathematics, Jung recognized parallels between Einstein's relativity of spacetime and the relativities inherent in the psyche as a "process", a "multiplicity within a unity", an affective and meaning-making organism. In Jung's model, the individuating psyche aimed for an equilibrium between consciousness and the unconscious, ego and its charged complexes. It lived on a continuum of current states of mind amidst memory and anticipation.

While still young, Pauli had reviewed Einstein's theories with an acumen that surprised the respected genius. Before he met Jung, Pauli's "exclusion principle" had unpacked the structure of the periodic table, altering the knowledge base of both physics and chemistry. He had also worked with Neils Bohr, whose idea of a complementarity of particle and wave pictures became important in quantum physics. Through Pauli's and Jung's commitment, between 1932 and Pauli's death in 1958, the Jamesian notion of complementarity met Bohr's concept of complementarity.

Jung guided Pauli toward an embrace of his emotions, a romance with his inner life, a fidelity to psyche. Through Jung, Pauli entered both his personal lifetime and the intensive domain of his dreams. For his part, Pauli applied his acute intellect to their mutual pre-occupation with mind's place in nature. Their dialogues focused on the interplay between the mental and material, the stretch of the mind-body continuum, originating from the earliest differentiations of energy at the birth of the universe to the human impact in our contemporary anthropocene age.

In his analytical psychology, Jung had long left a reductive view of the past as a singular cause for one's present and future. The unconscious was not packed with one's earlier repressions, but was a subliminal source of ongoing commentary aiming toward integration and resolution. For Jung, the psyche consisted of emergent, vibrational points from generational pre-dispositions, one's evolving personal identity, responses to immediate experience, and an ever-amplifying capacity for reflection. Its internal interactions were re-enacted in relationships, and intensified in his model for depth analysis as a two-person interactive field within and between patient and analyst. While transference of the patient on the analyst first carries the energies of earlier relations, it is a psychic vector toward more inclusive and expansive realization of a microcosmic self in a macrocosmic domain.

When Pauli contacted Jung, he had fallen into the cracks of his uneven and asymmetrical development. His desire to penetrate and understand was undone by a tendency to unravel and destroy. In Pauli, Jung saw intellectual acumen and emotional turbulence, thinking superiority and feeling inferiority. Pauli could not contain his emotionality. His affects manifested in an agitated mind, compulsive actions, dark obsessions, disturbing images, unwelcome nightmares. His presence had an odd affect on his environment. In a passage applicable to the "Pauli effect", the reported incidents of objects shattering in Pauli's proximity, Jung (1970a, par. 660) once wrote that archetypes

> seem to belong as much to society as to the individual; they are therefore numinous and contagious in their effects. (It is the emotional person who emotionalizes others.) In certain cases this transgressiveness also produces meaningful coincidences, i.e. acausal, synchronistic phenomena.

Jung also recognized an authentic suffering person. As he wrote in *Mysterium Coniunctionis* (Jung 1970b, par. 772):

> For no one who is one himself needs oneness as a medicine – nor we might add does anyone who is unconscious of his dissociation, for a conscious situation of distress is needed in order to activate the archetype of unity.

As he moved beyond his negativity, Pauli developed a capacity for dialogue with his own and others' feelings, symbols, philosophies. His personal statements and welcomed dreams revealed an extraordinary facility to extract, blend, and synthesize Western philosophies and Eastern wisdom traditions.

While towering intellects of their own eras, neither Pauli nor Jung were culturally provincial. They studied holistic symbol systems informed by binary pairings of all kinds of opposites, and pondered historical images and intuitions about the attraction/coagulation and repulsion/acceleration of forces, intimations of such current hypotheses as dark matter and dark energy.

Each engaged the mantic process of the *I Ching*, based on explicit analogies between the mental and environmental, to find and extract wisdom from outside themselves on themselves. In his physics, Pauli embraced Eastern models of symmetry, while Jung responded to the goal of equilibrium between alternating forces of Yin and Yang emerging from the Tao.

Jung and Pauli did not dismiss pre-enlightenment, pre-Cartesian perspectives. Foucault (1973, p. 17) reminds us that

> up to the end of the sixteenth century, resemblance played a constructive role in the knowledge of Western culture. It was resemblance that largely guided exegesis and interpretation of texts, organized the play of symbols, made knowledge of things visible and invisible, and controlled the art of representing them.

He lists the era's prevailing forms of similitude as *convenientia* – adjancy of place, juxtaposition, *aemulatio*, analogy, and sympathies (Foucault 1973, pp. 18–25). Jung and Pauli's consensus on synchronicity added the inconstant connections of contingency, equivalence, and meaning for events not explicable by temporal and linear causality, the constant, repeatable connections of cause and effect.

Pauli and Jung worked on and beyond professional frontiers. They accepted the co-existence of the impersonal and the subjective, the interpenetrations and mutual mirroring of matter and mind. Trusting that ancestral minds were engaging and changing their worlds, they did not dismiss the musings and wondering of the past. Aware that the psychologies and sciences of their day were in their beginnings, they did not cling to what was secure in the present. In their speculations on synchronicity, Pauli was expanding the causal rigidities of traditional science to non-causal probabilities, while Jung was releasing interpretation and identity from magical thinking about cause and effect.

2   Surprise and Synchronicity

In 1925, Pauli had theorized that, in addition to the known quantum numbers, an electron possessed an invisible spin as a fourth property. Then, in 1930, two years before contacting Jung, he surprised himself: "I have done a terrible thing. I have postulated a particle that cannot be detected" (Hirsch *et al.*, p. 42). Pauli's neutrino, an elusive, "ghostly" fundamental particle, was detected almost 30 years later.

Meanwhile, in his field, Jung had posited four primary functions in the ego's apprehension of reality. He had also described what could be seen only through their effects, the transgressive, non-subjective, trans-ego, archetypal dominants of a collective unconscious operating in a psychoid realm beyond mind and matter.

The Jung-Pauli study on synchronicity posited realities as yet beyond human ken, made manifest through shock and surprise. Surprise is a stimulus, an imprinting for the sake of action. Surprise is also a reaction to an unexpected conjunction in space at an unexpected time. Surprise is a prime and universal emotion, essential for survival. Surprise is the human, emotional concomitant of symmetry breaking.

In synchronistic experience, from the flint of a physical encounter and/or mental recognition, comes the spark which mandates a meaning-making move. Then synchronicities may emerge from an in-between: between the intrapsychic and the interpersonal, between inner states and outer events, between different minds, between dream and symptom and other mind-body interactions, between a misty dream figure and a beloved later met, between the now and the then. Synchronicities require a capacity for granting value, for reflection, evoked as psyche, and for making meaning. They depend on analogy, receptivity and openness to multi-dimensional realities.

While the arousing effect of surprise is an immediate emotional/physical reaction, response and reflection unfold over time. It is as if time slows, allowing mind to adjust to the matter at hand and psyche to find relevance through making the connection significant. When surprise transforms into a synchronicity, there is a relativization of one's world view and place within it. As in meta-cognition, one observes oneself as observer, breaking down barriers between an isolated self and a surround which has "happened" to one, and so demands to be observed. The psyche's granting of meaningful relevance comes with unfolding associations, as an "act of creation in time".

Since time immemorial, time exists in mythic, philosophical and scientific domains. Just as the "problem of time" in astrophysics and cosmology arises from the absence of time as a fundamental property in a universe of timeless laws, the theoretical symmetry between Pauli and Jung broke around the issue of time in the synchronistic moment. This invites us to consider variations of the concept of time.

3 Since Time Immemorial

Time immemorial is a singular state, before recounting and recording history began marking time, before memory offered the prime material for "the remembered present" of consciousness. Time immemorial is akin to empty space, paradoxically buzzing with potential. For time immemorial evokes a vague sense of implicit recollection.

In immemorial time, all was innate and implicit, present and complete. In this pre-initial state, not yet limited and specific, there were no initial conditions, no fundamental laws, no frozen accidents, no forces or gravities, no perturbations, no imagined parameters of time and space, and certainly no living observer. The cohesive intensity of all possible matter and energy was itself the one moment and place, an *unus mundus* which preceded the on-going separation of fields and forces.

In Taoist thinking, time immemorial is akin to the void, the not yet existent, wherein no connectedness is needed, since all is one. In his essay "Synchronicity: An Acausal Connecting Principle", Jung (1978, par. 918) quotes Lao-Tze:

> There is something formless yet complete
> That existed before heaven and earth.
> How still! How empty!
> Dependent on nothing, unchanging,
> All pervading, unfailing.
> One may think of it as the mother of all things under heaven.
> I do not know its name,
> But I call it "Meaning"
> If I had to give it a name, I should call it "The Great".

During his *Visions Seminars*, Jung (1997), p. 1025) described the Tao this way:

> Tao is the void, it is the utter emptiness and silence; therefore it is immortality because it is being forever. It is timeless, it has no attribute of time, and it is free from the pairs of opposites because it has no quality.

And Pauli wrote (Meier 2001, p. 92):

> Nonbeing is that which cannot be thought about, which cannot be grasped by thinking reason, which cannot be reduced to notions and concepts and cannot be defined. It was along these lines, as I see it, that the ancient philosophers discussed the question of being or nonbeing.

Immemorial time is the vacuum which also contains the fullness of the pleroma, those fleeting, "filled" moments in incarnated time, as elusive and quick as the latest quark. Immemorial time is also the realm of Nonbeing, from which all linked forms emerge.

Time immemorial is a paradoxical human conceit, a liminal notion of remembered time before memory. Past, present, and future were bundled in timelessness, and occupied spacelessness. To imagine it fills one with yearning for the potential before the beginning. Between all that was not, was, and was to come, an all-containing mindfulness had an omniscience embracing infinity and ubiquity.

Von Franz (1992, p. 252) notes that Jung spoke of an initial "luminosity" or "cloud of cognition". For her, the "Jungian concept of a single energy that manifests itself in lower frequencies as matter and in more intense frequencies as psyche in many ways resembles the Chinese idea of ch'i". Those aware of absence of self or of "ch'i" seek its return. Such a search brought Pauli to consult Jung and his own psyche. In *Keeping Things Whole*, the Pulitzer-Prize winner Mark Strand (1979) muses:

> In a field
> I am the absence
> of field.
> This is
> always the case.
> Wherever I am I am
> what is missing.
>
> When I walk
> I part the air
> and always
> the air moves in
> to fill the spaces
> where my body's been.
>
> We all have reasons
> for moving.
> I move
> to keep things whole.

4 Memorial Time And Cosmogonic Moments

Astrophysics describes the world unfolding from an infinitesimal speck of energy. The "before" ends with the turbulent rotations of the first fractions of a second, while in the "after" greater spaces expand the universe into infinity. As intensity extends, time is brought into space, incarnation becomes interaction through separations and multiplication, collisions and mergings, pairings and symmetries, breaking out and breaking through, moving through and moving out.

In mythic cosmogonies, at a critical point with critical mass, once before time breaks into once upon a time. The ancient Egyptians imagined that instinct and the longing for contact converged to create various aspects of the world. In one tale, the High God's ejaculation spewed his seminal force into the primeval Ben-Ben mound. Above and below, the dynamic and receptive, spirit and matter meet as the phoenix Ben-Ben bird alights on the new ground emerging from soundless, indistinguishable, dark waters. The sound of its piercing cry calls forth the world.

Close to the source of original unity, simple creation appears with symmetry, harmony, and balance. One's emotional response places one on the continuum of nature. Jung experienced such a moment in Africa (Jung 1989, p. 255):

> To the very brink of the horizon, we saw gigantic herds of animals; gazelle, antelope, gnu, zebra, warthogs, and so on. Grazing, heads nodding, the herds moved forward like slow rivers. There was scarcely any sound save the melancholy cry of a bird of prey. This was the stillness of the eternal beginning, the world as it had always been in the state of non-being; for until then no one had been present to know that it was this world. There I was now, as if the first human being to recognize that this was the world, but who did not know that in this moment it was as if he had first really created it. There the cosmic meaning of consciousness became overwhelmingly clear to me.

When creation tales mimic emotional processes, the time before time is emotionally empty, without affect, or feeling. Felt emotion brings realization. When the Hellenic high god laughs, the gods come forth. When his laughter breaks into a sob, the soul emerges from the crack between laughter and tears (von Franz 1972, p. 136). In another Egyptian myth, an androgynous god notes his loneliness, feels longing, and takes his member in his hand. While the world flows from his semen, he speaks to his shadow. His word is made flesh in humanity. Image and word shape the hieroglyphs, emotion creates art and dance. Above and below re-meet in each "thought of the heart" and "the tongue that speaks the heart's word". Von Franz (1972, p. 138) writes:

> We should not forget the original meaning of the word emotion – *emotio*
> – as that which moves one out of something, which makes one move.
> Emotion seems to be the absolutely basic factor in all creation myths,
> together with its concommitant psychological feelings and physical re-
> actions.

Ernst Cassirer (1946, p. 33) offers this description:

> Mythical thinking comes to rest in the immediate experience: the im-
> mediate content commands his religious interest so completely fills his
> consciousness that nothing else can exist beside and apart from it. ...
> Focusing of force on a single point is the prerequisite for all mythical
> thinking. When, on the one hand, the entire self is given up to a single
> impression, is possessed by it, and on the other hand, there is the ut-
> most tension between the subject and its object, the outer world; when
> external reality is not merely viewed and contemplated, but overcomes
> a man in sheer immediacy, with emotions of fear or hope, terror or wish
> fulfillment: then the spark jumps somehow across, the tension finds re-
> lease, as the subjective excitement becomes objectified, and confronts
> the mind as a god or a daemon.

And he continues (Cassirer 1946, p. 34):

> It is as though the isolated occurrence of an impression, its separation
> from the totality of ordinary, commonplace experience produced not
> only a tremendous intensification, but also the highest degree of con-
> densation, and as though by virtue of this condensation the objective
> form of the god were created so that it veritably burst forth from the
> experience.

Such intense and condensed moments mark the instant of discovery in experiment, a creative breakthrough, contact with a new sense of self, falling
in love, a meeting of minds, a numinous or synchronistic realization.

From his earliest days of tracking emotion in the "word association experiment", of formulating complex theory, and of identifying the archetypal
underpinnings of human predispositions, Jung followed William James in
positing the primacy of emotion before understanding. For Jung, emotion
was central in the unfolding of psyche via instinct and affect, action, image
and narrative. Emotions are the core of the complex, the force in dreams,
the fuel of transference.

In Jung's individuation model, one recognizes reactions, sorts responses,
moves from affect to images, from emotion to refined feeling. For Jung,

intense emotion comes with archetypal activity, allowing an experience. Interpretation and intellectual clarification become "meaningful and helpful when the road to original experience is blocked" (Jung 1997, p. x).

On Jung's scale, when emotion is dysfunctional, and original experience is blocked, the ego and its four functions of thinking feeling, intuition and sensation are imbalanced. Emotionality overwhelms rationality, process is disrupted, one becomes disassociated from the spacetime of the affective-cognitive continuum of mind and body.

In science, emotion may be such an initiating force as well as it is in myth, art, psychology and life. The following words of Einstein's suggest that his conversations with Jung exerted a two-way influence (quoted after Hadamard 1945, p. 142):

> Words or the language, as they are written or spoken, do not seem to play any role in my mechanism of thought. The psychical entities which seem to serve as elements in thought are certain signs and more or less clear images which can be "voluntarily" reproduced and combined.

Einstein notes that these were originally of a "visual and ... muscular type". He recognized (again in Hadamard 1945, p. 142)

> the desire to arrive finally at logically connected concepts is the emotional basis of this rather vague play ... before there is any connection with logical construction in words or other kinds of signs which can be communicated to others.

Pauli often remarked on the emotional dynamic which impels science. Echoing Nietzsche, and intimating current thinking in neuroscience: we feel, and thus we think, rather than the Cartesian we think and so we are. He once stated that "feeling is as deep as thinking and that *amo ergo sum* would be as justified as the *cogito ergo sum* by Avicenna-Descartes" (Pauli 1949). Like many who seek themselves through the recovery of feeling, reintegration of emotion, and acceptance of interdependence, Pauli had an intimation of integrity, an image of self as it might have been contained in the One Mind of time immemorial. To save his life, and possibly his science, Pauli turned to Jung, to the images of his dreams, and to the Taoist wisdom text *I Ching*.

5   Out of and in Tao

Surprise is relative – there is the sharp surprise of those synchronicities which crash into consciousness from beyond the margins, and those which are elicited and sought in the mantic. A 3500 year old Chinese text, the

I Ching, valorizes and functions from a perceived Tao of interrelatedness, equilibrium, and reciprocity among its images, and between its texts and the reader. These are connected through analogies provided in the text, and meaning extracted by the reader.

To consult the *I Ching* for pertinent frames of psychological reference implies psychoid archetypes (Jung 1978, par. 964), constellating patterns "in circumstances that are not psychic (equivalence of an outward physical process with a psychic one)", and continually go beyond their frame, an infringement to which "I would give the name transgressivity". Challenging the given categories of experience and distinctions between matter and mind can lead back to the empty state described as the Tao, and hence forward into a new beginning.

Finding a parallel to his notion that only conscious disintegration activates the search for unity, Jung (1997, p. 1025) cites Lao-Tze: "They are all so clear, only I am troubled". Troubled lack of clarity leads to "the great void, the positive nothing, the being non-being". With its mathematical structure and philosophical assumptions of multiple levels of existence, the *I Ching* was a reflective instrument for Pauli, an amplification in interpreting his dreams. The "scourge" and "whip" of physics, whose disruptive intensity produced the famous glass-shattering "Pauli effect", often constellated hexagram 51, The Arousing, Shock, Thunder. Aptly, it reads: "the setting of the sun suggests that the foundation is to be laid on chaos". We read (Wilhelm and Baynes 1950, p. 197): "The shock terrifies for a hundred miles. He does not let fall the sacrificial spoon and chalice." And the text then comments (Wilhelm and Baynes 1950, p. 198):

> when a man has learned within his heart what fear and trembling mean,
> he is safeguarded against any terror produced by outside influences...
> he remains so composed and reverent in spirit that the sacrificial rite
> is not interrupted. The aim is profound inner seriousness from which
> all outer terrors glance off harmlessly.

Pauli turned to his dreams as vectors toward psychic equilibrium with profound inner seriousness, and devotion: "I made a point of stressing the difference between the spontaneous appearance of the phenomenon ... and the induced phenomenon ... as is the case with the Mantic (the I Ching)" (Meier 2001, p. 44). He wrote to Jung (Meier 2001, p 40): "what you call a conjunction process is generally conducive to the appearance of the synchronistic phenomenon". In the *I Ching*, "this moment is depicted by the sign Chen".

Pauli's interest in synchronicities became an aspect of meaning-making in individuation rather than acausal connections being observed or registered within a specific time frame (Meier 2001, p. 44):

> In the matter of the mind, I preferred to use the term "meaning-correspondence" rather than "synchronicity", so as to place more emphasis on meaning rather than on simultaneousness and to link up with the old "correspondentia".

In his scientific work, Pauli referred to a fundamental feature of quantum theory which states that an observation or measurement changes the object observed and thus introduces an aspect of subjectivity. While the scientist seeks objectivity, as Jung wrote, the participant in a synchronistic occurrence has an emotional predisposition to become involved (Jung 1975, Vol. 2, p. 318):

> What is the psychological condition in which a synchronistic phenomenon may be expected? ... A certain affective condition seems to be indispensable. One has therefore to look for emotional conditions ... they permit a certain insight into the underlying unconscious constellations and their archetypal structure. I have observed personally quite a number of synchronistic events where I could establish the nature of the underlying archetype.

Pauli differentiated the spirit of mathematics and of his conscious attitude: "everything that is part of the counter-position of the sciences is a private matter" (Meier 2001, p. 89). He noted that the observing physicist, in contrast to the serious alchemist, is not himself transformed, because the "gift of sacrificing" is not a part of himself, but a portion of the external world. Insofar as the scientist must opt to know "which aspect of nature we want to make visible ... we simultaneously make a sacrifice ... a coupling of choice and sacrifice" (Card 1991, pp. 35f).

When one cannot bend to change, and be open to surprise, reality may indeed enter as an arousing shock. Pauli was unable to "outgrow" his fascination with symmetry, to "sacrifice" its primacy as the structure of the basic forces of nature. In 1956, it was proven that the weak interactive force violates left-right symmetry. Pauli declared (quoted from Stewart and Golubitsky 1992, p. 181):

> I am shocked not so much by the fact that the Lord prefers the left hand as by the fact that he still appears to be left-right symmetric when he expresses himself strongly. ... Why are strong interactions right-and left-symmetric?

6 Physical Time And Emotional Timing

In the fundaments of physical laws, time has no distinguished direction, nor are past, present and future absolutes. Breaking the symmetry of time reversal, time becomes a one-way parameter with a direction toward the future. The cracked egg, the shattered cup, the dead body cannot be made whole nor alive as before. As an incarnated organ, the brain shares in the chronology and biology of aging body and slowing synapses. The incarnated brain notes, measures, counts, and keeps the time made and marked by the clocks of human craft.

Biological time moves from past to present and future. Our inhabited bodies carry markers from both our own pasts and the transmissions of generations. In genetics, we carry our ancestors. In epigenetics, we bear marks of their constraints and choices, just as future generations will carry ours. In living matter, we keep time with the hands of nature's biological clock. Our inherited instincts, affects and reactions are aimed toward survival. Choice and decision assess risk and probability. Receptivity, interest, and emotions speed up and slow time. As Rosalind says in *As You Like It* (Shakespeare 1599/1954, p. 56):

> Time travels in divers paces with
> divers persons. I'll tell you who Time ambles
> withal, who Time trots withal, who Time gallops
> withal and who he stands still withal.

When humans are gripped with fear and terror, time seems to pause or slow down, as if the brain must notice the when and whereabouts while taking fast action. Like surprise, our other basic survival emotions, such as anger, fear, disgust, mobilize in the seconds between the imprints of various sense perceptions. The timing mechanisms of the brain are crucial operatives in mood states, psychoses, and autism.

Emotions also fashion the telling of our histories. While our collective time-keeping appears to be counted in a sequence of numbered years, their qualities are assigned by cultural, national and religious agendas. Certain epochs are named after dominant powers, eras are titled according to an empire's territories. Decades are linked to wars. National identities are counted in the years since a battle was won or lost. Millennia manifest systems of belief. Our many calendars number the years since Abraham, since Buddha, since Christ, since Mohammed.

Our personal time-keeping marks rites of passage – birth, initiation, marriage, the sequential anniversaries and birthdays until the day of death. We live in an envelope of time as an inexorable reality, guaranteed by our categories of before, now, later, and after.

Dream time adds other temporal dimensions. Our lived lifetime is augmented when we sleep to dream. And in our dreams, time sequence is often reversed between the dreaming and the telling. The dreaming mind follows a different clock than the waking mind. Neuroscientists such as Llinas describe waking consciousness as a dream modified by sensory input and motor output. Wakefulness is regarded as a dreamlike state modulated by sensory perceptions (Llinas and Churchland 1979, p. 6). In the same vein, von Franz (1992, p. 72) writes:

> It is a remarkable coincidence that, at approximately the same time as physicists discovered the relativity of time in their field, C.G. Jung came across the same fact in his explorations of the human unconscious. In the world of dreams, time also appears as relative and the categories of "before" and "after" seem to lose their meaning.

The time of the unconscious and a consciousness organized by sense perception may flow in different directions. The time of the psyche may wax while body wanes.

7 Psychic Time And Timelessness

Our incarnated time is both cyclic and linear. The cycles of nature inspire comparative analogies, of spring following winter, summer preceding autumn. Our mythic dying and reviving nature gods exemplify the many mortifications and rejuvenations of life phases. Our revealed religions freeze time in a moment of revelation, inscribed in stone. The dead gods who come back to life on earth and precede the chosen to another dimension are attempts to push back and reverse time, and to create an after lifetime time through salvation and reunion.

In the personal psyche, we combine the recall of past occurrences, the immediacy of present impacts, and time-independent imagination, as we recast identity, re-invent self, and re-envision what might come. When psyche is a process, it is released from entrapment in a memory that refuses recontextualization, a traumatic past episode, a cast in stone interpretation. Rather, time is a flow between once upon, then, there, here and now, when and where. The past is continually reshaped and future re-imagined as recollection and imagination come together in the "remembered present" of consciousness.

Jung described time "as a mere *modus cogtandi*; what we perceive in fact is a stream of inner and outer experiences; time is the flow of outwardly perceived events and the inwardly experienced train of thoughts, feelings, and emotion" (von Franz 1992, p. 123). Mind and psyche exist both in and out of the spacetime of personal incarnation and ego's education. When archetypal patterns or givens are dominant, ego time seems not to exist as a framing category. On the one hand, this can extend one's life sense through generations, while on the other, one may lose the personal in one's lifetime.

Time wears many guises in an analytic process. Time is a matter of timing, and untimeliness. Mind and psyche move in and out of time, as myth, memory, perception, projection, construct, and immediate experience converge, condense, dissolve, and merge. There are un-sequential reversals to the "frozen accidents" of embedded complexes, Rashoman relativities of perspective and focus, or the strange loops of intertwined psyches. In the transferential field, there are multiple instances of complementarities, synchronicities, symmetry breaking and entanglements. For the suffering, there are black holes and event horizons. In a personal mythopoeisis of apt analogy and fitting association, fact is relative to one's truth. What is formula for the physicist is narrative for the analyst.

Several collective and personal neuroses express themselves through time factors: a narcissistic race with time, a pretense that certain times do not exist, that age is unreal; an obsessive need to repeat actions rather than let time move on; the dour resentments that will not be released from injurious events in private or national histories.

Personal growth follows overlapping and intricate timelines. As a multiplicity within a unity, in different contexts, one may be infantile, adolescent, youthful, ripe, mature, wise, senile. As each complex is of different age, our struggles resemble inter-generational conflicts. Such temporal disjunctions constellate a longing for when we were of one age, at the beginning of the potential. Being out of time can constellate that search for internal unity imagined to beckon in time immemorial.

8 The Time-Timeless Totality

Ancient Egyptians oriented their lives to live on the universal grid of Maat. Alchemists intervened with matter so they might approach the mysterious conjunction between individual and world and so re-unify matter and re-integrate mind within an *unus mundus*. Pauli dreamt of a world clock, three different time pulses on horizontal and vertical discs within a rotating golden ring. In its image of diversity and totality, this dream was especially numinous for both Jung and Pauli.

Neuroscientists are driven to solve the binding problem, the homunculus in the brain which synthesizes and creates. Psychologists engage dissociations in the hope of repair after rupture. Jungian analysts foster the *lux moderna*, the light of insight when conscious and unconscious are mutual and compatible.

Pauli and Jung debated the context of time in synchronicity. At times, they seemed to speak past each other. Pauli was clear – synchronicity did not demand simultaneity. Jung slid around the temporal issue, arguing for simultaneity while giving examples of synchronicity as a retrodiction. Possibly the physicist was clocking outer time, while the analyst was locating the inner click, when the act of creation in time emerged from a rolling process of observation, surprise, emotional significance, mindful recognition, emergent realization. Ultimately, both released themselves from time. Toward the end of his life, Jung (1989, p. 305) said:

> Our world, with its time, space, and causality, relates to another order of things lying behind or beneath it, in which neither "here and there" nor "earlier and later" are of importance. I have been convinced that at least a part of our psychic existence is characterized by a relativity of space and time.

And Pauli declared in a letter to Fierz of March 30, 1947 (von Meyenn 1993, pp. 435f):

> I am particularly dissatisfied with the way in which the spacetime continuum is introduced at present. (Of course it is ingenious to disband time from ordering causal sequences and – "as once in May" – use it as a romping place for probabilities. But if one replaces ingenious by impudent, this is not less true. In fact, something happens only during an observation, where – as Bohr and Stern finally convinced me – entropy increases necessarily. Between observations nothing happens at all, only time has reversibly proceeded on our mathematical papers!). This spacetime continuum has now become a Nessus shirt which we cannot take off again! (Instead of "Nessus shirt" you can also say "prejudice", but this would, first, sound too harmless and, second, shift the mistake too much from a mere conception to a judgment.)

In 1952, while completing his essay on synchronicity, Jung (1973, p. 45) wrote this comment in a letter:

> It might be that psyche should be understood as unextended intensity, not as a body moving with time. One might assume the psyche gradually rising from minute extensity to infinite intensity, transcending for

instance the velocity of and thus irrealizing the body ... The brain might be a transformer station, where the relatively infinite tension or intensity of the psyche proper is transformed into perceptible frequencies or extensions. Conversely, the fading of introspective perception of the body explains itself as due to a gradual "psychification", i.e. intensification at the expense of extension. Psyche = the highest intensity in the smallest space. (But in itself the psyche would have no dimension in space and time at all ...).

Does this mean that ultimately mind must be included in our theories of reality? The neuroscientist Gerald Edelman (1992, p. 11) notes:

Einsteinian and Heisenbergian observers, while embedded in their own measurements, are still psychologically transparent. Their consciousness and motives do not have to be taken into account to practice physics. The mind remains well removed from nature.

And he continues (Edelman 1992, p. 15): "There must be ways to put the mind back into nature that are concordant with how it got there in the first place."

9 Epilogue: Magic, Beauty, Happiness, and Glamour

Synchronicities are experiential reminders of one's discrete, dynamic position in a network of existence and the mirrorings of one's surround. They comprise impersonal manifestations of interactive connection made personal and meaningful by the human mind's creative response and participation. Depending on one's tolerance for surprise they may be heartening or deflating. They initiate those who engage them into the mysteries of life both in and out of time, into more dimensions than either the Taoists of old or today's superstring theorists yet imagine.

The British physicist and author Clarke (1962) suggested that any sufficiently advanced technology is indistinguishable from magic. Jung and Pauli might see magic in the many instances when, for good or ill, the analogies from human imagination have resulted in the products of our sciences; when minds emerging from matter can discern probable correspondence and intervene in the patterns of nature and cosmos.

Jung looked at the theories of 20th century physics as amplifications and mirrors of the dynamics of psychic energy. He recognized the unfolding of alchemical intuitions, especially in the quantum physics of his era. He would evoke the language of complementarity, uncertainty, nonlocality, entanglement, symmetry, dark energy and dark matter, sometimes as equivalents, sometimes as analogies for psychic energy.

Pauli was called the conscience of physics. Jung thought that the attitude toward material and spiritual categories in synchronicity would influence "ethical, aesthetic, intellectual, social, and religious systems of value". He quotes the alchemist Gerhard Dorn: "No man can truly know himself unless first he see and know by zealous meditation ... what rather than who he is, on whom he depends" (Jung 1970b, par. 685).

The perspectives from the mind-matter continuum move us from who we are to what we are, in the process of individuals seeking their role in a grand, mysterious conjunction with a spacetime of existence. The ethic of the *what* connects us to all that exists, crucial in our anthropocene age of a universe at the effect of human behavior and intervention

In life and in practice as a Jungian analyst, an experienced synchronicity is a humbling call toward an I-other, I-you, and I-Thou continuum, a "showing up" for a moment of meeting, and for a mythopoetic, cosmogonic consciousness that allows an "act of creation in time". One's context becomes larger, with more complexity and mystery, and paradoxically with a greater clarity. One's sensibility extends to the spectrum of existence. Such contextualization layers life with the more than personal. Amidst global conflicts and personal suffering, this nonetheless can have beauty, allow happiness, and grant glamour to the human experience.

Werner Heisenberg believed that beauty is involved with the age-old problem of the "one" and the "many" – in close connection with the problem of "being" and "becoming". Pauli suggested "a cosmic order", noting that in the unconscious the place of concepts is taken by images with strong emotional content. For Pauli, images of emotionally determined archetypes which link sense perception and mental concepts emerge from the collective unconscious via "the happiness that man feels in understanding nature".

Astrophysicists now speak of "the anthropic principle", proposing that the human brain is capable of deciphering the contents and dynamics of the universe because it is just a special case of the universe's design. And, toward the end of his life, Jung (1989, p. 300) reflected:

> We cannot visualize another world ruled by quite other laws, the reason being that we live in a specific world which has helped to shape our minds and establish our basic psychic conditions. We are strictly limited by our innate structure and therefore bound by our whole being and thinking to this world of ours. Mythic man, to be sure, demands a "going beyond all that" but scientific man cannot permit this. To the

intellect, all my mythologizing is futile speculation. To the emotions, however, it is a healing and valid activity; it gives existence a glamour which we would not like to do without. Nor is there any good reason why we should.

References

Borges J.L. (1962): The garden of forking paths. In *Labyrinths*, ed. by D.A. Yates and J.E. Irby, New Directions Books, New York.

Card C.R. (1991): The archetypal view of C.G. Jung and Wolfgang Pauli. *Psychological Perspectives* **24**, 19–33 and **25**, 52–69.

Cassirer E. (1946): *Language and Myth*, transl. by S. Langer, Harper & Row, New York.

Clark A.C. (1962): Hazards of prophecy: The failure of imagination. In *Profiles of the Future*, Harper & Row, New York.

Edelman G. (1992): *Bright Air, Brilliant Fire. On the Matter of the Mind*, Basic Books, New York.

Foucault M. (1973): *The Order of Things, An Archeology of the Human Sciences*, Vintage Books, New York.

Hadamard J. (1945): *The Psychology of Invention in the Mathematical Field*, Princeton University Press, Princeton.

Hirsch M., Pas H. and Porod W. (2013): Ghostly beacons of new physics. *Scientific American* **308**(4), 42–47.

James W. (1890): *The Principles of Psychology Vol. One*, Holt, New York.

Llinas R. and Churchland P. (1997): *The Mind-Brain Continuum*, Carfax Publishing, New York.

Jung C.G. (1970a): *Civilization in Transition, Collected Works Vol. 10*, Princeton University Press, Princeton.

Jung C.G. (1970b): Mysterium coniunctionis. *Collected Works Vol. 14*, Princeton University Press, Princeton.

Jung C.G. (1973): *Letters, Vol. 1, 1906–1950*, Princeton University Press, Princeton.

Jung C.G. (1975): *Letters, Vol. 2, 1951–1961*, Princeton University Press, Princeton.

Jung C. G. (1978): Synchronicity, an acausal connecting principle. In *The Structure and Dynamics of the Psyche, Collected Works Vol. 8*, Princeton University Press, Princeton, pp. 417–519.

Jung C.G. (1989): *Memories, Dreams, Reflections*, ed. by A. Jaffe, Vintage Books, New York.

Jung C.G. (1997): *Visions Seminars*, Princeton University Press, Princeton.

Meier C.A., ed. (2001): *Atom and Archetype: The Pauli/Jung Letters 1932–1958*, Princeton University Press, Princeton.

Pauli (1949): Letter to Goldschmidt of February 19, 1949. In *Nochmals Dialogik*, ETH Stiftung Dialogik, Zürich, p. 25.

Shakespeare W. (1599/1954): As you like it. In *The Yale Shakespeare*, ed. by S.C. Burchell, Oxford University Press, London.

Strand M. (1979): *Keeping Things Whole. From Selected Poems*, Alfred A. Knopf, New York.

Stewart I. and Golubitsky M. (1992): *Fearful Symmetry: Is God a Geometer?*, Penguin Books, London.

Wilhelm R. and Baynes C.F. (1950): *The I Ching*, Princeton University Press, Princeton.

von Franz M.L. (1972): *Creation Myths*, Spring Publications, New York.

von Franz M.L. (1992): *Psyche and Matter*, Shambhala, Boston.

Contributors

Marcus Appleby
Whitchurch, Hampshire, UK
marcus.appleby@gmail.com

Harald Atmanspacher
Collegium Helveticum
University and ETH Zurich
Switzerland,
Institute for Frontier Areas of
Psychology and Mental Health
Freiburg, Germany
atmanspacher@collegium.ethz.ch

Hans Christian von Baeyer
Physics Department
College of William and Mary
Williamsburg, Virginia, USA
henrikritter@gmail.com

Joseph Cambray
Providence, Rhode Island, USA
cambrayj@earthlink.net

Wolfgang Fach
Institute for Frontier Areas of
Psychology and Mental Health
Freiburg, Germany
fach@igpp.de

Thomas Filk
Department of Physics
University of Freiburg, Germany
Thomas.Filk@t-online.de

Christopher Fuchs
Quantum Information Processing
Technology Group
Raytheon BBN Technologies
Cambridge, Massachusetts, USA
qbism.fuchs@gmail.com

Suzanne Gieser
Stockholm, Sweden
sg@gieser.se

George Hogenson
Oak Park, Illinois, USA
hogenson@icloud.com

Dave Laveman
Laveman & Associates
New Hope, Pennsylvania, USA
dalaveman@aol.com

Roderick Main
Center for Psychoanalytic Studies
University of Essex, UK
main@essex.ac.uk

Christian Roesler
Clinical Psychology
Catholic University Freiburg
Germany
roesler@kh-freiburg.de

Rüdiger Schack
Department of Mathematics
Royal Holloway

University of London, UK
r.schack@rhul.ac.uk

William Seager
Department of Philosophy
University of Toronto, Canada
seager@utsc.utoronto.ca

Beverley Zabriskie
New York City, New York, USA
BevZab@aol.com

Index